INTRODUCTION TO RENORMALIZATION GROUP METHODS IN PHYSICS

INTRODUCTION TO RENORMALIZATION GROUP METHODS IN PHYSICS

R. J. CRESWICK

H. A. FARACH

C. P. POOLE, JR.
Department of Physics and Astronomy
University of South Carolina
Columbia, SC

A Wiley-Interscience Publication
JOHN WILEY & SONS, INC.
New York • Chichester • Brisbane • Toronto • Singapore

In recognition of the importance of preserving what has been
written, it is a policy of John Wiley & Sons, Inc., to have books
of enduring value published in the United States printed on
acid-free paper, and we exert our best efforts to that end.

Copyright ©1992 by John Wiley & Sons, Inc.

All rights reserved. Published simultaneously in Canada.

Reproduction or translation of any part of this work
beyond that permitted by Section 107 or 108 of the
1976 United States Copyright Act without the permission
of the copyright owner is unlawful. Requests for
permission or further information should be addressed to
the Permissions Department, John Wiley & Sons, Inc.

Library of Congress Cataloging in Publication Data:
Creswick, R. J.
    Introduction to renormalization group methods in physics/R. J.
Creswick, H. A. Farach, C. P. Poole, Jr.
      p. cm.
    "A Wiley Interscience publication."
    Includes bibliographical references.
    1. Renormalization (Physics)   2. Mathematical physics.
I. Farach, Horacio, A.   II. Poole, Charles P., Jr.   III. Title.
QC20.7.R43C74 1991
530.1'5—dc20   91-19221
                                                                    CIP

ISBN 0-471-60013-X

Printed in the United States of America

10 9 8 7 6 5 4 3 2 1

We dedicate this book to our wives

R. J. C. to Victoria

H. A. F. to Blanca

C. P. P. to Kathleen

CONTENTS

Preface   xiii

Chapter 1   **Self-Similarity and Scale Invariance**   1

     1.0   Introduction, 1
     1.1   Self-Similarity and Dimension of Cantor Sets, 3
     1.2   Fractals, 5
     1.3   Fractals with More than One Scale Factor, 8
     1.4   Dimension of the Random Walk, 9
     1.5   Anatomy of a Renormalization Group Transformation, 11
     1.6   Renormalization Group Analysis of the Random Walk, 14
     1.7   Universality Class of the Random Walk, 20
     1.8   Real-Space Renormalization Group for the Self-Avoiding Random Walk, 22
          Problems, 30
          References, 32

Chapter 2   **Renormalization Group Approach to Chaos**   34

     2.0   Introduction, 34
     2.1   One-Dimensional Maps and the Period-Doubling Route to Chaos, 39

viii CONTENTS

      2.2 Renormalization Theory of the Universality of Period-Doubling in One-Dimensional Maps, 47
      2.3 Global Structure of the Aperiodic Attractor, 54
      2.4 Renormalization Theory of Two-Dimensional Maps, 56
      2.5 Renormalization Group for the Period-Doubling Route to Chaos, 59
      2.6 Universal Properties of Conservative Maps, 65
      Problems, 66
      References, 68

Chapter 3 Renormalization Approach to Percolation   69

      3.0 Introduction, 69
      3.1 Percolation in One Dimension, 81
      3.2 Percolation on a Cayley Tree, 83
      3.3 Renormalization Group for Bond Percolation on a Hierarchical Lattice, 86
      3.4 Majority-Rule Renormalization Group for the Triangular Lattice, 91
      3.5 Site Percolation on the Square Lattice: Cell Renormalization, 94
      3.6 Real-Space Renormalization Group for Bond Percolation, 96
      3.7 Site—Bond Dilution: The Sol–Gel Transition, 99
      Problems, 105
      References, 107

Chapter 4 Renormalization Group and Critical Phenomena   108

      4.0 Introduction, 108
      4.1 Construction of the Renormalization Group Transformation, 110
      4.2 Renormalization Group in the Neighborhood of a Fixed Point, 117
      4.3 Correlation Length at a Fixed Point, 120
      4.4 Homogeneity Properties of the Free Energy, 121

4.5 Scaling Form of the Correlation Function, 124
4.6 Scaling Laws, Hyperscaling, and Universality, 126
4.7 Fixed Points, and First-Order Phase Transitions, 128
4.8 Crossover Phenomena, 129
4.9 Finite-Size Scaling, 131
4.10 Invariance of Critical Exponents with Respect to Change of Variable, 134
4.11 Goldstone Modes and the Lower Critical Dimension, 137
Problems, 142
References, 143

Chapter 5 The Ising Model — 144

5.0 Introduction, 144
5.1 The One-Dimensional Ising Model, 147
5.2 The One-Dimensional Ising Model in a Magnetic Field, 152
5.3 The Ferromagnetic Ising Model in Two Dimensions, 155
5.4 Exact Solution of the Ising Model on the Square-Lattice: The Partition Function, 162
5.5 Exact Solution for the Ising Model on the Square Lattice: Correlation Function, 168
Problems, 180
References, 181

Chapter 6 Renormalization Group for the Ising Model — 183

6.0 Introduction, 183
6.1 Ising Model on the Triangular Lattice, 183
6.2 The First Cumulant, 187
6.3 The Second Cumulant, 188
6.4 Cumulants Revisited, 191
6.5 Fixed-Point Analysis, 192
6.6 Renormalization Equations in the Presence of a Magnetic Field, 196

- 6.7 The First Cumulant with $h \neq 0$, 197
- 6.8 Renormalization Group Equations to Second Order, 198
- 6.9 Antiferromagnetic Ising Model on the Triangular Lattice, 205
- Problems, 213
- References, 216

Chapter 7 Other Real-Space Renormalization Group Methods for the Ising Model 217

- 7.0 Introduction, 217
- 7.1 Decimation Transformation on a Hierarchical Lattice, 218
- 7.2 The Random-Bond Ising Model, 219
- 7.3 The Migdal-Kadanoff Transformation, 223
- 7.4 The Migdal-Kadanoff Transformation in a Weak Magnetic Field, 227
- 7.5 The Monte Carlo Renormalization Group, 230
- Problems, 237
- References, 237

Chapter 8 Mean Field Theory and the Gaussian Fixed Point 238

- 8.0 Introduction, 238
- 8.1 The Mean Field Theory for the Ising Model, 239
- 8.2 Mean Field Theory as a Variational Theory, 243
- 8.3 Cluster Approximations, 246
- 8.4 The Hubbard-Stratonovich Transformation, 247
- 8.5 Landau-Ginzburg Form for the Free-Energy Functional and Spontaneous Symmetry Breaking, 255
- 8.6 The Gaussian Model, 259
- 8.7 Renormalization Group Analysis of the Gaussian Model, 261
- 8.8 Stability of the Gaussian Fixed Point, 264
- Problems, 266
- References, 266

Chapter 9 The ε Expansion   **267**

     9.0   Introduction,   267
     9.1   Construction of the RG Transformation, 269
     9.2   The Cumulant Expansion and Feynman Diagrams,   272
     9.3   Evaluation of the Second Cumulant,   277
     9.4   Renormalization Equations to Order ε, 282
     9.5   The Wegner–Houghton Infinitesimal Renormalization Group Generator,   291
     9.6   Solution of the Renormalization Equations to $O(\varepsilon)$,   296
     9.7   Solution of the RG Equations to $O(\varepsilon^2)$: Evaluation of η,   300
     9.8   Field-Theoretic Approach to the ε Expansion,   307
     9.9   Critical Behavior for $d = 4$: The Upper Critical Dimension,   317
     Problems,   320
     References,   321

Chapter 10 The Spherical Model and the $1/n$ Expansion   **322**

     10.0   Introduction,   322
     10.1   The Equation of State and Critical Exponents for the Spherical Model,   326
     10.2   Critical Properties of the n-Vector Model in the Limit $n \to \infty$,   329
     10.3   Renormalization Group in the Limit $n \to \infty$,   332
     10.4   Field-Theoretic Approach to the $1/n$ Expansion,   336
     Problems,   342
     References,   343

Chapter 11 The Two-Dimensional X–Y Model and the Kosterlitz–Thouless Transition   **344**

     11.0   Introduction,   344
     11.1   Real-Space Renormalization for the Vortex Gas,   350

- 11.2 Analysis of the Renormalization Group Equations, 356
- 11.3 The Spin–Spin Correlation Function, 357
- 11.4 Universal Behavior of Superfluid Helium Films, 359
  Problems, 362
  References, 362

Appendix A The Cumulant Expansion — 364

Appendix B Feynman Diagrams — 371

Appendix C Combinatorial Solution to the Ising Model — 381

Author Index — 395

Subject Index — 399

PREFACE

This text is intended as an introduction to the renormalization group (RG) in physics. The term "group," when used in this context, is somewhat misleading because the renormalization transformations do not form a mathematical group (there is no inverse transformation), and one rarely needs the machinery of group theory in applying RG methods. It is also misleading to refer to the RG as if it is a well-defined mathematical procedure. Although a great deal of formalism has been developed for the RG in field theory and in statistical mechanics, RG approaches have been applied successfully to a large variety of problems in physics, ranging from classical mechanics to the ground states of quantum many-body systems.

By treating examples taken from these diverse fields which contain all of the characteristics of more complex RG calculations, but which are mathematically transparent enough so that one does not get bogged down in formalism, we hope to bring out the essential features they share in common. For this reason no attempt has been made to treat the RG in field theory, a discussion of which is beyond the scope of this text.

The underlying philosophy of the book is that one learns best by doing, and therefore problems are included at the end of each chapter, which hopefully illuminate and extend the material covered in the text. Some of the problems require simple numerical calculations for which a microcomputer should prove adequate.

The text is intended for use in a one-semester second- or third-year graduate course. Although we have done everything possible to make the book as self-contained as possible, due to the diverse nature of the material covered, we must rely on the student to have a good undergraduate background in classical mechanics, statistical physics, and quantum mechanics.

If imitation is the greatest flattery, then this text flatters many. By design we cover a wide range of fields and can claim expertise in none of them. Therefore, we are forced to borrow heavily from the literature.

Several texts and review articles exist that treat the RG within the more restricted context of field theory and statistical physics, and they are highly recommended as supplements to this book. The standard against which any text on RG theory is measured is Ma's excellent text *Modern Theory of Critical Phenomena*. More recently, and at a more introductory level, is *Introduction to the Renormalization Group and to Critical Phenomena* by Pfeuty and Toulouse. For those who have more of an inclination toward field-theoretical methods, we recommend *Field Theory, the Renormalization Group and Critical Phenomena* by Amit. We should also mention the review article by M. N. Barber, "An Introduction to the Fundamentals of Renormalization Group in Critical Phenomena" [*Phys. Rep.* **29C** 1 (1977)] and of course the standard review of Wilson and Kogut, "The Renormalization Group and the ε-Expansion" [*Phys. Rep.* **12C** 75 (1974)].

We would like to thank the Fereydoon Family for many helpful suggestions and Michael Schick and Per Hemmer for elucidating their work on RSRG calculations for two-dimensional Ising models. We would also like to thank M. Mesa, C. Sisson, and J. Lane for help in preparation of the text and figures.

R. J. Creswick
H. A. Farach
C. P. Poole, Jr.

1

SELF-SIMILARITY AND SCALE INVARIANCE

1.0 INTRODUCTION

Although the renormalization group (RG) is not restricted to the study of systems at their critical points, it is here that it has achieved its greatest success. This is fortunate because in the critical region the traditional tools of the physicist, principally perturbation theory, fail completely. The critical behavior of a system is characterized by a loss of scale; the system fluctuates strongly at all wavelengths and there is no natural "small parameter" on which to base a perturbation expansion. The critical point is also characterized by nonanalytic behavior in the various correlation functions of the system; so again perturbation theory cannot be expected to yield good results. In this chapter the concepts of scale invariance and self-similarity are introduced through the fascinating class of geometric objects called fractals [5, 8].

An object is self-similar if every part of the object, when magnified by a suitable scale factor, is identical to the original object. A related and stronger property is scale invariance, which can be thought of as self-similarity on all scales.

The idea of dimension plays an important role in RG theory, and the structures that one encounters at the critical point of a system do not always fit into our intuitive idea of dimension. To begin, we define the "Euclidean" dimension d as the dimension of the space in which the system resides; this will always be an integer, 1, 2, 3, and so on.

2 SELF-SIMILARITY AND SCALE INVARIANCE

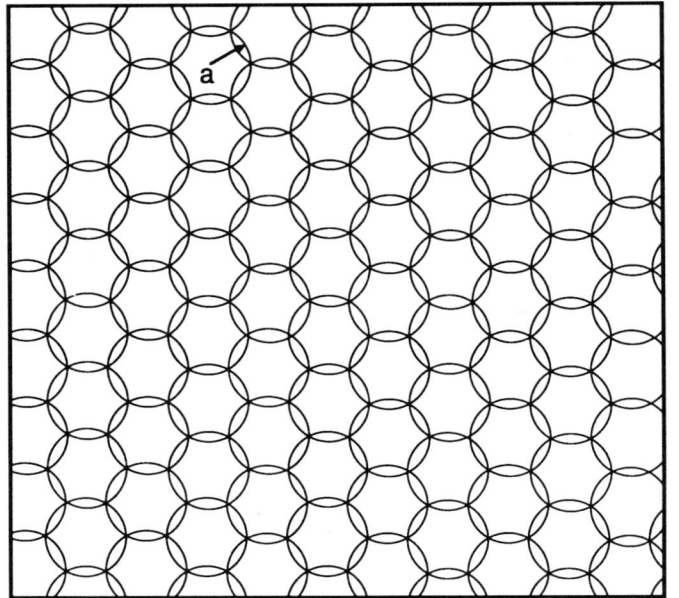

Fig. 1.0.1 Covering the plane by circles of radius a. The minimum number needed is $N(a) \sim a^{-2}$.

The "topological dimension" d_T of a curve or surface conforms to our intuitive idea of its dimension. A continuous curve has topological dimension 1, a continuous surface has topological dimension 2, and so on.

Because a curve or surface cannot have a larger dimension than the space in which it is embedded, we have $d_T \leq d$. We now define a third dimension, which we denote by D, called the fractal (or Hausdorff) dimension. A fractal is an object with a nonintegral dimension. This term was coined by Mandelbrot [5] who has more or less single handedly brought fractals to the attention of physicists.

The fractal dimension may be defined as follows: Cover the object whose dimension you wish to find by d-dimensional balls[†] of radius a, and let the minimum number of balls needed be $N(a)$. Then the fractal dimension D is defined by

$$N(a) \sim a^{-D} \quad a \to 0 \qquad (1.0.1)$$

[†]A ball in a space of dimension d is the generalization of a ball in three-dimensional space (the interior of a sphere). The 1-ball of radius a is a line segment of length $2a$, a 2-ball is a disk of radius a, and so on.

Equation (1.0.1) is an example of a scaling relation. We say that the number of balls scales with the radius, and the scaling exponent, which in this case is D, is the fractal dimension.

The fractal dimension D satisfies the inequality

$$d_T \le D \le d \qquad (1.0.2)$$

That $D \le d$ is clear; no curve, no matter how convoluted, can do more than fill the space in which it is embedded. The lower limit is also clear; an object must at least have a dimension equal to that of its elements.

Clearly, (1.0.1) gives us the answer we should expect for a smooth curve or surface: This is shown in Fig. 1.0.1 for the case of a square ($d_T = 2$) in which the fractal dimension is the same as the topological dimension, $d_T = D = 2$. To see that this is not always identical to our usual notion of dimension, we now consider the case of Cantor sets.

1.1 SELF-SIMILARITY AND DIMENSION OF CANTOR SETS

Many of the self-similar geometric structures that we will consider are constructed recursively, and this is nicely illustrated by the Cantor set in Fig. 1.1.1.

Take a line segment ($d = 1$) of length a_0 and divide it into thirds; remove the middle third. Repeat this process with the remaining two line segments and so on ad infinitum. In the limit one is left with a countable set of isolated points so d_T, the topological dimension, is 0 and the Euclidean dimension is 1. Now let us calculate the fractal dimension of the preceding Cantor set using the prescription (1.0.1).

The covering "balls" in this case are simply line segments, and as the length of these segments is reduced, more and more of the structure of the Cantor set is revealed. The simplest approach is to choose line segments of length $a = a_0 3^{-n}$, $n = 1, 2, 3, \ldots$. At the nth level of detail, we will need $N(a) = 2^n$ line segments to completely

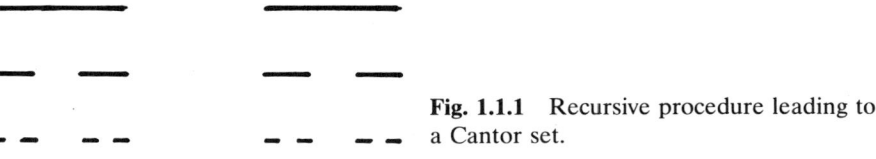

Fig. 1.1.1 Recursive procedure leading to a Cantor set.

cover the Cantor set. The dimension can then be determined by eliminating n:

$$n = -\frac{\ln(a/a_0)}{\ln 3} \quad (1.1.1)$$

We have $N(a) = 2^n$ or

$$\begin{aligned} N(a) &= 2^{\left[-\frac{\ln(a/a_0)}{\ln 3}\right]} \\ &= \left(\frac{a}{a_0}\right)^{-(\ln 2/\ln 3)} \end{aligned} \quad (1.1.2)$$

which by (1.0.1) gives

$$D = \frac{\ln 2}{\ln 3} = 0.6309\ldots \quad (1.1.3)$$

The Cantor set is self-similar in the following sense. Imagine taking a part of the set, for example, the first third, and magnifying, or rescaling, the unit of length by a factor of 3. The resulting set is again a Cantor set and indistinguishable from the original set. The Cantor set is also scale invariant because it looks the same on any length scale. There is no smallest length scale beyond which the set looks "smooth," but rather one encounters structure on all scales, and this is just what is reflected in the scaling relation (1.0.1).

The calculation of the dimension of the Cantor set can be reformulated by making direct use of its self-similarity. Because each piece of the Cantor set is itself an identical Cantor set apart from a scale factor, we can write for the previous example,

$$N(a) = 2N(3a) \quad (1.1.4)$$

That the same function N appears on both sides of (1.1.4) reflects the self-similarity of the Cantor set.

Assuming that (1.0.1) holds, we have

$$a^{-D} = 2 \cdot 3^{-D} \cdot a^{-D} \quad (1.1.5)$$

or
$$1 = 2 \cdot 3^{-D} \tag{1.1.6}$$
from which (1.1.3) follows immediately.

For many years Cantor sets were familiar only to mathematicians, but recent work in the field of nonlinear dynamics has brought them to the attention of physicists. In Chapter 2 we will see that the trajectory of a simple two-dimensional map is a Cantor set in the chaotic region of parameter space.

1.2 FRACTALS

It is possible to construct self-similar structures like the Cantor set in higher Euclidean dimensions as well. Consider, for example, the "triadic Koch island" (Fig. 1.2.1).

Fig. 1.2.1 Triadic Koch island [5].

6 SELF-SIMILARITY AND SCALE INVARIANCE

The boundary is formed by taking each side of a unit equilateral triangle, dividing it in thirds, and erecting on the middle third another equilateral triangle of side 1/3. As with the Cantor set, this process is repeated for each element of the boundary ad infinitum, and the resulting curve has a dimension greater than 1 but less than 2.

The fractal dimension of the triadic Koch island can be calculated just as for the Cantor set. In this case the Euclidean dimension is 2, and we are interested in the number of 2-spheres (circles) needed to cover the boundary. After each iteration there are four times as many sides as in the previous iteration, each of which is one-third as long.

Fig. 1.2.2 Quadric Koch island [5].

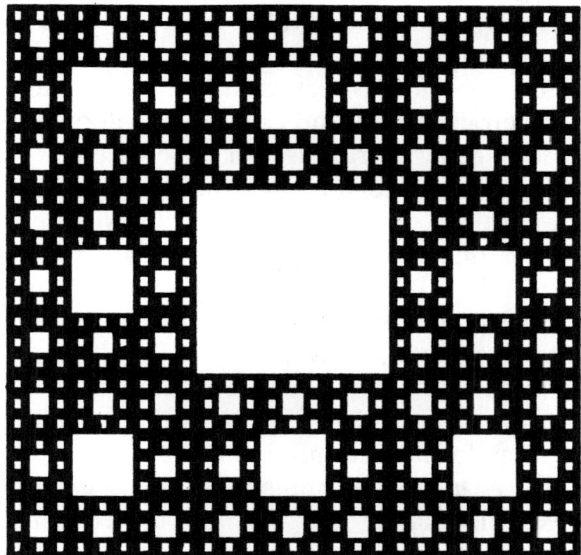

Fig. 1.2.3 The Sierpinski carpet.

Therefore, the number of circles needed to cover the boundary increases as

$$N(a) = 4N(3a) \tag{1.2.1}$$

from which we conclude that the fractal dimension is

$$D = \frac{\ln 4}{\ln 3} = 1.262\ldots \tag{1.2.2}$$

In Problem 1.2 you are asked to calculate the dimension of the quadric Koch island shown in Fig. 1.2.2.

As a final example, we consider the Sierpinski carpet (Fig. 1.2.3), which is a generalization to two-dimensional Euclidean space ($d = 2$) of the Cantor set. In this case the unit square is divided into nine squares, the middle square is removed, and the process is repeated for the remaining eight squares ad infinitum. In this case the topological dimension $d_T = 1$, and we have

$$N(a) = 8N(3a)$$
$$D = \frac{\ln 8}{\ln 3} = 1.8928\ldots \tag{1.2.3}$$

8 SELF-SIMILARITY AND SCALE INVARIANCE

1.3 FRACTALS WITH MORE THAN ONE SCALE FACTOR

Thus far we have considered fractals that are generated by taking an elementary geometrical figure (the generator) and scaling each part of it by $r < 1$. More general fractals can be created by allowing each element of the generator to be scaled independently by a factor r_i, $i = 1, \ldots, N$, where $\Sigma r_i < 1$ for a generator of N components.

The simplest example of this is again a Cantor set in which we introduce two scales r_1 and r_2. As before the Cantor set is constructed recursively by taking a line segment of unit length and dividing it into segments of length r_1, $1 - (r_1 + r_2)$, and r_2 and removing the middle segment. This procedure is illustrated in Fig. 1.3.1 for $r_1 = 0.4$ and $r_2 = 0.25$ [4]. Each of the remaining segments is treated in the same fashion ad infinitum.

As before, let $N(a)$ be the number of line segments of length a needed to cover the Cantor set. We now make explicit use of the self-similarity of the Cantor set by observing that the subsets that lie in the intervals $[0, r_1]$ and $[1 - r_2, 1]$ are identical to the whole set apart from a scale factor.

Imagine that we have covered the whole set with line segments of size a. Now take the left part of the set in the interval $[0, r_1]$ and magnify, or rescale, it by a factor of $1/r_1$. The covering line segments also become rescaled. Because this part of the set is similar to the whole set, on rescaling it actually becomes congruent, and so the number of covering line segments on the left part of the set is $N(a/r_1)$. The same argument can be made for the right part of the set with scale factor $1/r_2$. Therefore, we have

$$N(a) = N\left(\frac{a}{r_1}\right) + N\left(\frac{a}{r_2}\right) \qquad (1.3.1)$$

Because we expect $N(a) \sim a^{-D}$, we find immediately that

$$r_1^D + r_2^D = 1 \qquad (1.3.2)$$

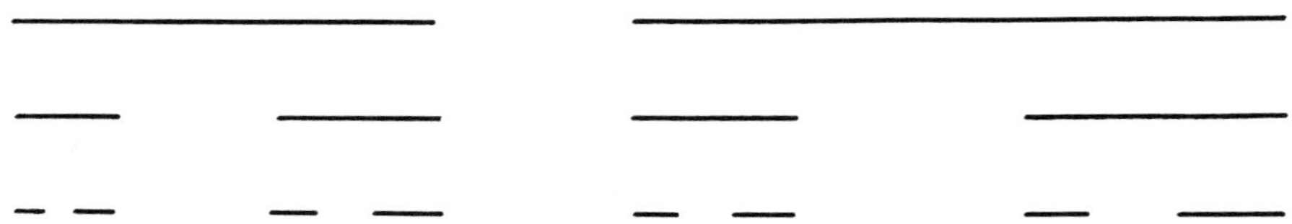

Fig. 1.3.1 Cantor set with two scales, $r_1 = 0.40$ and $r_2 = 0.25$. For further details see Halsey et al. [4].

In the general case where the generator is composed of n elements, the fractal dimension is determined by

$$\sum_{j=1}^{n} r_j^D = 1 \qquad (1.3.3)$$

For the Cantor set of Fig. 1.3.1, D is approximately 0.611.

1.4 DIMENSION OF THE RANDOM WALK

The scale-invariant objects we have studied thus far have been rather artificial, and here we shall consider a problem of a more physical nature, the random walk (RW). Random walks underlie the whole important field of stochastic processes, including Brownian motion and diffusion.

First, we will address the question of the dimension (in the fractal sense) of an RW in a space of Euclidean dimension $d \geq 2$. A random walk is made up of statistically independent steps $\mathbf{r}$ of fixed length a_0, with the direction of each step arbitrary or unrestricted. The mean distance from the starting point can be determined by calculating the mean square of the net displacement

$$\mathbf{R} = \sum_{i=1}^{M} \mathbf{r}_i \qquad (1.4.1)$$

By symmetry, after M steps the mean displacement is $\langle \mathbf{R} \rangle = 0$. Because the individual steps are statistically independent, $\langle \mathbf{r}_i \cdot \mathbf{r}_j \rangle = 0$ if $i \neq j$, and we have

$$\langle |\mathbf{R}|^2 \rangle = \sum_{i=1}^{M} \sum_{j=1}^{M} \langle \mathbf{r}_i \cdot \mathbf{r}_j \rangle = \sum_{i=1}^{M} \langle r_j^2 \rangle = M a_0^2 \qquad (1.4.2)$$

and the root mean square (RMS) displacement is

$$R = \sqrt{\langle |\mathbf{R}|^{2'} \rangle} = \sqrt{M}\, a_0 \qquad (1.4.3)$$

Now let us assume that $M \gg 1$ and that we divide the random walk of M steps into n subwalks of M/n steps. Taking $1 \ll n \ll M$, the preceding argument can also be made for each of the subwalks; the

RMS distance covered by each subwalk is

$$r(n) = \sqrt{n}\, a_0 \qquad (1.4.4)$$

Thus the random walk of M steps each of length a_0 is self-similar to the RW of M/n steps, each of length $\sqrt{n}a_0$.

The statistics of these "coarse-grained" steps is discussed in Sec. 1.6. Here we will assume that we can treat the coarse-grained RW as a new RW with step size $a = \sqrt{n}a_0$. For large n the relative error made in treating the step size as fixed is negligible. Let $N(a)$ be the number of balls of radius a needed to cover the walk. If we take $a = a_0$, then let us assume that in general we need

$$N(a_0) \sim M \qquad (1.4.5)$$

balls to cover the walk. However, if we subdivide the walk into M/n subwalks of n steps, then we can, as we have argued previously, treat this as a new random walk with step size $a = \sqrt{n}a_0$ of M/n steps. By (1.4.5) we then have

$$N(\sqrt{n}\, a_0) \sim \frac{M}{n} \qquad (1.4.6)$$

Putting these two relations together, we have

$$N(a) = nN(\sqrt{n}\, a) \qquad (1.4.7)$$

From which we see that

$$N(a) \sim a^{-2} \qquad (1.4.8)$$

and the fractal dimension of the random walk is $D = 2$. It is interesting to note that the dimension of an unrestricted random walk is independent of the Euclidean dimension of the space in which the walk takes place (so long as $d \geq 2$).

The preceding argument is only heuristic, and a more complete and rigorous treatment will be given in Sec. 1.6.

We really ought to call the assumption of (1.4.5) into doubt. Essentially what it says is that on the smallest scales, where the covering balls are just one step in size, we need one such ball for every step of the RW. This is a good assumption if the RW turns out to be fairly ramified, that is, if $D < d$. For $d = 2$, (1.4.5) is certainly not consistent with $D = 2$, although this turns out to be correct also in $d = 2$. In two dimensions the balls of radius a_0 do overlap frequently. We can, however, argue that even for $d = 2$, (1.4.5) is reasonable. The RW will essentially cover an area $\sim R^2$, and so the number of 2-balls needed to cover this area is

$$N(a_0) \sim \frac{R_2}{a_0^2} \sim M \qquad (1.4.9)$$

where we have used (1.4.3). In $d = 1$, the RW completely covers the line and so $D = 1$ as well.

1.5 ANATOMY OF A RENORMALIZATION GROUP TRANSFORMATION

Now that the ideas of scale invariance and self-similarity as applied to the RW have been discussed, we are ready to embark on our first RG calculation. Before we do, it is a good idea to try and formulate a general notion as to just what is an RG.

Although RG transformations on parameter space do not actually comprise a mathematical group, there is very often a kind of invariance principle at work. For example, we shall see in Chap. 4, when RGs for equilibrium critical phenomena are constructed, the partition function is invariant under the RG transformation. In the case of the RW to be considered presently, the RG is constructed in such a way as to keep the RMS length of the random walk invariant; in the case of the self-avoiding walk, Sec. 1.7, the RG keeps the generating function invariant.

RG calculations generally are composed of two steps. In the first step, which is sometimes called "decimation" or "coarse graining," one "averages out" a subset of the degrees of freedom of the system. Usually this set includes the degrees of freedom which vary on very short scales, the motivation being that at the critical point the behavior of the system is dominated by fluctuations on arbitrarily large scales.

The second part of the RG transformation, called rescaling, is to redefine the unit of length. The scale factor, which we shall denote by b, is the ratio of the coarse-grained unit of length to the original unit of length. In the first step, coarse graining has smeared out everything so that the smallest scale on which the system changes has been increased. In the rescaling part of the RG, the unit of length is redefined so that, in terms of the new length scale, this smallest distance is restored to its original value.

As a result of these two operations, the parameters of the system h_k, $k = 1, 2, \ldots$, for example, the coupling constants in a Hamiltonian, will be renormalized. The equations relating the new, renormalized, parameters h'_k to the original parameters h_k are called the RG equations, which we may formally express as

$$h'_k = R_k(\{h_j\}) \qquad (1.5.1)$$

Imagine for the moment that for an initial choice of parameters $h_j^{(c)}$ the renormalized parameters tend, under repeated applications of the RG, to a fixed point h_k^*,

$$\lim_{n \to \infty} h_k^{(n)} = R_k\big(R_k(\ldots n \text{ renormalizations } R_k(h_j^{(c)})\ldots)\big) \to h_k^* \qquad (1.5.2)$$

Equation (1.5.2) implies that in the limit of many renormalizations, corresponding to very large length scales, the system and its renormalized copy are identical. If we now recall that the unit of length has been rescaled by a factor b in the RG, we see that the system is in fact self-similar at a fixed point of the RG. The special set of values $h_j^{(c)}$ for which (1.5.2) holds defines the "critical surface" in parameter space, which includes, but should not be identified with, the fixed point. Another way of putting this is the critical surface is the "basin of attraction" of the fixed point of the RG transformation.

The scaling behavior characteristic of systems near a critical point follows in a very natural way from the properties of the RG equations in the neighborhood of a fixed point. Near the fixed point the RG equations can be linearized. Writing $h_k = h_k^* + \varepsilon_k$ with $|\varepsilon_k| \ll 1$,

$$\varepsilon'_k = \sum_j \frac{\partial h'_k}{\partial h_j} \varepsilon_j \qquad (1.5.3)$$

The general solution of this set of linear equations can be expressed in terms of the right eigenvectors of the linear operator $\partial h'_k / \partial h_j = R_{kj}$:

$$\sum_j R_{ij} \phi_j^{(n)} = \lambda^{(n)} \phi_i^n \qquad (1.5.4)$$

1.5 ANATOMY OF A RENORMALIZATION GROUP TRANSFORMATION

Generally, the operator R_{kj} is not symmetric, and a complete solution of the linearized RG equations also requires the left eigenvectors $\tilde{\phi}_i^{(n)}$, which obey the relation

$$\sum_k \tilde{\phi}_k^{(n)} R_{kj} = \lambda^{(n)} \tilde{\phi}_i^{(n)} \qquad (1.5.5)$$

The left and right eigenvectors are orthonormal

$$\sum_k \tilde{\phi}_k^{(n)} \phi_k^{(n')} = \delta_{nn'} \qquad (1.5.6)$$

and complete

$$\sum_n \tilde{\phi}_j^{(n)} \phi_k^{(n)} = \delta_{jk} \qquad (1.5.7)$$

If the initial point in parameter space is close to the fixed point, we can use (1.5.7) to write

$$\varepsilon_i = \sum_i u_n \phi_i^{(n)} \qquad (1.5.8)$$

where

$$u_n = \sum_k \tilde{\phi}_k^{(n)} \varepsilon_k \qquad (1.5.9)$$

is the amplitude of the scaling field $\tilde{\phi}^{(n)}$ at the point $h_k^* + \varepsilon_k$. If the eigenvalue $\lambda^{(n)}$ associated with the eigenvector $\phi_k^{(n)}$ satisfies $|\lambda^{(n)}| < 1$, then on repeated application of the RG, u_n tends to 0 and we say that $\phi^{(n)}$ is "irrelevant." On the other hand, if $|\lambda^{(n)}| > 1$, then u_n increases exponentially, and $\phi^{(n)}$ is said to be "relevant." If $|\lambda^{(n)}| = 1$, then $\phi^{(n)}$ is "marginal."

In order for the trajectory in parameter space to reach the fixed point, the amplitudes of all the relevant scaling fields must vanish. If there are r relevant scaling fields, this places r constraints on the initial set of parameters, and unless there are as many physically adjustable parameters, the fixed point will be inaccessible. Or, to put it another way, physically realizable Hamiltonians will comprise a subspace in parameter space which is generally much smaller in dimension than parameter space itself. The intersection of the subspace of physical Hamiltonians and the critical surface defines the set of physically accessible critical points. This set can be a single point, as

14   SELF-SIMILARITY AND SCALE INVARIANCE

in the case of an ordinary ferromagnet, a line as one finds in the superfluid transition in He^4, or possibly a set of even higher dimension.

To put this general discussion on a more concrete basis, consider the critical properties of a magnetic system. The parameter space in this case will consist, first of all, of the temperature and external magnetic field, both of which are "physically adjustable parameters" in the sense of the preceding paragraph. In addition, the Hamiltonian will depend on the exchange couplings between spins. For the sake of argument, let us assume that the original Hamiltonian contains a coupling between pairs of nearest-neighbor spins only. Under the action of the RG, couplings between second-neighbor spins and four-spin interactions may emerge. It is usually the case that an RG transformation produces a proliferation of new couplings, and one is generally forced to truncate parameter space at some point. For a more complete illustration of this point, see Chap. 6.

If any of these new couplings are relevant with respect to the fixed point of interest, then, because we have no control over their amplitude, the fixed point will be unstable.[†]

1.6 RENORMALIZATION GROUP ANALYSIS OF THE RANDOM WALK

The usefulness of the RG approach is well illustrated by the example of the RW. In order to make the calculation easier, we relax the constraint that each step in the RW is of length a_0 and let the length of the step be governed by the Gaussian probability distribution

$$p(\mathbf{r}) = \left(2\pi\sigma_0^2\right)^{-d/2} \exp\left[-\frac{|\mathbf{r}|^2}{2\sigma_0^2}\right] \qquad (1.6.1)$$

The direction of each step is still unrestricted. Typically, in RG calculations one is not interested in the microscopic behavior of the system, but rather its large-scale properties. The natural thing to try then is to "average out" the small-scale details. This process is also

---
[†]Of course properly speaking all nontrivial fixed points are unstable because there must be at least one relevant scaling field to yield critical behavior. However, within the context of the literature on the RG, we call a fixed point stable if there are as many physically adjustable parameters as there are relevant scaling fields. In many cases there is just a single relevant scaling field.

1.6 RENORMALIZATION GROUP ANALYSIS OF THE RANDOM WALK

sometimes referred to as "coarse graining." For the random walk, coarse graining, can be accomplished by constructing the probability distribution $P(\mathbf{r}')$, where $\mathbf{r}'$ is a new random variable related to the old variables by

$$\mathbf{r}' = \sum_{i=1}^{n} \mathbf{r}_i \tag{1.6.2}$$

Note that $\mathbf{r}'$ is a "rescaled step" in a new random walk. The probability distribution for $\mathbf{r}'$ is given by

$$P(\mathbf{r}') = \int d\mathbf{r}_1 \cdots d\mathbf{r}_n \, \delta\left(\mathbf{r}' - \sum_{i=1}^{n} \mathbf{r}_i\right) p(\mathbf{r}_1) \cdots p(\mathbf{r}_n) \tag{1.6.3}$$

If the Dirac delta function is represented by its Fourier transform, the integrals over the old variables can be performed, giving

$$P(\mathbf{r}) = \int \frac{d^d k}{(2\pi)^d} \exp\left[-n|\mathbf{k}|^2 \frac{\sigma_0^2}{2} - i\mathbf{k} \cdot \mathbf{r}\right] \tag{1.6.4}$$

The integral over k can also be performed, which yields for the coarse-grained probability distribution

$$P(\mathbf{r}') = \frac{1}{(2\pi n \sigma_0^2)^{d/2}} \exp\left[-\frac{|\mathbf{r}'|^2}{n\sigma_0^2}\right] \tag{1.6.5}$$

Note that this distribution is of the same form as the original; the only difference is that the parameters (in this case there is only σ) are changed. The coarse-grained σ is given by

$$\sigma' = \sqrt{n}\, \sigma_0 \tag{1.6.6}$$

The second step in the RG procedure, called "rescaling," is to redefine the unit of length. We choose to construct the RG in such a way as to preserve the width of the Gaussian. This requires that we define rescaled random variables $\mathbf{r}''$:

$$\mathbf{r}' = \sqrt{n}\, \mathbf{r}'' \tag{1.6.7}$$

The rescaled probability distribution $P(r'')$ is related to $P(r')$ by

$$P(\mathbf{r}')\,d^d r' = P(\mathbf{r}'')\,d^d r'' \tag{1.6.8}$$

which gives for the renormalized probability distribution

$$P(\mathbf{r}'') = \frac{1}{(2\pi\sigma_0^2)^{d/2}} \exp\left[-\frac{(\mathbf{r}'')}{2\sigma_0^2}\right] \tag{1.6.9}$$

(Note that $d^d r' = n^{d/2} d^d r''$.) The renormalized probability distribution is identical in form to the original, and the full renormalization group equation for σ is

$$\sigma^{(R)} = \sigma_0 \tag{1.6.10}$$

Because σ remains unchanged by the RG procedure, it is a "marginal" variable.

A consequence of the identity of the original and renormalized distributions is that there is no way of distinguishing, by any of its statistical properties, a random walk from its renormalized copy, which is illustrated in Fig. 1.6.1. This is what is meant by self-similarity in a probabilistic context.

Returning to (1.6.6), consider the RMS distance $R(M)$ covered by a walk of M steps. $R(M)$ must be a function of M and the typical step size σ. Because σ is the only quantity in the problem with the dimension of length, we can assume that R is given by

$$R(M,\sigma) \sim \sigma M^\nu \tag{1.6.11}$$

Note that (1.6.11), like (1.0.1), is a scaling law and embodies the idea that, at least in a statistical sense, the trajectory of a random walk is self-similar. From (1.6.6) and the observation that the coarse-grained random walk is of $M' = M/n$ steps, we must have

$$\sigma M^\nu = \sqrt{n}\,\sigma \left(\frac{M}{n}\right)^\nu \tag{1.6.12}$$

from which we conclude that the scaling exponent ν must be $1/2$. The

1.6 RENORMALIZATION GROUP ANALYSIS OF THE RANDOM WALK

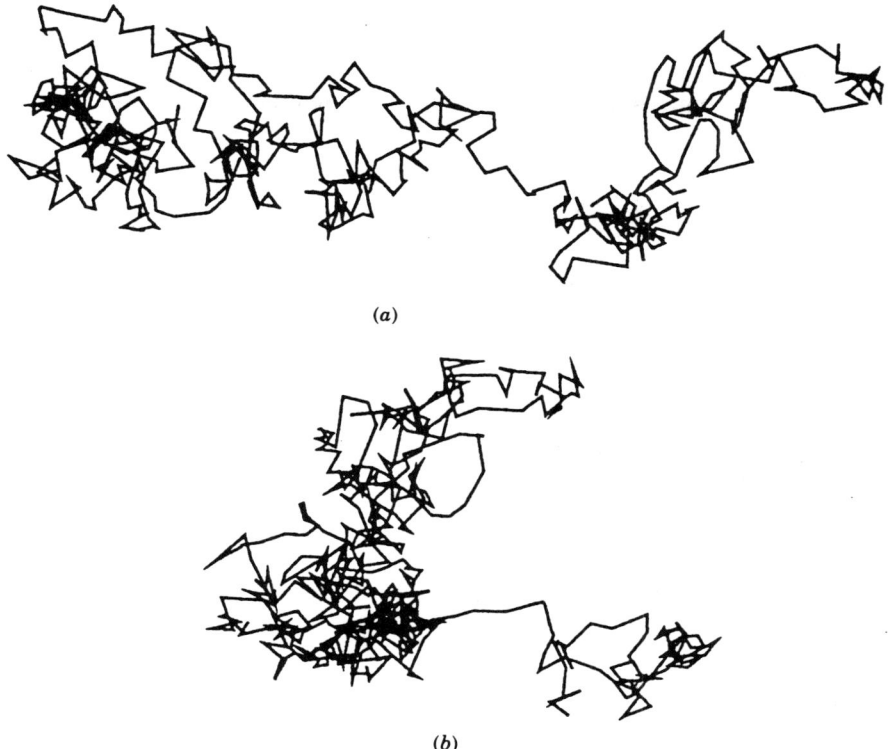

Fig. 1.6.1 (*a*) Random walk. (*b*) Renormalized copy of random walk shown in (*a*).

relationship between the scaling exponent ν and the fractal dimension of the random walk is evidently

$$\nu = \frac{1}{D} \tag{1.6.13}$$

This result is quite general (Prob. 1.4), although for other kinds of random walks, for example, the self-avoiding random walk, which does not permit self-intersections, the exponent ν will take on a different value.

The RG equation (1.6.10) given above for the random walk is typical of "Gaussian" models in that we have constructed it in such a way that the coefficient of the Gaussian term σ remains invariant under the RG transformation. The model is so simple, however, that most of the details of a general RG outlined in Sec. 1.5 are missing. Therefore, let us generalize the simple Gaussian model (1.6.1) to include a systematic drift in the direction $\mathbf{r}_0$, and take for $p(\mathbf{r})$:

$$p(\mathbf{r}) = \frac{1}{(2\pi\sigma^2)^{d/2}} \exp\left[\frac{(\mathbf{r} - \mathbf{r}_0)^2}{2\sigma^2}\right] \qquad (1.6.14)$$

Coarse graining just as in (1.6.2) to (1.6.5), we find

$$p(\mathbf{r}') = \frac{1}{(2\pi n\sigma^2)^{d/2}} \exp\left[-\frac{(\mathbf{r} - n\mathbf{r}_0)^2}{2n\sigma^2}\right] \qquad (1.6.15)$$

The coarse-grained parameters σ' and $\mathbf{r}_0'$ are therefore

$$\sigma' = \sqrt{n}\,\sigma \qquad (1.6.16)$$

$$\mathbf{r}_0' = n\mathbf{r}_0 \qquad (1.6.17)$$

If we now rescale as in (1.6.7), the RG equations are

$$\sigma^{(R)} = \sigma \qquad (1.6.18)$$

$$\mathbf{r}_0^{(R)} = \sqrt{n}\,\mathbf{r}_0 \qquad (1.6.19)$$

Again by construction σ remains unchanged under renormalization, whereas $\mathbf{r}$ scales with n as $\sqrt{n}$. In the language of the RG, the systematic drift $\mathbf{r}_0$ is a relevant variable and the scaling exponent for $\mathbf{r}$ is $1/2$. For a given value of σ, the RG equations (1.6.18) and (1.6.19) have fixed points $(\sigma, \mathbf{r} = 0)$ and $(\sigma, \mathbf{r}_0 = \infty)$. The first fixed point, which we call the Brownian fixed point, corresponds to the pure or unbiased random walk. To interpret the second fixed point, let us calculate the dimension of the walk. The RMS distance covered by a walk of M steps governed by (1.6.14) is

$$\sqrt{\langle |R|^2 \rangle} = \sqrt{M^2 r_0^2 + M\sigma_0^2} \qquad (1.6.20)$$

For M large the first term dominates and the number of spheres of

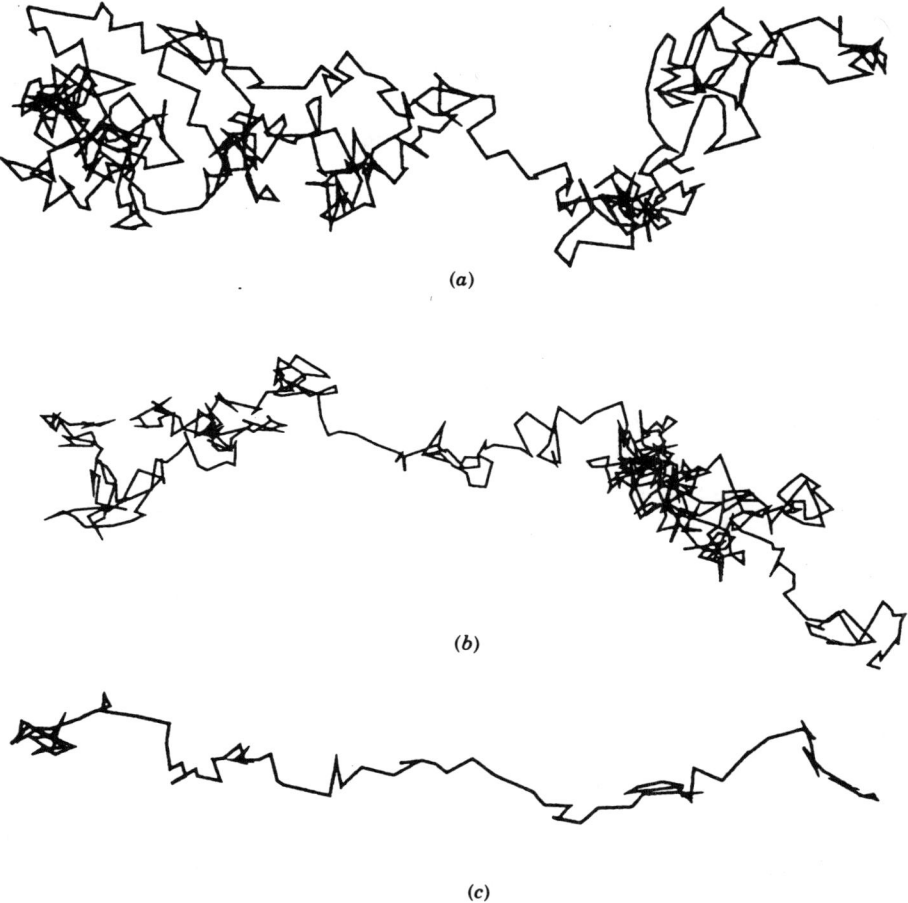

Fig. 1.6.2 Random walk with a drift term. (*a*) A small segment of the walk is shown and should be compared with the pure random walk, Fig. 1.6.1*a*. (*b*) A larger scale view of the same walk shows one-dimensional behavior beginning to emerge. (*c*) At even larger scales the random component of the walk is much less noticeable and its ultimate one-dimensional nature emerges.

size *a* needed to cover the walk is just

$$N(a) \sim \frac{1}{a} \tag{1.6.21}$$

and the dimension of the walk is $D = 1$.

The Brownian fixed point is unstable with respect to inclusion of a drift term in the distribution function. For $|\mathbf{r}|/\sigma \ll 1$ the walk looks Brownian, or two dimensional, on small scales, but as one examines the walk on larger and larger scales it crosses over from being two dimensional to one dimensional. This is illustrated in Fig. 1.6.2.

1.7 UNIVERSALITY CLASS OF THE RANDOM WALK

Here we consider a simple extension of the probability distribution (1.6.1) to include any distribution that is a function of the magnitude of **r** only

$$p(\mathbf{r}) = p(|\mathbf{r}|) \tag{1.7.1}$$

In addition to the Gaussian distribution considered in the previous section, the family of distributions we are now considering also includes the case of a random walk of fixed step size a, for which the distribution is (in three dimensions)

$$p(\mathbf{r}) = \frac{1}{4\pi a^2}\delta(|\mathbf{r}| - a) \tag{1.7.2}$$

The only restriction, besides (1.7.1), which we place on the probability distribution is that all the moments of the distribution be finite.

We again want to construct the probability distribution for the coarse-grained step

$$\mathbf{r}' = \sum_{i=1}^{n} \mathbf{r}_i \tag{1.7.3}$$

which, with the help of the Dirac delta function, can be written as

$$P(\mathbf{r}) = \int d^d r_1, \ldots, d^d r_n\, \delta\!\left(\mathbf{r} - \sum_{i=1}^{n} \mathbf{r}_i\right) p(\mathbf{r}_1), \ldots, p(\mathbf{r}_n) \tag{1.7.4}$$

Introducing the Fourier transform of the delta function, (1.7.4) becomes

$$P(\mathbf{r}') = \int \frac{d^d k}{(2\pi)^d} e^{-i\mathbf{k}\cdot\mathbf{r}'}\left[\langle e^{i\mathbf{k}\cdot\mathbf{r}}\rangle\right]^n \tag{1.7.5}$$

1.7 UNIVERSALITY CLASS OF THE RANDOM WALK

where

$$\langle e^{i\mathbf{k}\cdot\mathbf{r}}\rangle = \int d^d r\, e^{i\mathbf{k}\cdot\mathbf{r}} p(\mathbf{r})$$
$$\equiv g(\mathbf{k}) \qquad (1.7.6)$$

The function $g(\mathbf{k})$ is sometimes called the characteristic function of the probability distribution $p(\mathbf{r})$. It can be thought of as a generating function for the moments of $p(\mathbf{r})$ because any moment of p can be calculated by taking the appropriate derivatives of g. In order to evaluate $g(\mathbf{k})$, we invoke the cumulant expansion (see App. A), which tells us that

$$\langle e^{i\mathbf{k}\cdot\mathbf{r}}\rangle = \exp\left(i\mathbf{k}\cdot\langle\mathbf{r}\rangle - \tfrac{1}{2}\sum k_\mu k_\nu (\langle r_\mu r_\nu\rangle - \langle r_\mu\rangle\langle r_\nu\rangle) + \cdots\right) \qquad (1.7.7)$$

By symmetry $\langle\mathbf{r}\rangle = 0$, and defining the matrix of second moments,

$$\sigma^2_{\mu\nu} = \langle r_\mu r_\nu\rangle - \langle r_\mu\rangle\langle r_\nu\rangle = \sigma_0^2 \delta_{\mu\nu} \qquad (1.7.8)$$

the coarse-gained distribution now becomes

$$P(\mathbf{r}') = \int \frac{d^d k}{(2\pi)^d} e^{-i\mathbf{k}\cdot\mathbf{r}'} e^{-(n/2)\sigma_0^2 k^2 + \cdots} \qquad (1.7.9)$$

The contribution of the higher cumulants to the integral in (1.7.9) can be neglected for large n. To see this, let us assume that the Gaussian part of the exponent dominates the integrand. Then the integrand will be appreciable only for

$$k \sim \frac{1}{\sqrt{n\sigma_0^2}} \qquad (1.7.10)$$

In the mth cumulant there are m factors of k, and so the mth cumulant is of order $n^{1-m/2}$ and can be neglected for large n and $m > 2$. The remaining integration is Gaussian and we finally find for the coarse-grained distribution

$$P(\mathbf{r}') = \frac{1}{(2\pi n\sigma_0)^{d/2}} \exp\left(-\frac{1}{2}\frac{|\mathbf{r}'|^2}{n\sigma_0^2}\right) \qquad (1.7.11)$$

We have arrived back at the starting point of the previous section, with a Gaussian distribution, and we can now simply follow the same steps that led to the scaling behavior (1.6.12).

The observation that coarse-grained random variables are governed by Gaussian distributions was first proved by Bernoulli and is usually referred to in probability theory as the "law of large numbers." In the context of RG theory, we would put things a little differently and say that all probability distributions of the form (1.7.1) belong to the same universality class. By this we mean that in the limit of very long random walks (very many independent statistical events), all random walks of this class have the same statistical properties. In particular, the RMS distance R covered by such a walk will scale with N as

$$R \sim N^{1/2} \qquad (1.7.12)$$

1.8 REAL-SPACE RENORMALIZATION GROUP FOR THE SELF-AVOIDING RANDOM WALK[†]

The unrestricted random walk is physically interesting as a microscopic model of the diffusion of noninteracting particles and Brownian motion. A type of restricted RW called the self-avoiding random walk (SAW) is also interesting physically as a model of the behavior of long chain molecules.

The probability distribution (1.6.1) is not a very realistic model of a real linear polymer, and so it is interesting to explore models that embody some of the properties of real polymers. A very detailed account of these various models can be found in the text by Freed [3].

A SAW is any connected path on the lattice that does not intersect itself, Fig. 1.8.1. This restriction models in a simple way the short-range repulsive force between monomers in a long polymer, and therefore the SAW has been considered a good statistical model for the config-

---

[†]In the literature one generally distinguishes between RG transformations formulated in terms of the degrees of freedom at each point in real space, and those formulated in terms of the degrees of freedom in momentum space. The former is called the real-space renormalization group (RSRG).

1.8 REAL-SPACE RENORMALIZATION GROUP FOR THE SAW

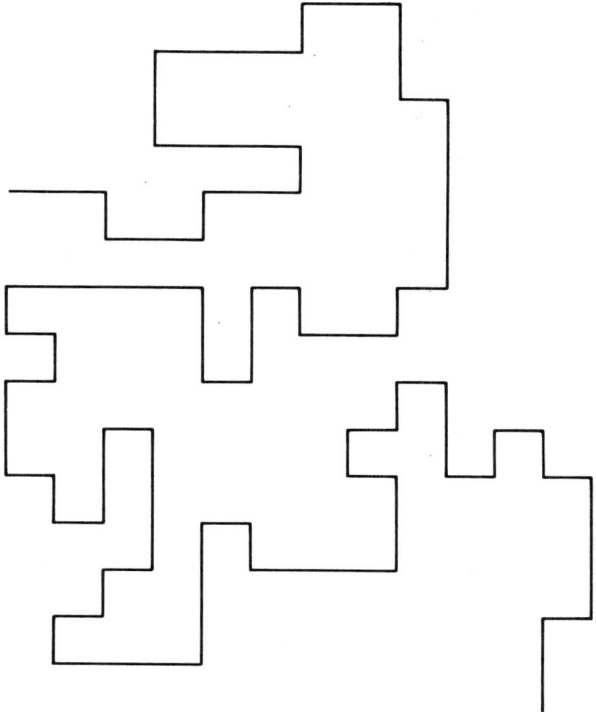

Fig. 1.8.1 A SAW. Note the absence of self-intersections.

urations of this important class of molecules. Just as for the unrestricted random walk, we expect that the characteristic length of a SAW will depend on the number of steps (monomers) according to a scaling law of the form

$$R \sim N^\nu \qquad (1.8.1)$$

Unfortunately, because of the complicated geometrical constraint on self-intersection, we cannot construct the renormalized probability distribution for the SAW in closed form as we did for the RW.

Define $M_n(\mathbf{R})$ to be the number of SAWs of n steps that connect the origin to the point $\mathbf{R}$. $M_n(\mathbf{R})$ is a rapidly increasing function of n, and for fixed n these walks will cover a typical distance R_n as given in (1.8.1). Then the mean square displacement for all SAWs of n steps is

$$\langle R_n^2 \rangle = \frac{\sum_{\mathbf{R}} |\mathbf{R}|^2 M_n(\mathbf{R})}{\sum_{\mathbf{R}} M_n(\mathbf{R})} \qquad (1.8.2)$$

One possible approach to solving this problem is to explicitly enumerate all the walks of n steps and evaluate (1.8.2) directly, but this program can only be carried out for $n < 100$ even on large computers. We can make some progress if we borrow a trick from statistical mechanics and introduce a "grand canonical" ensemble in which walks of all values of n are included.

In going from the canonical (fixed number of particles) to the grand canonical (variable number of particles) ensemble, one takes the canonical partition function Z_n, weights it with z^n, where $z = e^{\beta \mu}$ is the fugacity (μ is the chemical potential), and sums over n to give the grand canonical partition function

$$Q(z) = \sum_n z^n Z_n \qquad (1.8.3)$$

In the same way let us assign to each step in the walk a weight K so that walks of n steps receive a weight K^n and form the "generating function" $G(K)$, where

$$G(K) = \sum_n K^n \sum_R M_n(\mathbf{R}) \qquad (1.8.4)$$

The generating function allows us to study the whole range of SAWs rather than SAWs with a fixed number of steps. The correlation length $\xi(K)$ is defined in a way analogous to (1.8.2)

$$\xi^2(K) = \frac{\sum_R \sum_n |\mathbf{R}|^2 K^n M_n(\mathbf{R})}{G(K)} \qquad (1.8.5)$$

The correlation length may be thought of as the typical size of a walk for given fugacity K. The number of SAWs of n steps, $M_n = \sum_\mathbf{R} M_n(\mathbf{R})$, is a rapidly increasing function of n, whereas, for $K < 1$, K^n is rapidly decreasing. Assuming M_n does not increase faster than exponentially, the product of these two factors in the sum (1.8.4) will therefore have a sharp maximum for a particular value $n = n_0(K)$. By varying K we can therefore study the entire family of SAWs. For small K only SAWs of finite length contribute to the generating function. As K increases so does $n_0(K)$, and finally as K approaches the critical value K_c, $n_0(K)$ diverges and the generating function exhibits singular behavior. It is the region $K \lesssim K_c$ that will yield information about the incipient infinite SAW.

1.8 REAL-SPACE RENORMALIZATION GROUP FOR THE SAW

In order to clearly establish the connection between the critical behavior of the generating function and the scaling law (1.8.1), let us assume that for large n, M_n is given by (see [6] for details)

$$M_n \sim K_c^{-n} n^{\gamma-1} \tag{1.8.6}$$

Each term in the sum (1.8.5) is then

$$K^n M_n = \exp\{n \ln K/K_c + (\gamma - 1)\ln n\} \tag{1.8.7}$$

The value of $n_0(K)$ is determined by the condition that the exponent in (1.8.7) be maximal. Differentiating with respect to n, we see that $n_0(K)$ is given by

$$n_0(K) = \frac{\gamma - 1}{\ln(K_c/K)} \tag{1.8.8}$$

As $K \to K_c$, $n_0(K)$ does indeed diverge as

$$\lim_{K \to K_c} n_0(K) \to (\gamma - 1)\left(\frac{K_c - K}{K_c}\right)^{-1} \tag{1.8.9}$$

The peak in the function $K^n M_n$ is very sharp, and the sum (1.8.5) will therefore be dominated by terms with $n \cong n_0$. Assuming that R_n scales as in (1.8.1), we have

$$\lim_{K \to K_c} \xi(K) \to n_0^\nu(K) \sim \left(\frac{K_c - K}{K_c}\right)^{-\nu} \tag{1.8.10}$$

We see from (1.8.10) that the class of incipient infinite SAWs corresponds to the critical value of the weight factor $K = K_c$, and by (1.8.10) as the critical value is approached the typical length of a walk diverges with the critical exponent ν defined in (1.8.1). This is the connection we were looking for which allows us to transcribe the scaling law $R \sim N^\nu$ into the language of critical behavior and the RG. The problem now is to formulate the RG for the generating function $G(K)$.

In (1.8.4) the weight associated with each SAW depends only on the number of nearest-neighbor steps or "bonds." It is, however, easy to imagine generalizing the generating function by introducing a weight K_2 for each pair of next-neighbor sites visited by the walk, K_3 for third neighbors, and so on. In this way the generating function can be formulated in such a way as to contain whatever statistical information about the SAWs one is interested in. Let us denote this general set of parameters by $\{K_i\}$ and the generalized statistical weight function by $W(\Gamma, \{K_i\})$, where Γ stands for a particular SAW (or "graph").

The object of the RG transformation is to map all SAWs Γ on the original lattice which connect the origin to a point **R**, to a single SAW Γ' on the coarse-grained lattice. In order to preserve the statistical properties of the SAW, the statistical weight associated with Γ' must equal the sum of the statistical weights of the walks that map into Γ',

$$W(\Gamma', \{K_i'\}) = \sum_{\Gamma \in \Gamma'} W(\Gamma, \{K_i\}) \qquad (1.8.11)$$

The symbol $\Gamma \in \Gamma'$ indicates that the sum is over all SAWs Γ that map under the RG into Γ'.

Following Shapiro [10], we construct an RG that maps a segment of a SAW with fugacity K within a region of size b on the original lattice to a single step in a SAW with fugacity K' on the coarse-grained lattice. It then follows that the renormalized correlation length $\xi'(K')$ is related to the original correlation length by

$$\xi(K) = b\xi'(K') \qquad (1.8.12)$$

where b, the scale factor, is the ratio between the coarse-grained lattice spacing and the original lattice spacing. If K is the only relevant parameter near the fixed point, then $\xi'(K)$ is the same function of K as $\xi(K)$.

Consider the cell of bonds shown in Fig. 1.8.2 for which $b = 2$ [2, 9, 11]. A vertical step in the renormalized SAW (with renormalized fugacity K') is given by all SAWs in the cell that start at the lower-left corner of the cell and exit the top edge. Similarly, all SAWs in the cell that start at the lower-left corner and leave the cell by the right edge map into a single horizontal step on the coarse-grained lattice. By symmetry the sum over the weights of all SAWs in either case is the same, so it is sufficient to consider only the SAWs that map into a vertical step.

Note that not all SAWs on the original lattice map unambiguously to a SAW on the coarse-grained lattice if we use the simple one-parameter weight function of the corner rule. Figure 1.8.3 shows two SAWs on the square lattice. In Fig. 1.8.3a the corner rule with the nearest-neighbor fugacity K gives the proper renormalized weight on

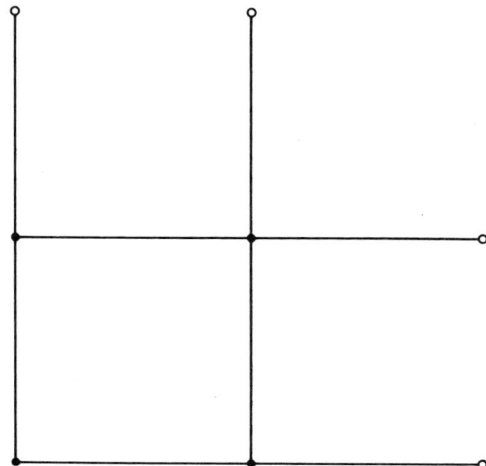

Fig. 1.8.2 Cell of bonds ($b = 2$) for the RSRG for the self-avoiding walk.

the coarse-grained lattice, whereas in Fig. 1.8.3b the corner rule does not give the weight associated with the renormalized walk. To obtain the proper result, one needs to introduce weights associated with second- and third-neighbor sites. This approach has been explored by Napiorkowski, Hauge, and Hemmer [7], and a similar calculation for the Ising model is described in detail in Chap. 6.

Figure 1.8.4 shows the SAWs that contribute to a vertical step in the renormalized SAW together with their proper statistical weights. The renormalization transformation is therefore

$$K' = K^4 + 2K^3 + K^2 \qquad (1.8.13)$$

The fixed points of this transformation are the two trivial fixed points $K^* = 0$ (finite SAWs), $K^* = \infty$ (lattice filling SAW), and the nontrivial fixed point $K^* = K_c = 0.4655\ldots$, which is the fixed point of interest and corresponds to the incipient infinite SAW.

In order to determine the exponent ν, we must examine the renormalization equations in the neighborhood of the nontrivial fixed point K_c. For K slightly smaller than K_c, the renormalization transformation will drive K to 0, whereas for K greater than K_c, repeated applications of (1.8.13) drive K to ∞. For $K < K_c$, we assume that $\xi(K)$ diverges as $K \to K_c$ according to

$$\xi(K) \sim |K - K_c|^{-\nu} \qquad (1.8.14)$$

28 SELF-SIMILARITY AND SCALE INVARIANCE

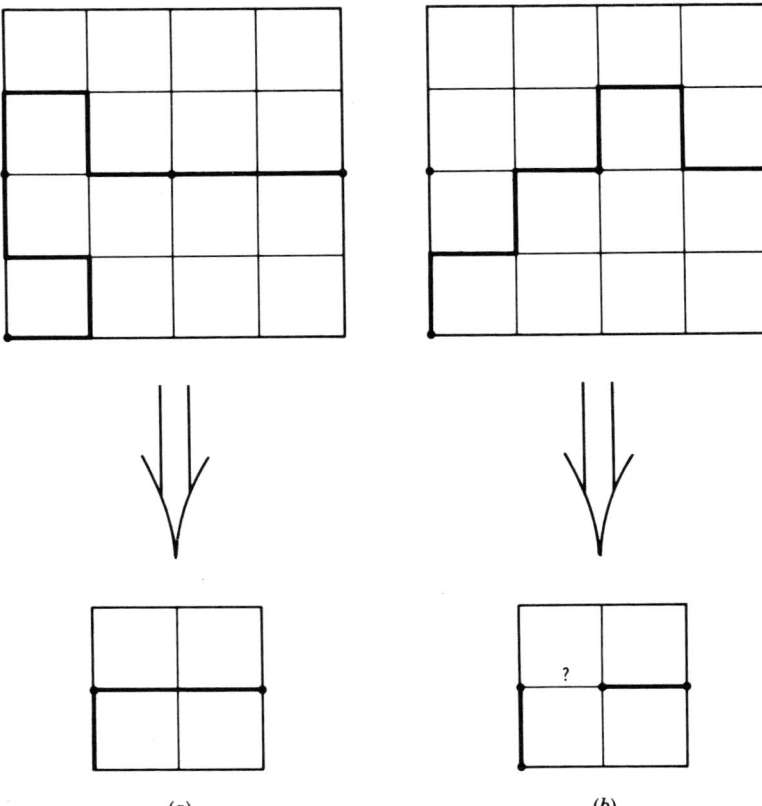

Fig. 1.8.3 (*a*) A SAW on the square lattice that maps properly under the corner rule to a SAW on the coarse-grained lattice. (*b*) A SAW that does not transform properly under the simple corner rule with nearest-neighbor fugacity only. After Redner and Reynolds [9].

Because K' is an analytical function of K, we have

$$K' - K_c = \left| \frac{\partial K'}{\partial K}(K_c) \right| (K - K_c) \qquad (1.8.15)$$

Putting this together with (1.8.12) gives

$$|K - K_c|^{-\nu} = \left[\left| \frac{\partial K'}{\partial K}(K_c) \right| \right]^{-\nu} |K - K_c|^{-\nu} \qquad (1.8.16)$$

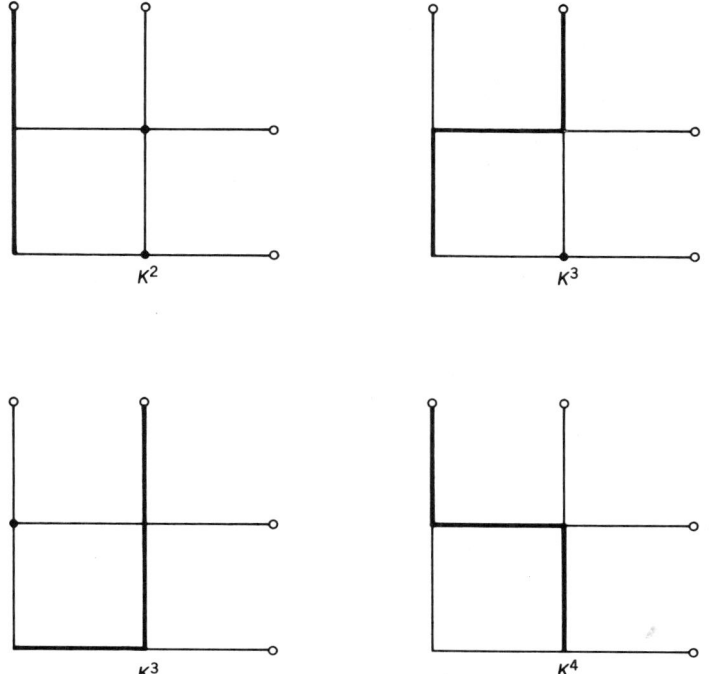

Fig. 1.8.4 SAWs in the cell of Fig. 1.8.2 that contribute to the renormalized fugacity.

Solving for ν, we find

$$\nu = \frac{\ln b}{\ln \left| \frac{\partial K'}{\partial K}(Kc) \right|} = 0.715 \qquad (1.8.17)$$

The values $K_c = 0.4655$ and $\nu = 0.715$ should be compared with the best numerical estimates [10]

$$\begin{aligned} K_c &= 0.379 \\ \nu &= 0.74\text{--}0.75 \end{aligned} \qquad (1.8.18)$$

Typically, the accuracy of RGs formulated on small cells is difficult to establish a priori, but can be systematically improved by considering larger cells (see Prob. 1.5) [9].

Note that because ν is larger for the SAW than for the RW, a SAW of N steps has a greater RMS size than the RW of the same number

30 SELF-SIMILARITY AND SCALE INVARIANCE

of steps. This is reasonable as the SAW cannot double-back on itself. Also the fractal dimension $d = 1/\nu = 1.34$ is smaller than that of the RW, reflecting the greater ramification of the SAW.

When the scaling properties of two RWs are the same, as in the examples considered in Sec. 1.7, we say they belong to the same universality class. On a microscopic level the two types of walks may look very different—for example, one may be made up of steps with fixed length, whereas the other may have steps of all possible length—and yet on a sufficiently large scale they are indistinguishable.

On the other hand, if we consider the SAW, because it scales differently from an RW, as reflected in the exponent ν, we say that the SAW belongs to a different universality class than the RW.

PROBLEMS

1.1 Calculate the dimension of the Cantor set shown in Fig. 1.P.1.

1.2 Calculate the dimension of the quadric Koch island shown in Fig. 1.2.2.

1.3 As an alternate definition of the fractal dimension, consider a hypersphere of radius R, and let the number of points of a set within R be $N(R)$. Then $N(R)$ will increase as

$$N(R) \sim R^D$$

Use this definition to derive the relationship between the exponent ν and the fractal dimension (1.0.1).

1.4 As a variation of the RG transformation, (1.6.16) to (1.6.19), choose the scale factor so that the RMS distance covered by the walk is invariant (instead of the coefficient of the Gaussian

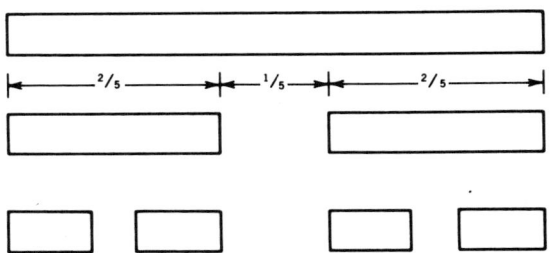

Fig. 1.P.1 Recursive procedure for Cantor set for Prob. 1.1.

Fig. 1.P.2 A Siepinski carpet [5].

term). Rederive the RG equations. In this formulation are σ and **r** relevant, irrelevant, or marginal?

1.5 Find the RG transformation for SAWs on a 3×3 cell. Write a simple computer program to find the fixed point and evaluate ν.

1.6 Calculate the fractal dimension of the Sierpinski carpet shown in Fig. 1.P.2.

1.7 What is the dimension of the two-scale Cantor set if $r_1 = 0.5$ and $r_2 = 0.25$?

1.8 Calculate $g(\mathbf{k})$ explicitly for the distribution (1.7.2). Find the first two nonvanishing moments and estimate the value of n for which the Gaussian approximation is valid.

1.9 Calculate the fractal dimension of the Sierpinski sponge shown in Fig. 1.P.3 [1]. Note that in each iteration a six-pointed star, something like a child's "jacks," is removed from each elementary cube. What is the topological dimension of the sponge?

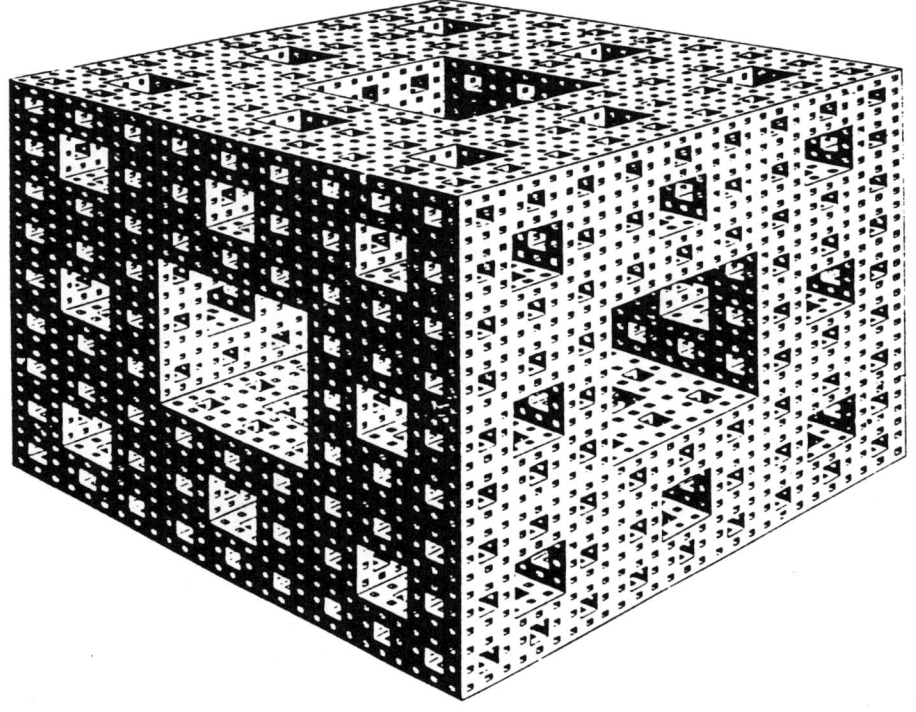

Fig. 1.P.3 The Menger sponge [1, 5].

1.10 Using the same corner rules of Sec. 1.8, construct the RG for the simple cubic lattice.

REFERENCES

1. Blumenthal, L. M. and Menger, K. (1970) *Studies in Geometry*. W. H. Freeman, San Francisco.
2. Family, F. (1980) *J. Phys. A* **13** L325.
3. Freed, K. F. (1987) *Renormalization Group Theory of Macromolecules*. Wiley, New York.
4. Halsey, T. C., Jensen, M. H., Kadanoff, L. P., Procaccia, I., and Shraiman, B. I. (1986) *Phys. Rev. A* **33** 1141.
5. Mandelbrot, B. B. (1977) *Fractals: Form, Chance, and Dimension*. W. H. Freeman, San Francisco; *The Fractal Geometry of Nature*. W. H. Freeman, San Francisco (1983).

6. McKenzie, D. S. (1976) *Phys. Rep.* **27** 35.
7. Napiorkowski, M., Hauge, E. H., and Hemmer, P. C. (1979) *Phys. Lett. A* **72** 193.
8. Pietronero, L. and Tosatti, E. (eds.) (1986) *Fractals in Physics*. North-Holland, Amsterdam.
9. Redner, S. and Reynolds, P. J. (1981) *J. Phys. A* **14** 2679.
10. Shapiro, B. (1978) *J. Phys. C* **11** 2829.
11. Stanley, H. E., Reynolds, P. J., Redner, S., and Family, F. (1982) In *Real Space Renormalization* (T. W. Burkhardt and J. M. J. van Leeuwen, eds.). Springer, Berlin.

2

RENORMALIZATION GROUP APPROACH TO CHAOS

2.0 INTRODUCTION

We continue our study of RG methods in physics with an example taken from the literature on chaos in nonlinear dynamics. Good reviews of the subject are Helleman [15] and Hao [14]. Once the basic concepts of chaotic behavior have been described and the motivation for studying mappings is established, we will examine the theory of the onset of chaos through a sequence of period-doubling bifurcations due to Feigenbaum [6–10]. The universal behavior of one-dimensional maps is demonstrated by Feigenbaum's functional renormalization method. In section 2.3 we discuss the structure of the aperiodic attractor. We then present an approximate RG due to Helleman for treating two-dimensional maps [12], which is very close in spirit to Feigenbaum's method.

One of the most active areas of physics in the past few years has been the study of nonlinear dynamical systems. Nonlinear dynamics is at least as old as Newton and one may well ask: Why is this area so active today? The reason is that, except for a few visionaries such as Poincaré, it was not until recently that the implications of "nonintegrability" were fully appreciated. Much of the renaissance of interest in problems in nonlinear dynamics can be ascribed to the extraordinary growth in computer technology in recent years.

A typical course in mechanics deals in problems that are essentially one dimensional; either there is a single degree of freedom, or there are enough conservation laws to reduce the problem to one dimension by quadrature. Problems that do not fall in this category, such as the three-body problem, are "unsolved," although one has the vague notion that, given sufficient computational resources, a solution of arbitrary accuracy can be constructed for any set of initial conditions. The situation in nonlinear dynamics is actually much worse than this. It is not that the solutions to most of the problems of interest are known, or at least known to exist. Rather, the set of problems we know how to solve is of measure zero; the rest are nonintegrable.

For the moment, consider Hamiltonian systems. One might take the point of view that even if it is not possible to construct an analytic solution, or express the solution in terms of integrals of the motion, one can always construct a "numerical solution." To qualify as a solution of Hamilton's equations, the result of any numerical process must be stable; otherwise there is no way to estimate the accuracy of the method. For most Hamiltonians, no such guarantee of stability can be made and so in this sense the numerical solution also may not exist.

The equations of motion of a system are integrable if there exists a sufficient number of conserved analytic functions of the coordinates to completely specify the orbit. Put another way, it is possible to construct, either analytically or by a convergent canonical perturbation series, the action-angle variables of Hamilton–Jacobi theory. When this program fails, when the action integrals are nonanalytic functions and the canonical perturbation series does not converge, the system is called nonintegrable.

A good way to contrast integrable and nonintegrable systems is to look at the behavior of two nearby orbits. If the system is integrable, then each orbit will be specified by the values of the integrals of the motion. Because these are analytic functions of the coordinates and momenta, the difference between the values taken by the integrals of the motion for two nearby orbits will also be small. If we follow the evolution of these two orbits, they will move apart for short times at a rate proportional to the time (Prob. 2.1). On the other hand, if the system is nonintegrable these integrals of the motion do not exist and two nearby orbits will generally diverge exponentially.

If all of the phase space is this pathological, that is, if every orbit diverges exponentially from nearby orbits, then a single orbit comes

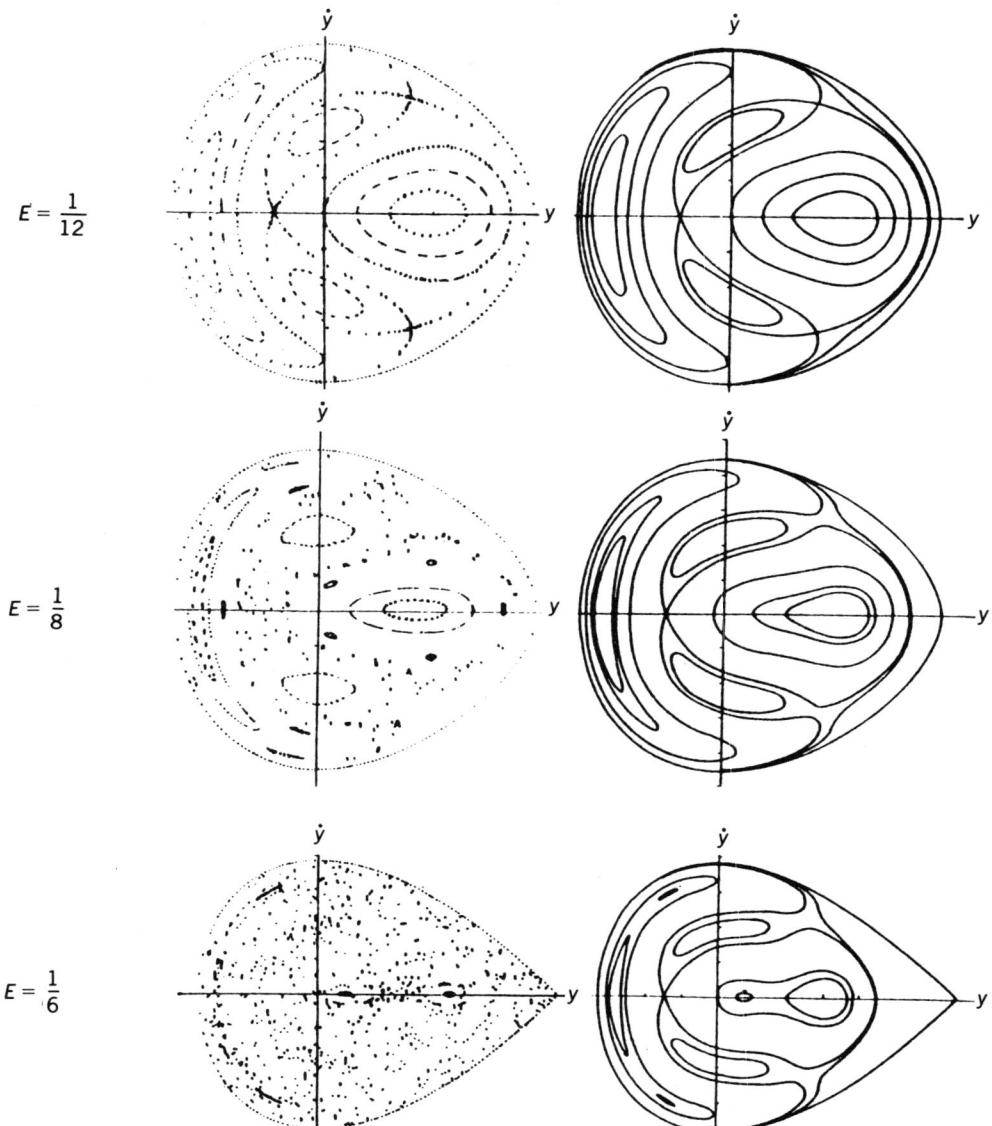

Fig. 2.0.1 Poincaré map for the Henon–Heiles Hamiltonian. Figures on the left are several orbits calculated by computer. Figures on the right are orbits of the map calculated by canonical perturbation theory. Note that the scattered points for $E = 1/6$ are intersections of a single chaotic orbit [10, 17]. See also [13].

arbitrarily close to every point in phase space: The system is ergodic and mixing. Clearly, there is little interest in constructing such an orbit because a small change in the initial conditions leads to a completely different (also phase space-filling) orbit. However, the ergodicity of the system leads to a new level of simplicity: We know from statistical mechanics that when the system is ergodic time averages over a particular orbit can be replaced by phase space averages, and so the statistical properties of the system can be calculated.

The structure of phase space in the generic case is much more interesting than either the completely integrable or ergodic cases. According to the KAM [14, 1, 16] (Kolmogorov, Arnold, and Moser) theorem, there are regions of phase space where the system is integrable as well as regions of nonintegrability, and what is worse, the regions of integrability and nonintegrability are nested one inside the other on all scales in phase space. This situation is illustrated in Fig. 2.0.1 for the famous Henon–Heiles Hamiltonian

$$H = \tfrac{1}{2}(\dot{x}^2 + \dot{y}^2) + \tfrac{1}{2}(x^2 + y^2) + x^2 y - \tfrac{1}{3}y^3 \qquad (2.0.1)$$

which was developed to model the motion of stars in the galactic disk. The figures shown [10, 13, 17] are called Poincaré maps, and are made by taking a cross section of the three-dimensional energy surface in phase space.

The orbit of the system is confined to the energy hypersurface by energy conservation, and every time the orbit crosses the plane of intersection a dot is drawn on the figure. Note that at low energy ($E = 1/12$ in dimensionless units), the true orbits lie very close to those calculated from canonical perturbation theory [13, 17]; the system appears to be regular and integrable. However, as the energy is raised, "chaotic" regions appear along with some islands of integrability. Note that in the chaotic region all the dots are made by a single orbit. Finally, at an energy of $1/6$, the chaos is widespread.

One might be tempted to say that for low energy the Henon–Heiles Hamiltonian is integrable. Certainly the perturbation calculation gives very good results. However, if one were to make an enlargement of the region around one of these supposedly integrable orbits, one would find a thin region called a stochastic layer in which the orbits are chaotic. As the energy increases, the width of the stochastic regions increases and eventually they merge into large-scale chaos. A similar situation is illustrated in Figs. 2.0.2a and b for the intersecting

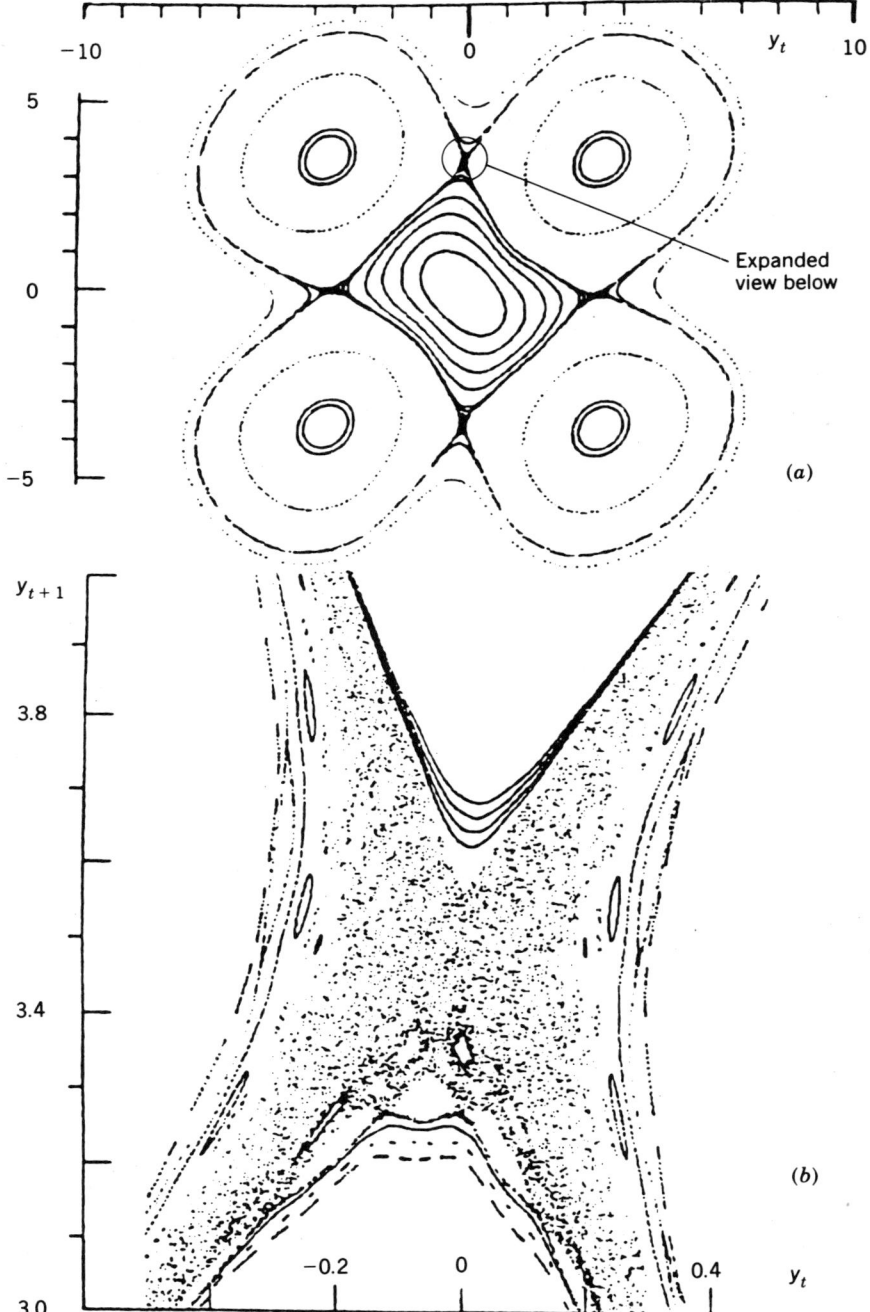

Fig. 2.0.2 (*a*) Stroboscopic map for the intersecting storage ring equations (2.0.2) [5]. (*b*) Magnified view of the circled region in Fig. 2.0.2*a* showing chaotic behavior. All points belong to a single chaotic orbit.

storage ring model [5]

$$Y_{t+1} = CY_t + \left(\frac{S}{Q}\right)P_t$$
$$P_{t+1} = -SQY_t + CP_t + BF(Y_{t+1}) \tag{2.0.2}$$
$$C = \cos(2\pi Q) \quad S = \sin(2\pi Q) \quad F(Y) = \frac{2}{Y}\left[1 - e^{-Y^2/2}\right]$$

From the upper half of Fig. 2.0.2a you would guess that the system is integrable, but a closer examination, as in the blowup of the circled region of Fig. 2.0.2b, reveals chaotic behavior. Again remember that all the dots in the chaotic region are created by a single orbit.

One of the consequences of nonintegrability is the difficulty in predicting the long-time behavior of dynamical systems. If a small error is made in the initial conditions, this error will grow exponentially with time, and eventually one will have no idea what the system is doing. To extend the time over which the orbit can be followed with confidence, the error in the initial conditions must be reduced by others of magnitude. One of the central problems in nonlinear dynamics is estimating the degree of chaotic behavior and the time scale over which an orbit can be followed with confidence.

A related problem, which is of importance to statistical physics, is the transition to chaos that occurs when the stochastic layers merge and the last vestiges of integrability disappear. As stated by Boltzmann, in order to apply statistical mechanics to an isolated system, it must be ergodic. The assumption is usually made that any system with a large number of degrees of freedom is ergodic, but this has been proved in only a very small number of cases. We observed for the Henon–Heiles Hamiltonian that the degree of chaotic behavior increased with the energy. If the "transition to chaos" for a large system occurs at an energy $E(N)$, where N is the number of particles, then the behavior of $E(N)$ in the thermodynamic limit determines whether the system is properly ergodic or not.

2.1 ONE-DIMENSIONAL MAPS AND THE PERIOD-DOUBLING ROUTE TO CHAOS

For many purposes it is more convenient to study discrete dynamical systems, or dynamical maps, rather than their underlying differential equations. This approach is convenient because it lends itself to numerical studies carried out by computer. Indeed, the work of Metropolis, Stein, and Stein [15] laid the foundation for the discovery by Feigenbaum that the period-doubling route to chaos is universal [2, 4–8].

The simplest dynamical map is the one-dimensional map, which we write in the form

$$x_{n+1} = f(x_n, C) \tag{2.1.1}$$

The constant C, called the control parameter, plays the role of an adjustable parameter such as the energy in the Henon–Heiles system, the Reynolds number in a flow, or the temperature difference in a Rayleigh–Bernard cell. By varying C we can control the degree of nonlinearity of the map. For definiteness and to make contact with the original papers of Feigenbaum [4–8], we will take for f the well-known logistic map

$$f(x, C) = 4Cx(1 - x) \tag{2.1.2}$$

We should emphasize, however, that the theory we are about to develop is completely independent of the particular choice of the function $f(x, C)$. For example, we might consider the map

$$f(x, C) = C \sin \pi x \tag{2.1.3}$$

which is of historical interest because it was the comparison of the period-doubling sequences of (2.1.2) and (2.1.3) that led Feigenbaum to the observation of universality in period doubling. The "fig tree" diagram of bifurcations of this map is shown in Fig. 2.1.1 and should be compared with that of logistic map (Fig. 2.1.2). If $f(x, C)$ is monotonic, then the behavior of the iterates is qualitatively simple. Nontrivial maps are "folded over," that is, they exhibit an extremum, and in the generic case the second derivative of f at the extremum does not vanish. Therefore, the theory of Feigenbaum covers most one-dimensional maps.

The simple graphical construction in Fig. 2.1.3 shows that the map has a fixed point or period-1 orbit, which is given by

$$x_0^* = f(x_0^*, C) \tag{2.1.4}$$

If we set $x_n = x_0$ in (2.1.1), we obtain $x_{n+1} = x_0$, so the point is "fixed." If we now consider small perturbations around the fixed

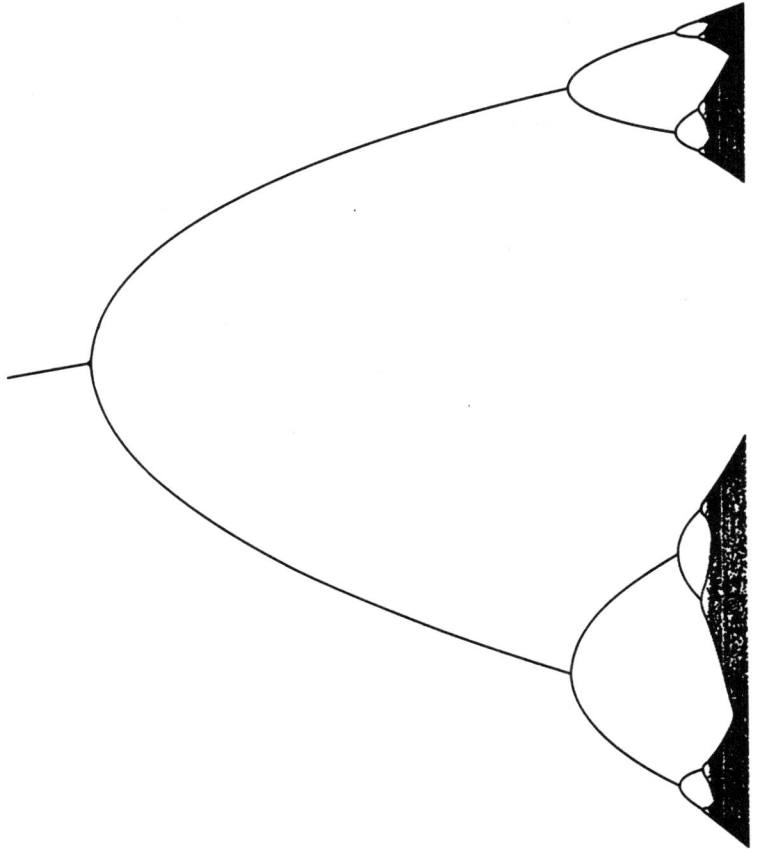

Fig. 2.1.1 Period-doubling sequence for the sine map, Eq. (2.1.3).

point, these perturbations will tend to 0 so long as

$$|f'(x_0^*, C)| < 1 \qquad (2.1.5)$$

In this situation we say the period-1 orbit is an "attractor." The stability of the period-1 orbit is illustrated in Fig. 2.1.4. Note how the iterates of the map rapidly converge on the fixed point.

As the control parameter C increases, the magnitude of the slope of the map at the fixed point also increases, and at some value $C = C_0$, condition (2.1.5) will be violated. In this case, the period-1 orbit is unstable, and a stable period-2 orbit emerges as shown in Fig. 2.1.5. The period-2 orbit is given by the pair of equations

$$\begin{aligned} x_1^* &= f(x_0^*, C) \\ x_0^* &= f(x_1^*, C) \end{aligned} \qquad (2.1.6)$$

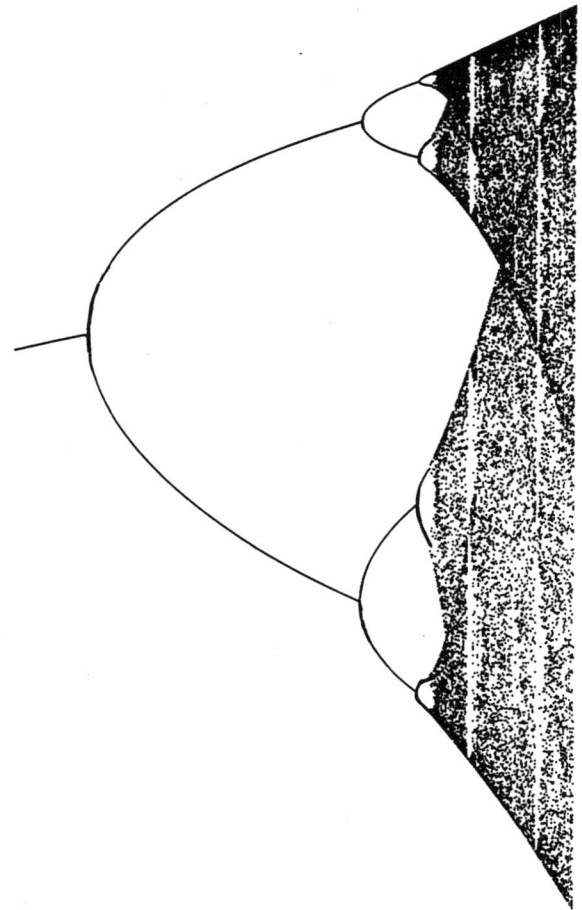

Fig. 2.1.2 The full sequence of period-doubling bifurcations for the logistic map leading up to the aperiodic or chaotic attractor at C_∞.

The phenomenon of a period-1 orbit becoming unstable and a stable period-2 emerging is called a "pitchfork bifurcation" or a period doubling. As C is increased still further, the separation between the period-2 iterates increases as shown in Fig. 2.1.6. At the value $C = C_1$, the period-2 orbit becomes unstable and bifurcates in turn into a stable period-4 orbit. The elements of the period-4 orbit are given by the set of four coupled equations

$$\begin{aligned} x_0^* &= f(x_1^*, C) & x_3^* &= f(x_2^*, C) \\ x_2^* &= f(x_1^*, C) & x_0^* &= f(x_3^*, C) \end{aligned} \quad (2.1.7)$$

2.1 MAPS AND THE PERIOD-DOUBLING ROUTE TO CHAOS

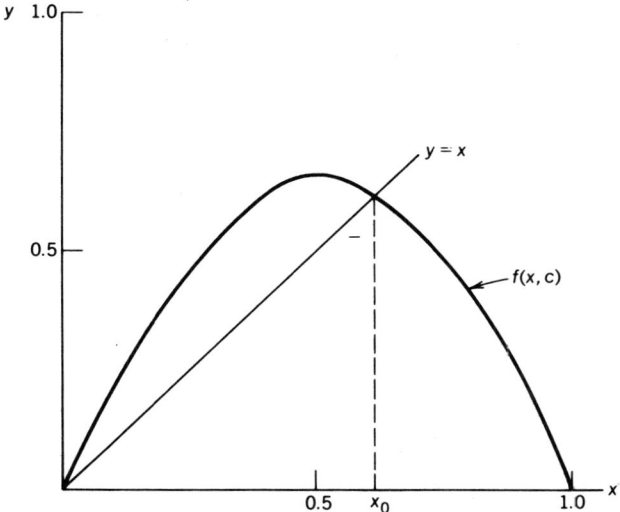

Fig. 2.1.3 The logistic map $f(x, C)$ for $C = 0.50$. The intersection of the graph of f with the line $y = x$ is the fixed point x_0.

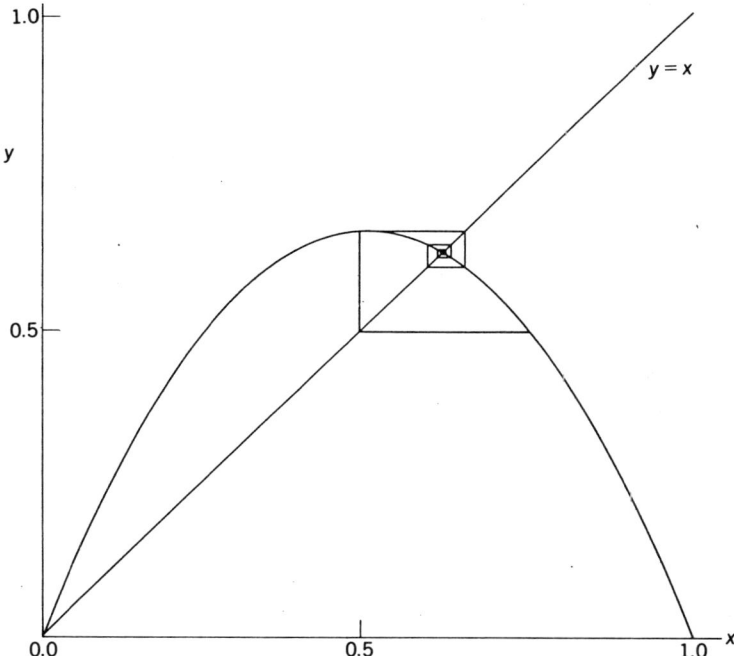

Fig. 2.1.4 Orbit of a point within the basin of attraction of the period-1 attractor. The iterates can be determined graphically as shown. Logistic map, $C = 0.75$.

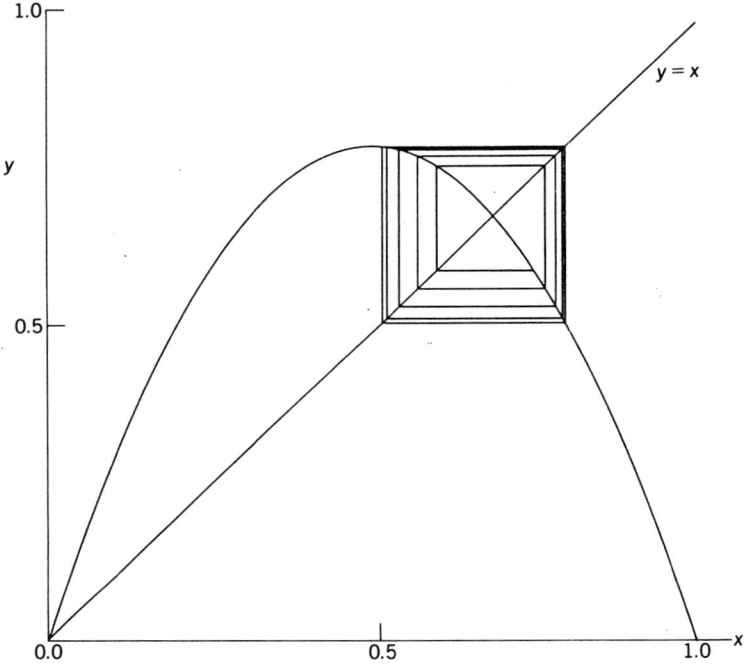

Fig. 2.1.5 Orbit attracted to the period-2 attractor. Logistic map, $C = 0.80$.

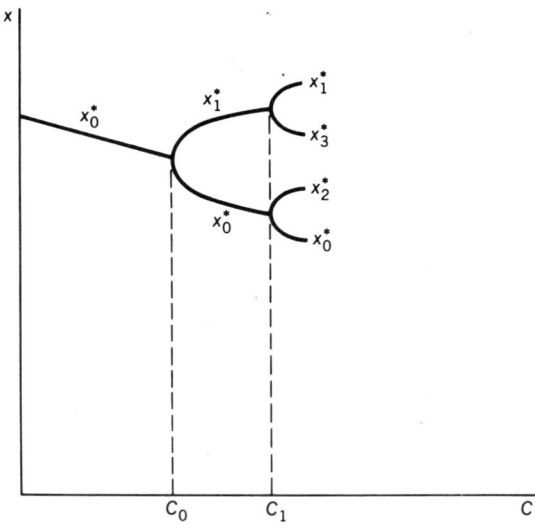

Fig. 2.1.6 The first two bifurcations of the logistic map, Eq. (2.1.2).

2.1 MAPS AND THE PERIOD-DOUBLING ROUTE TO CHAOS

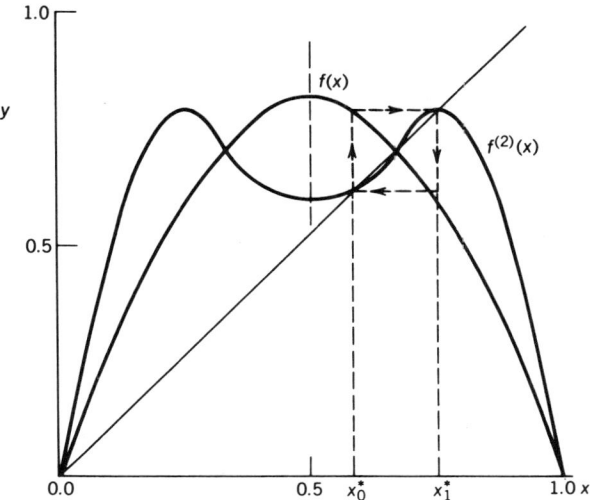

Fig. 2.1.7 The function $f^{(2)}(x)$ and $f(x)$. Note that for the period-2 orbit of $f(x)$, x_1, x_2 are fixed points, or period-1 orbits of $f^{(2)}(x)$.

In general, we define C_k to be the value of C at which the period-2^k orbit becomes unstable.

We now want to argue that this sequence of period doublings continues ad infinitum, and that the period of the attractor increases exponentially as the control parameter approaches a finite limit. The first step in this demonstration is to show that each successive bifurcation follows recursively from the previous one. Therefore, we begin by studying the first bifurcation.

As shown graphically in Fig. 2.1.7, each of the elements of the stable period-2 orbit of $f(x, C)$ is a period-1 orbit, or fixed point, of the map $f^{(2)}(x, C) = f(f(x, C), C)$ formed by composing f with itself. The function $f^{(2)}(x, C)$ possesses a simple quadratic extremum and each of its period-1 orbits will in turn become unstable and bifurcate into a pair of stable period-2 orbits as the control parameter is increased beyond the value C_1. This simple observation is essential to understanding the infinite sequence of period doublings because now we can take $f^{(2)}(x, C)$ as our map.

The elements of these two stable period-2 orbits of $f^{(2)}(x, C)$ comprise a stable period-4 orbit of the original map. Following the same line of reasoning, each of the elements of the period-2 orbits of $f^{(2)}(x, C)$ are period-1 orbits of $f^{(4)}(x, C) = f^{(2)}(f^{(2)}(x, C), C)$. Note that although $f^{(4)}(x, C)$ can be regarded as the fourth iteration of the original map $f(x, C)$, we can build it from $f^{(2)}(x, C)$ and do not need to make reference to $f(x, C)$ itself.

One might think that as this line of reasoning is followed the proliferation of period-1 orbits will quickly render the problem intractable, but it turns out the behavior of the map at each fixed point is identical, so we only need to focus our attention on one of them.

To see this, consider the map $f^{(2^n)}(x, C)$, which in the range $C_{n-1} < C < C_n$ will have 2^n stable period-1 orbits. The set of fixed points of $f^{(2^n)}(x, C)$ are the elements of the period-2^n orbit of f. The stability of each of these orbits is governed by the condition that the magnitude of the slope of $f^{(2^n)}(x, C)$ at the fixed point be less than unity, (2.1.4). Differentiating the map, one finds by the chain rule that the slope at one of these period-1 orbits is

$$\frac{d}{dx} f^{(2^n)}(x, C) \bigg|_{x=x^*} = \prod_{k=0}^{2^n-1} f'(x_k^*, C) \qquad (2.1.8)$$

Thus the slope of the map $f^{(2^n)}(x, C)$ is identical at each of its fixed points, and all 2^n period-1 orbits will simultaneously bifurcate at the value C_n.

To summarize, we have first demonstrated the mechanism by which a stable period-1 orbit bifurcates into a stable period-2 orbit. The recursive nature of the period-doubling sequence then follows from the observation that all of the 2^n period-1 orbits of $f^{(2^n)}(x, C)$ bifurcate simultaneously, that is, for the same value of $C = C_n$.

Several successive period doublings are illustrated in Fig. 2.1.8. There are two important observations to make with regard to this figure. The first is that the sequence of values $C_0, C_1, \ldots$ at which the bifurcations take place continually become more closely spaced, and in fact they converge geometrically to a finite value C_∞,

$$\lim_{k \to \infty} C_\infty - C_k \sim \delta^{-k} \qquad (2.1.9)$$

In addition, the ratio of the spacing Δ_k between the kth and $k + 1$st branches of the "fig tree" tends to a finite limit

$$\lim_{k \to \infty} \frac{\Delta_k}{\Delta_{k+1}} \to \alpha \qquad (2.1.10)$$

The remarkable discovery of Feigenbaum is that the values of the parameters α and δ are universal for the broad class of one-dimen-

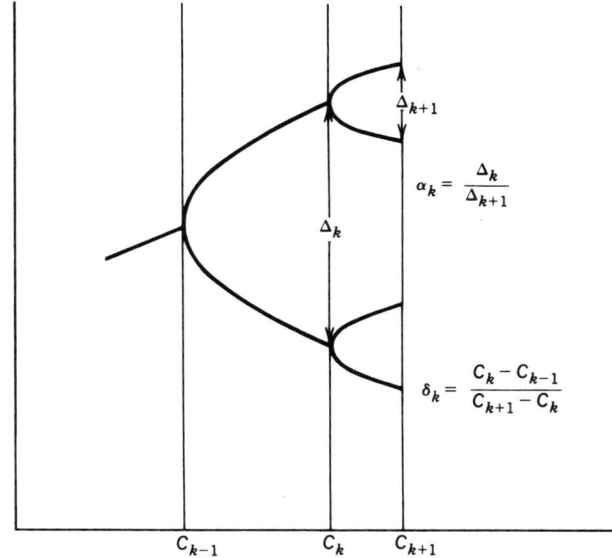

Fig. 2.1.8 Scaling properties of the period-doubling sequence.

sional maps that we have defined, and they take on the values

$$\alpha = 2.50290787\ldots$$
$$\delta = 4.6692016\ldots$$
(2.1.11)

The scaling properties of the orbits reflected in (2.1.9) and (2.1.10) and the universality of the period-doubling scheme are a strong indication that the recursive nature of successive bifurcations becomes, in the asymptotic limit, self-similar. The universal nature of the period-doubling, or Feigenbaum, sequence can be understood by taking the recursive procedure outlined previously and using it as the basis of an RG on the space of functions with a quadratic maximum.

2.2 RENORMALIZATION THEORY OF THE UNIVERSALITY OF PERIOD DOUBLING IN ONE-DIMENSIONAL MAPS

As we have argued previously, to clarify its general properties, we need to consider the behavior of the map in the neighborhood of only one fixed point, and for the logistic map it is convenient to focus our attention on the fixed point closest to $x = 1/2$.

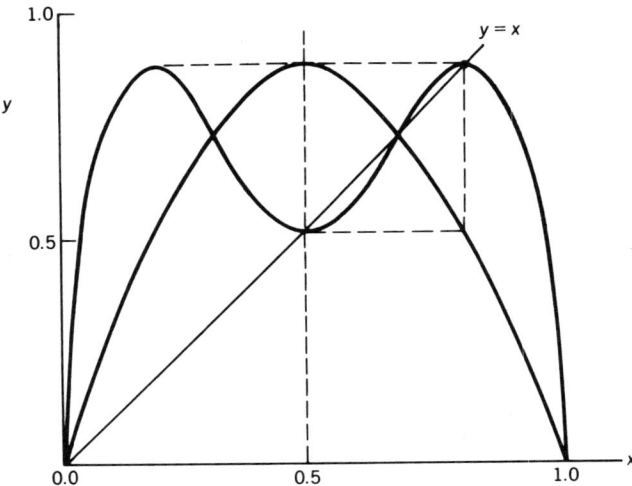

Fig. 2.2.1 The maps $f(x, \lambda_1)$ and $f^{(2)}(x, \lambda_1)$ for the period-2 superstable orbit.

At $C = C_0$ the slope of $f(x_0^*, C_0)$ is -1 and the slope of $f^{(2)}(x_0^*, C_0)$ is therefore $+1$. As C is increased the elements of the period-2 orbit separate until at the value $C = \lambda_1$ one of the fixed points is at $x = 1/2$ and the slope vanishes. This situation is shown in Fig. 2.2.1. [By symmetry the point $x = 1/2$ is an extremum of $f^{(n)}(x, C)$.] Because the magnitude of the slope determines the rate at which iterates converge to the attractor, when the slope vanishes the attractor is said to be superstable. In fact, one can show (Prob. 2.3) that the iterates converge faster than exponentially.

In general, we define λ_n to be the value of C at which one of the fixed points of $f^{(2^n)}(x, C)$ lies at $x = 1/2$,

$$f^{(2^n)}(\tfrac{1}{2}, \lambda_n) = \tfrac{1}{2} \qquad (2.2.1)$$

Figure 2.2.2 shows in a similar way $f^{(2)}(x, \lambda_2)$ and $f^{(4)}(x, \lambda_2)$ for the superstable period-4 orbit. If we compare these two figures in the neighborhood of $x = 1/2$, the first thing we notice is that in Fig. 2.2.2 the "circulating square," that is, the area containing $x = 1/2$ and its image under the map, is reduced in size by a factor α_n. Defining the distance of the fixed point from $x = 1/2$,

$$d_n = f^{(2^{n-1})}(\tfrac{1}{2}, \lambda_n) - \tfrac{1}{2} \qquad (2.2.2)$$

2.2 UNIVERSALITY OF PERIOD DOUBLING   49

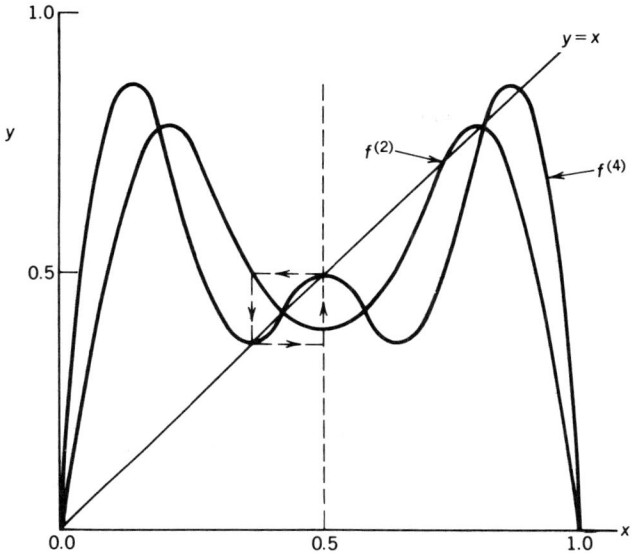

Fig. 2.2.2 The maps $f^{(2)}(x, \lambda_2)$ and $f^{(4)}(x, \lambda_2)$ for the period-4 superstable orbit.

the ratio α_n is given by

$$\alpha_n = -\frac{d_n}{d_{n+1}} \qquad (2.2.3)$$

The minus sign in (2.2.3) is introduced because the fixed point closest to $x = 1/2$ shifts from left to right of $1/2$ in successive self-compositions, as can be seen by comparing Figs. 2.2.1 and 2.2.2

If we now take the region enclosed in the dotted square in Fig. 2.2.1 and first reflect $f^{(2)}(x, \lambda_1)$ in the point $x = 1/2$, $y = 1/2$ and then rescale by the factor α, we see, as shown in Fig. 2.2.1, that this "renormalized" map resembles quite closely the original map. Similarly, applying the renormalization transformation (reflection and rescaling) twice to $f^{(4)}(x, \lambda_2)$, we get a map that is nearly indistinguishable from the renormalized $f^{(2)}$, as shown in Fig. 2.2.3.

If the renormalized and original maps were exactly equal, then the sequence of bifurcations would be exactly self-similar. Once we had worked out the first period doubling, all the rest would follow automatically. Although the first few renormalized maps differ slightly, as we consider maps of higher and higher degree of composition and renormalization, the transformation converges rapidly to a unique,

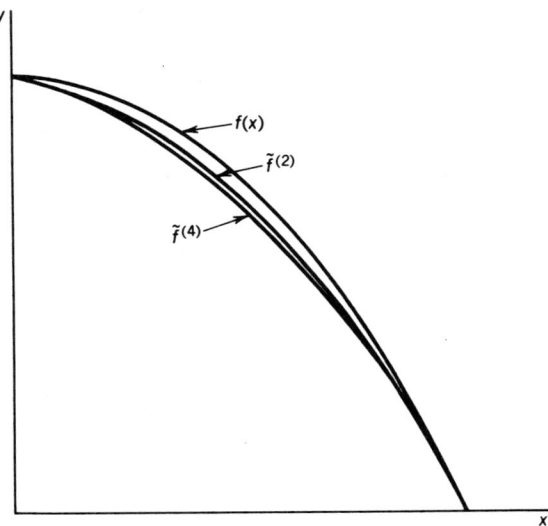

Fig. 2.2.3 Comparison of the original map and the renormalized versions of the maps $f^{(2)}(x)$ and $f^{(4)}(x)$ after shifting the origin of the coordinates.

universal limiting function, which Feigenbaum [4–8] calls $g_1(x)$. At this point it is convenient to shift the origin of coordinates by $x \to x - 1/2$. Then $g_1(x)$ is given by

$$g_1(x) = \lim_{n \to \infty} (-\alpha)^n f^{(2^n)}\left(\frac{x}{\alpha^n}, \lambda_{n+1}\right) \qquad (2.2.4)$$

In fact, one can define a family of universal functions $g_r(x)$ by

$$g_r(x) = \lim_{n \to \infty} (-\alpha)^n f^{(2^n)}\left(\frac{x}{\alpha^n}, \lambda_{n+r}\right) \qquad (2.2.5)$$

It then follows (Prob. 2.4) under composition and rescaling that

$$g_{r-1}(x) = -\alpha g_r\left(g_r\left(\frac{x}{\alpha}\right)\right) \qquad (2.2.6)$$

Taking the limit $r \to \infty$, or in (2.2.5) keeping λ fixed at the special value λ_∞, leads to the Feigenbaum–Cvitanovic universal function $g(x)$, which is a fixed function of the RG, and which, by (2.2.6), satisfies

$$g(x) = -\alpha g\left(g\left(\frac{x}{\alpha}\right)\right) \qquad (2.2.7)$$

The solution to the Feigenbaum–Cvitanovic equation (2.2.7) is unique if we specify the order of the maximum (quadratic in this case) at $x = 0$ and the scale of $g(x)$ by requiring that $g(0) = 1$. One can in fact show that if $g(x)$ is a solution of (2.2.7), then $g_\mu(x) = \mu g(x/\mu)$ is also a solution (Prob. 2.5). The first scaling parameter α is then given simply by

$$g(1) = -\frac{1}{\alpha} \qquad (2.2.8)$$

The RG transformation, which we call T, acts on a function $\psi(x)$ according to the rule

$$T[\psi](x) = -\alpha\psi\left(\psi\left(\frac{x}{\alpha}\right)\right) \qquad (2.2.9)$$

The universal function $g(x)$ is a fixed point (in the space of functions) of T,

$$g(x) = T[g](x) \qquad (2.2.10)$$

whereas the action of T on the sequence of universal functions $g_r(x)$ is given by

$$g_{r-1}(x) = T[g_r](x) \qquad (2.2.11)$$

In order to complete our analysis of the asymptotic properties of the period-doubling sequence, we must extract the second universal scaling parameter δ from the RG theory.

As with any RG calculation, we are interested in the behavior of the transformation not only at the fixed point, but also in the neighborhood of the fixed point. This is where we discover scaling behavior, and so it should not come as any great surprise that δ, which governs how the sequence of universal functions g_r converges to g, is the single relevant eigenvalue of the RG transformation linearized about the fixed point.

The function $g_r(x)$ for large r is a good approximation to $g(x)$, but by (2.2.11) we see that on repeated applications of T we move away from the fixed point in the sequence $g_r, g_{r-1}, g_{r-2}, \ldots$. Therefore, the fixed point is unstable and the asymptotic properties of the bifurcation sequence are determined by a relevant scaling field, which we call $h(x)$, and its associated eigenvalue δ.

First, we linearize the transformation T about the fixed point $g(x)$, defining the linear operator L by

$$L[\psi](x) = \lim_{\varepsilon \to 0} \frac{1}{\varepsilon}[T[g + \varepsilon\psi](x) - g(x)] \qquad (2.2.12)$$

Explicitly, by (2.2.9), L is given by

$$L[\psi](x) = -\alpha\left[g'\left(g\left(\frac{x}{\alpha}\right)\right)\psi\left(\frac{x}{\alpha}\right) + \psi\left(g\left(\frac{x}{\alpha}\right)\right)\right] \qquad (2.2.13)$$

Because $\Delta g_r(x) = g(x) - g_r(x)$ is small for large r, we can apply L to Δg_r, which gives

$$L[\Delta g_r] = -T[g - \Delta g_r] + g$$
$$= -T[g_r] + g = \Delta g_{r-1} \qquad (2.2.14)$$

by (2.2.11). Using the explicit form of L, (2.2.13), this becomes

$$-\alpha\left[g'\left(g\left(\frac{x}{\alpha}\right)\right)\Delta g_r\left(\frac{x}{\alpha}\right) + \Delta g_r\left(g\left(\frac{x}{\alpha}\right)\right)\right] = \Delta g_{r-1} \qquad (2.2.15)$$

The operator L is independent of the index r, and therefore we can solve (2.2.15) by a "separation of variables" and assume that the solution can be written in the form

$$\Delta g_r(x) = p_r h(x) \qquad (2.2.16)$$

Substitution of (2.2.16) into (2.2.15) then yields

$$p_r L[h](x) = p_{r-1} h(x) \qquad (2.2.17)$$

For this equation to hold for all r and x, we must have

$$L[h](x) = \delta h(x) \qquad (2.2.18)$$

2.2 UNIVERSALITY OF PERIOD DOUBLING   53

where δ is a constant and therefore

$$\delta p_r = p_{r-1}$$

which has the solution

$$p_r = \delta^{-r} \tag{2.2.19}$$

To finally identify the eigenvalue δ of the linearized RG transformation with Feigenbaum's universal parameter, we must show that in fact the values of the control parameter for superstable orbits λ_n converge geometrically to λ_∞ according to (2.1.9). The careful reader will have noticed that in (2.1.9) δ is defined in terms of the values C_k at which the bifurcations take place, not the values λ_k. Because the λ_k are bounded by C_k and C_{k+1} and because we have firmly established the smooth scaling properties of highly iterated maps, we are safe in assuming that if the λ_k converge geometrically, so will the C_k. Returning to the definition of $\Delta g_r(x)$, we see that asymptotically, that is, for large r, $g_r(x)$ is given by

$$g_r(x) \cong g(x) - \delta^{-r} h(x) \tag{2.2.20}$$

Let us now define $\tilde{\Delta} g_r(x) = g_{r+1}(x) - g_r(x)$. From the definition of the g_r, (2.2.5), we have

$$\tilde{\Delta} g_r(x) = \lim_{n \to \infty} (-\alpha)^n \left[f^{(2^n)}\left(\frac{x}{\alpha^n}, \lambda_{n+r}\right) - f^{(2^n)}\left(\frac{x}{\alpha^n}, \lambda_{n+r+1}\right) \right] \tag{2.2.21}$$

and by (2.2.20)

$$\tilde{\Delta} g_r(x) = \delta^{-1} \tilde{\Delta} g_{r-1}(x) \tag{2.2.22}$$

which can be written as

$$\lim_{n \to \infty} (-\alpha)^n \left[f^{(2^n)}\left(\frac{x}{\alpha^n}, \lambda_{n+r}\right) - f^{(2^n)}\left(\frac{x}{\alpha^n}, \lambda_{n+r+1}\right) \right]$$
$$= \frac{1}{\delta} \lim_{n \to \infty} \left[f^{(2^{n'})}\left(\frac{x}{\alpha^{n'}}, \lambda_{n'+r-1}\right) - f^{(2^{n'})}\left(\frac{x}{\alpha^{n'}}, \lambda_{n'+r}\right) \right] \tag{2.2.23}$$

Taking $n' = n + 1$ on the right-hand side then gives, for large n,

$$f^{(2^n)}\left(\frac{x}{\alpha^n}, \lambda_{n+r}\right) - f^{(2^n)}\left(\frac{x}{\alpha^n}, \lambda_{n+r+1}\right)$$
$$\cong -\left(\frac{\alpha}{\delta}\right)\left[f^{(2^{n+1})}\left(\frac{x}{\alpha^{n+1}}, \lambda_{n+r}\right) - f^{(2^{n+1})}\left(\frac{x}{\alpha^{n+1}}, \lambda_{n+r+1}\right)\right] \quad (2.2.24)$$

For convenience we now set $x = 0$, $r = 0$, and by (2.2.24) we have

$$f^{(2^n)}(0, \lambda_n) - f^{(2^n)}(0, \lambda_{n+1}) \cong -\left(\frac{\alpha}{\delta}\right)\left[f^{(2^{n+1})}(0, \lambda_n) - f^{(2^{n+1})}(0, \lambda_{n+1})\right] \quad (2.2.25)$$

Because we are only interested in asymptotic properties, we can relax the restriction that n is large and extrapolate (2.2.25) back to $n - 1$,

$$f^{(2^n)}(0, \lambda_n) - f^{(2^n)}(0, \lambda_{n+1}) \sim \left(-\frac{\delta}{\alpha}\right)^n \left[f^{(1)}(0, \lambda_n) - f^{(1)}(0, \lambda_{n+1})\right] \quad (2.2.26)$$

Multiplying both sides by $(-\alpha)^n$ and using the fact that $f^{(1)}(1/2, C) = C$ yields

$$g_1(0) - g_0(0) \cong \delta^n(\lambda_{n+1} - \lambda_n) \quad (2.2.27)$$

Because $g_1(0) - g_0(0)$ is a constant of order unity, we have

$$\lim_{n \to \infty} \frac{\lambda_n - \lambda_{n-1}}{\lambda_{n+1} - \lambda_n} = \delta \quad (2.2.28)$$

from which it follows that $\lambda_\infty - \lambda_n \sim \delta^{-n}$ in agreement with (2.1.9).

2.3 GLOBAL STRUCTURE OF THE APERIODIC ATTRACTOR

By construction, the region around the maximum of $g_1(x)$ accommodates a period-2 orbit, $g_2(x)$ accommodates a period-4 orbit, and in general $g_r(x)$ will accommodate a period-2^r orbit in the same region. The limiting function $g(x)$ accommodates the entire aperiodic (or chaotic) attractor. What is perhaps more surprising is that, apart from a trivial translation and scale factor, this attractor is universal for all

2.3 GLOBAL STRUCTURE OF THE APERIODIC ATTRACTOR

quadratic maps. That is to say the sequence of iterates of the point $x = 0$ for the special value $C = C_\infty$ of any map will be identical, and in particular it is equal to the sequence generated by the universal function $g(x)$ itself.

Let us denote the elements of the attractor by $x_0, x_1, x_2, \ldots,$ where $x_0 = 0$ and

$$x_{n+1} = g(x_n) \qquad (2.3.1)$$

By (2.2.6) it follows that the subset of even iterates $x_0, x_2, x_4 \ldots$ reproduces the entire sequence scaled by $-1/\alpha$,

$$x_k = -\alpha x_{2k} \qquad (2.3.2)$$

To see this, first observe that, because $g(0) = 1$, $x_1 = 1$ and $x_2 = -1/\alpha$, so (2.3.2) is correct for $k = 1$. Substituting (2.3.2) into (2.2.7) then gives

$$\begin{aligned} x_k &= g(x_{k-1}) \\ &= -\alpha g\left(g\left(\frac{\alpha x_{2k-2}}{\alpha}\right)\right) \\ &= \alpha x_{2k} \end{aligned} \qquad (2.3.3)$$

which proves (2.3.2) by induction. Therefore, the even elements of the aperiodic attractor reproduce the entire attractor scaled by α, the set of every fourth element reproduces the entire set scaled by α^2, and so on. This self-similarity is just what we have come to recognize as the hallmark of a fractal, in this case, a Cantor set. Unfortunately, the odd iterates are not so well behaved, and although they resemble the original set scaled down by α^2, there is some "distortion," and the aperiodic attractor is only approximately described as a Cantor set with two scales, α and α^2. In fact, the aperiodic attractor contains an infinite number of scales. However, within the "two scales" approximation the dimension of the aperiodic attractor is given by

$$\frac{1}{\alpha^D} + \frac{1}{\alpha^{2D}} = 1 \qquad (2.3.4)$$

This gives $D = 0.524$, which is close to the exact value $D = 0.538$ [9].

2.4 RENORMALIZATION THEORY OF TWO-DIMENSIONAL MAPS

Now that we have established the universal nature of the period-doubling route to chaos for one-dimensional maps, it is important to extend these insights to higher-dimensional maps. First, very few dynamical systems are one dimensional, and yet many physical systems have been found that exhibit period-doubling routes to chaos. Also, all one-dimensional conservative systems are, as was mentioned in Sec. 2.0, integrable and therefore incapable of complicated behavior. Numerical experiments, however, show that conservative systems in higher dimensions are indeed capable of complicated behavior, and in particular period doubling to chaos. In this section we present an approximate RG treatment of the period-doubling route to chaos due to Helleman [13].

A general two-dimensional mapping can be expressed as

$$x_{t+1} = F(x_t, y_t)$$
$$y_{t+1} = G(x_t, y_t)$$
(2.4.1)

Two-dimensional maps can be classified as conservative or nonconservative (dissipative) depending on whether they are area preserving or not. By area preserving we mean that the set of points that occupy an area $\Delta x \Delta y$ before a mapping occupy the same area, $\Delta x' \Delta y' = \Delta x \Delta y$, after the mapping. The same distinction holds for higher-dimensional maps. The area-preserving quality of a conservative map is an example of Liouville's theorem.

Numerical experiments on conservative and nonconservative maps show that both types of maps exhibit period doubling to chaos, and one can define parameters α and δ just as for the one-dimensional map. What is perhaps surprising is that the values for α and δ found for two-dimensional nonconservative maps, (2.1.11), are identical to those for one-dimensional maps, whereas the conservative two-dimensional map yields distinct values as shown in Table 2.4.1. To see why all two-dimensional maps might exhibit universal behavior, recall from Feigenbaum's theory that at each level of bifurcation the region of interest around the fixed point is reduced by a factor of α, so that asymptotically all maps with the same local properties, that is, a nonvanishing second derivative, will follow the same route to chaos.

The difference in behavior between conservative, or area-preserving, maps and nonconservative maps has to do with the manner in

2.4 RENORMALIZATION THEORY OF TWO-DIMENSIONAL MAPS

TABLE 2.4.1 Comparison of Exact and RG Values for Feigenbaum's Universal Parameters for Conservative and Dissipative (Nonconservative) Maps [12]

Universal Parameter	Conservative		Dissipative	
	Exact	RG	Exact	RG
C_∞	-1.2663	-1.2656	-0.78497	-0.78078
δ	-4.1081	-4.0955	-2.5029	-2.2399
α	8.7211	9.06	4.6692	5.1231

which areas contract under a nonconservative mapping. Consider a region around the point $\mathbf{x}_0$ and the image of this region about the image point x_1, as shown in Fig. 2.4.1. Linearizing the map within this small region and defining $x = x_0 + \Delta x_0$, $x' = x_1 + \Delta x_1$, we have

$$\begin{bmatrix} \Delta y_1 \\ \Delta x_2 \end{bmatrix} = \begin{bmatrix} \dfrac{\partial x_1}{\partial x_0} & \dfrac{\partial x_1}{\partial y_0} \\ \dfrac{\partial y_1}{\partial x_0} & \dfrac{\partial y_1}{\partial y_0} \end{bmatrix} \begin{bmatrix} \Delta x_0 \\ \Delta y_0 \end{bmatrix} \qquad (2.4.2)$$

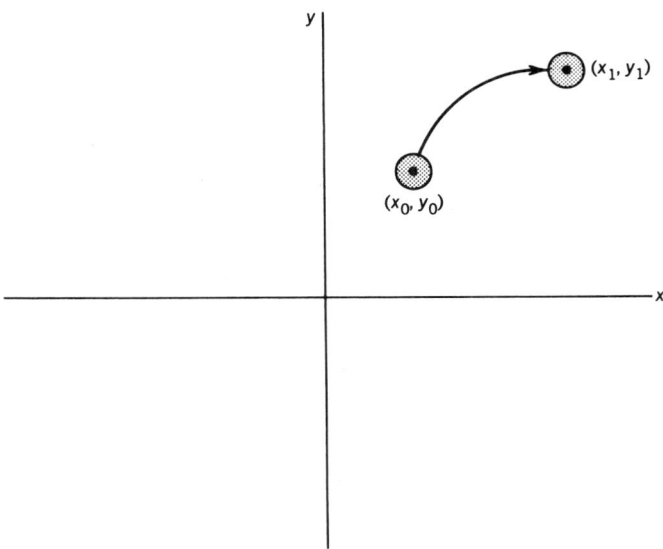

Fig. 2.4.1 Image of the neighborhood of the point $\mathbf{x}_0$ under a general two-dimensional map.

The area around the image point is

$$dA_1 = \det\left[\frac{\partial(x_1, y_1)}{\partial(x_0, y_0)}\right] dA_0 \qquad (2.4.3)$$

If the Jacobian in (2.4.3) is unity, the map is area preserving. If it is less than unity, it is nonconservative. If the Jacobian is greater than unity, we simply take the iterates of the map in reverse order and the map is again nonconservative.

One can define eigenvectors of the linearized map, $\psi^{(1)}$ and $\psi^{(2)}$, which satisfy

$$\begin{bmatrix} \dfrac{\partial x_1}{\partial x_0} & \dfrac{\partial x_1}{\partial y_0} \\ \dfrac{\partial y_1}{\partial x_0} & \dfrac{\partial y_1}{\partial y_0} \end{bmatrix} \begin{bmatrix} \psi_1^{(m)} \\ \psi_2^{(m)} \end{bmatrix} = \lambda^{(m)} \begin{bmatrix} \psi_1^{(m)} \\ \psi_2^{(m)} \end{bmatrix} \qquad (2.4.4)$$

Then the Jacobian is

$$J = \lambda^{(1)}\lambda^{(2)} \qquad (2.4.5)$$

Typically, the two eigenvectors will not be equal, and so it follows that an element of phase space will contract the fastest along the eigenvector with the smallest eigenvalue, call it $\lambda^{(1)}$. Therefore, after only a few iterations the map is essentially one dimensional, the $\lambda^{(1)}$ dimension being, in the language of the RG, irrelevant.

The same cannot, however, be argued for conservative maps, which retain their true two-dimensional behavior no matter how many iterations one performs. This then is the heuristic argument for why the nonconservative two-dimensional maps share the same universality class with one-dimensional maps, and why conservative two-dimensional maps fall into a separate universality class.

As we saw in the case of the one-dimensional map, the scale on which bifurcations take place is reduced by a factor of α in each successive bifurcation, and therefore only the local properties of the map in the neighborhood of a fixed point are relevant. To lowest order then most two-dimensional nonlinear maps can be replaced by a pair of first-difference equations. This pair of first-difference equations can then be further reduced to a single, second-difference equation called the "standard map,"

$$y_{t+1} + By_{t-1} = 2Cy_t + 2y_t^2 \qquad (2.4.6)$$

2.5 RENORMALIZATION GROUP FOR THE PERIOD-DOUBLING ROUTE TO CHAOS

The RG analysis of the standard map using the Feigenbaum sequence proceeds as follows. First, we find the value of C for which the trivial period-1 orbit, $y_0 = 0$, becomes unstable. We then construct an exact, stable, period-2 orbit. Using the original map T and expanding about the stable period-2 orbit, a new map T' is constructed with twice the time step. By rescaling the dynamical variable and the time step, this new map can also be cast in the standard form, but with renormalized values for the parameters B and C. Note that because the time step has doubled, a period-2 orbit of T is a period-1 orbit of T'.

The origin, $y_0 = 0$, is clearly a fixed point of the standard map. To find out for what range of values this orbit is stable, let

$$y = y_0 + y' \tag{2.5.1}$$

Linearizing the map (2.4.6) about the period-1 orbit, that is, simply dropping the quadratic term, we have

$$y'_{t+1} + By'_{t-1} \cong 2Cy'_t \tag{2.5.2}$$

As with a linear differential equation, this linear difference equation has exponential solutions

$$y'_t \sim \lambda^t \tag{2.5.3}$$

Substitution of (2.5.3) into (2.5.2) yields the exponents λ,

$$\lambda_\pm = C \pm \sqrt{C^2 - B} \tag{2.5.4}$$

Both values of λ are less than 1, and therefore the orbit is stable, if

$$|C| \leq \frac{1 + B}{2} \tag{2.5.5}$$

Beyond this value of the control parameter C, the period-1 orbit is unstable. For smaller values of C there is, however, a stable period-2 orbit, which we can assume takes the form

$$y_t^* = a - (-1)^t b \tag{2.5.6}$$

so $y_{t+1} = y_{t-1}$. Substituting this expression into (2.4.6) gives for a and b,

$$2a = -\frac{1+B}{2} - C$$
$$4b^2 = \left(\frac{1+B}{2} + C\right)\left\{C - \frac{3}{2}(1+B)\right\}$$
(2.5.7)

The solution is real if

$$C < -\left(\frac{1+B}{2}\right)$$
(2.5.8)

which is just the value of C for which the period-1 orbit becomes unstable. Thus we have constructed exactly the first bifurcation of the standard map.

The next step in the renormalization procedure is to construct the renormalized map T' from T, (2.4.6), and cast it in the standard form. We can then, by the previous calculation, construct the first bifurcation of T', which is equivalent to the second bifurcation of the original map T. Because T' takes the iterate y_t into y_{t+2}, the "time scale" for T' is doubled; to regain the standard form for the map T', the time will be rescaled.

Let y_t^* be the period-2 orbit constructed previously. Defining y_t' by

$$y_t = y_t^* + y_t'$$
(2.5.9)

the mapping T in terms of y_t' takes the form

$$y_{t+1}' + By_{t-1}' = 2Cy_t' + 4y_t^* y_t' + 2(y_t')^2$$
(2.5.10)

Now let $t = 2\tau + 1$; by (2.5.10) one has

$$y_{2t+2}' + By_{2t}' = (2C + 4y_{2\tau+1}^*)y_{2t+1}' + 2(y_{2t+1}')^2$$
(2.5.11)

and similarly for $t = 2\tau - 1$ one has

$$y_{2t}' + By_{2t-2}' = (2C + 4y_{2\tau-1}^*)y_{2t-1}' + 2(y_{2t-1}')^2$$
(2.5.12)

Using the fact that we are expanding about the period-2 solution y_t^*,

2.5 RENORMALIZATION GROUP FOR THE ROUTE TO CHAOS 61

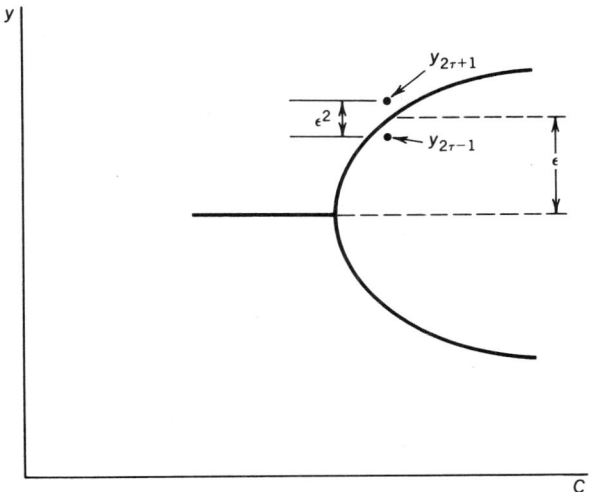

Fig. 2.5.1 Second iterates $y'_{2\tau+1} - y'_{2\tau-1}$ differ in $O(\varepsilon^2)$.

one finds, after a little algebra,

$$y'_{2t+2} + B^2 y'_{2t-2} = [(2C + 4y_1^*)(2C + 4y_0^*) - 2B] y'_{2t}$$
$$+ 2(2C + 4y_1^*)(y'_{2t})^2 + 2\left[(y'_{2t+1})^2 + B(y'_{2t-1})^2\right]$$
(2.5.13)

Except for the last term on the right-hand side of (2.5.13), which contains the iterate at times that are not even multiples of the original time step, this is nearly in the form of the standard map. We can get rid of this term (and here is where we make an approximation), by the following argument.

Let us assume that y'_t is of $O(\varepsilon)$, where $\varepsilon \ll 1$. Because we are expanding about a stable orbit, we will show that

$$y'_{2t+1} - y'_{2t-1} \sim O(\varepsilon^2) \qquad (2.5.14)$$

This idea is illustrated in Fig. 2.5.1. Accepting (2.5.14) for the moment, we can then write

$$y'_{2t+1} = (1 + A\varepsilon) y'_{2t-1} \qquad (2.5.15)$$

where A is a constant of order unity. Then

$$(1 + B + A\varepsilon) y'_{2\tau-1} = (2C + 4y_0^*) y'_{2\tau} + 2(y'_{2\tau})^2 \qquad (2.5.16)$$

Comparing terms to lowest order in ε, we must require

$$y'_{2\tau+1} \cong y'_{2\tau-1} \cong \left[\frac{2C + 4y_0^*}{1 + B}\right] y'_{2\tau} + O(\varepsilon^2) \qquad (2.5.17)$$

which verifies the consistency of (2.5.14). From these results we can then express the term involving y_t at odd times as

$$(y'_{2\tau+1})^2 + B(y'_{2\tau-1})^2 \cong \frac{(2C + 4y_0^*)^2}{1 + B}(y'_{2\tau})^2 + O(\varepsilon^3) \qquad (2.5.18)$$

Neglecting terms of $O(\varepsilon^3)$, one has for the new mapping

$$y'_{2\tau+2} + B'y'_{2\tau} = 2C'y'_{2\tau} + 2\alpha(y''_{2\tau})^2 \qquad (2.5.19)$$

where the scaling parameter α is

$$\alpha = 2C + 4y_1^* + \frac{(2C + 4y_0^*)^2}{1 + B} \qquad (2.5.20)$$

As is typical with RG transformations, in order to regain a map in the standard form we must perform some scale transformations. The coefficient of the quadratic term α can be eliminated by rescaling the dynamic variable y. Defining

$$z = \alpha y' \qquad (2.5.21)$$

and a new time variable

$$t' = 2\tau \qquad (2.5.22)$$

the new mapping is of the standard form

$$z_{t+1} + B'z_{t-1} = 2C'z_t + 2z_t^2 \qquad (2.5.23)$$

The renormalized values for B and C are

$$B' = B^2 \qquad (2.5.24a)$$
$$C' = 2(1 + B)^2 + 2C(1 + B) - 2C^2 - B \qquad (2.5.24b)$$

If our renormalization transformation were exact, then the sequence of bifurcations of the map (2.5.23) would correspond exactly to the bifurcations of the original map. In Prob. 2.7 you are asked to compare the RG results for the values C_n where the nth bifurcations

2.5 RENORMALIZATION GROUP FOR THE ROUTE TO CHAOS

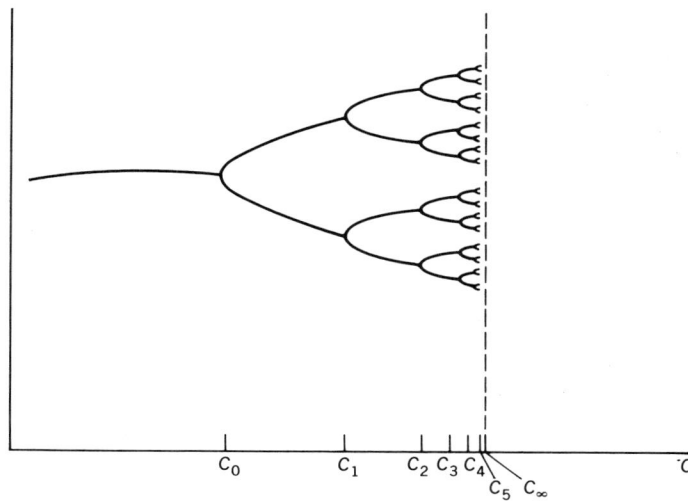

Fig. 2.5.2 Original bifurcation sequence for the logistic map.

occur with exact results. The approximation can be improved, at the cost of a great deal of effort, by finding the stable period-4 orbit of T and constructing a map with four times the original time step.

The scaling relations, (2.1.9) and (2.1.10), imply that as C approaches C^*, the pattern of bifurcations becomes self-similar. This is illustrated in Figs. 2.5.2 and 2.5.3.

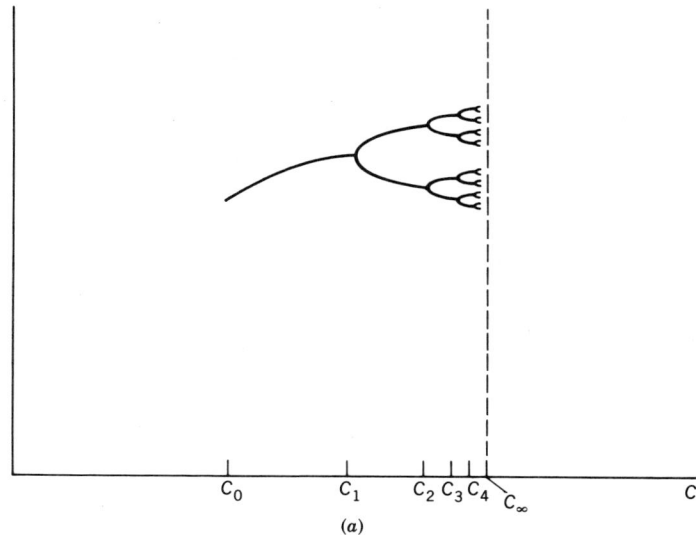

Fig. 2.5.3 The three steps in renormalizing the logistic map. (*a*) Bifurcation sequence when the time step is scaled by a factor of 2.

64   RENORMALIZATION GROUP APPROACH TO CHAOS

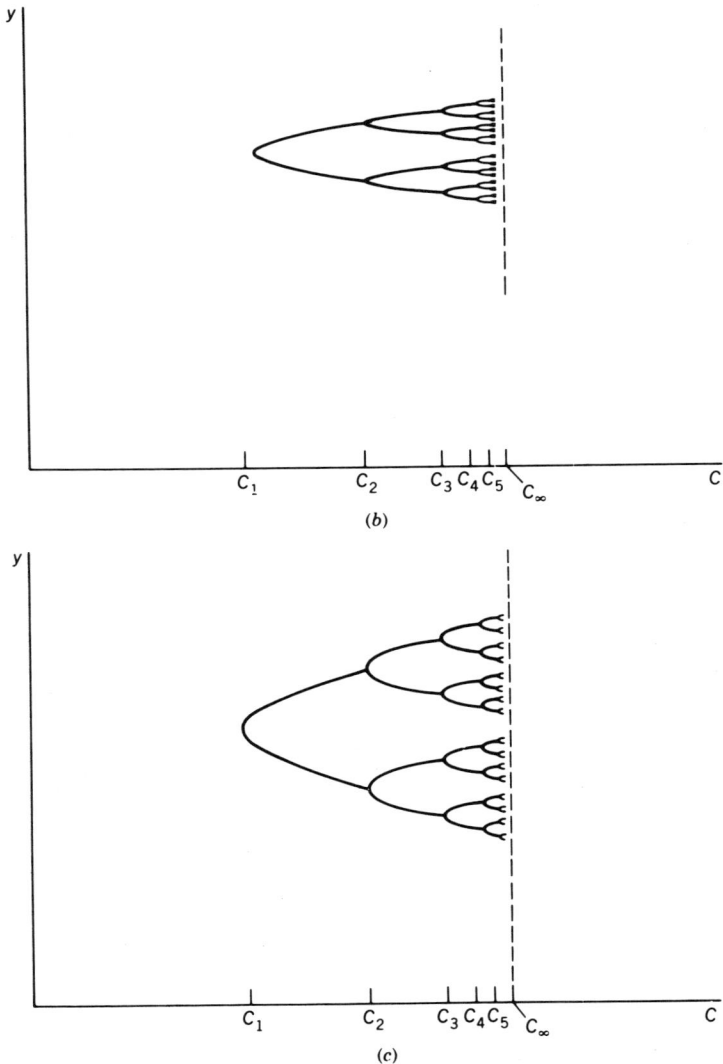

Fig. 2.5.3 (continued) (b) x_t is scaled by α. (c) The control parameter C is scaled by δ.

In Fig. 2.5.2 one sees the Feigenbaum sequence showing several levels of bifurcation. In Fig. 2.5.3 three scaling operations have been performed: The time step has been rescaled by a factor of 2, the dynamical variable has been enlarged by a factor α, and the horizontal scale has been enlarged by a factor of δ. The congruence of the two figures is clear. It is also apparent that the mapping for $C = C_\infty$ is invariant under this procedure. This invariance is reflected in the RG approach by the existence of a fixed point of the RG equation, (2.5.24),

$$B' = B = B^* \tag{2.5.25a}$$
$$C' = C = C^* \tag{2.5.25b}$$

In order to avoid confusion between the exact value C_∞ for the standard map and the fixed-point value obtained by solving (2.5.25), we denote the fixed-point value by C^*. We will now show that the existence of a fixed point of the RG equations implies self-similarity and leads in a natural way to the scaling properties of the Feigenbaum sequence.

Let $C = C^*$ and let $\{y_k, k = 1, \infty\}$ be the trajectory of the map for this value of C. Under the three operations outlined previously, the time step is doubled so half, say all the odd k, of the points in the orbit are retained and the spacing between iterates is reduced by a factor of α. Let this new set be $\{z_k\}$. Because the map is invariant under renormalization the two sets are equal. Therefore, the orbit is self-similar on arbitrarily long time scales.

Equations (2.5.24a) and (2.5.25a) have two possible solutions, $B^* = 1$ and $B^* = 0$. Note that for any value of $B < 1$ (dissipative map), B tends to 0 under repeated applications of the RG transformation. Therefore, all dissipative maps share the same fixed point, although the particular values of C at which the bifurcations take place differ. This explains the universal behavior of all dissipative maps in the neighborhood of the onset of chaos. Here we also find the reason why the conservative case, $B = 1$, is distinct. In the language of the RG we say that conservative and dissipative maps fall into different universality classes.

2.6 UNIVERSAL PROPERTIES OF CONSERVATIVE MAPS

It will be constructive to evaluate the Feigenbaum universal parameters C_∞, α, and β for conservative maps and to compare the RG values with their exact values.

Setting $B = 1$, (2.5.24b) and (2.5.25b) can be solved for C^*. Because we are interested in the parameter range $C < 0$, we have by (2.5.24b)

$$C^* = -2(C^*)^2 + 4C^* + 7 \tag{2.6.1}$$

$$C^* = \frac{3 - \sqrt{65}}{4} = -1.2656\ldots \tag{2.6.2}$$

which should be compared to the exact value [12], $C_\infty = -1.2663\ldots$.

The behavior of the map as C approaches C^* is governed by the RG in the neighborhood of the fixed point. According to Feigenbaum, the value C_k at which the kth bifurcation takes place approaches C^* geometrically

$$\lim_{n \to \infty} C_n \to C^* + \delta^{-n} \qquad (2.6.3)$$

The kth bifurcation of the renormalized map corresponds to the $k + 1$st bifurcation of the original map. Therefore, by (2.5.24b), with $B^2 = 1$,

$$C_k = -2C_{k+1}^2 + 4C_{k+1} + 7 \qquad (2.6.4)$$

Substituting the asymptotic form (2.6.2) in (2.6.3) and comparing lowest-order terms in δ gives

$$\delta = \sqrt{65} + 1 = 9.06\ldots \qquad (2.6.5)$$

whereas the exact value [15] is $\delta = 8.721$.

Finally, the ratio α, which is defined in (2.5.20), can be identified with the scaling factor (2.1.10). Substituting the fixed-point values for B and C gives

$$\alpha = -4.0955 \qquad (2.6.6)$$

whereas the exact value [15] is $\alpha = -4.1081$.

The results for the dissipative map are left as an exercise. The results of the RG analysis are compared to the exact numerical values in Table 2.4.1.

PROBLEMS

2.1 Show that if a system is integrable, nearby orbits diverge linearly in time. (You may assume that a transformation to action-angle variables is possible.)

2.2 Show that the slope of the function $f^{(2)}(x, C)$ is equal at each point of a period-2 orbit. Generalize this result for a period-2^k orbit.

2.3 Show that if the slope of the function $f(x)$ vanishes at the fixed point, then the iterates of the map in the neighborhood of the

PROBLEMS

2.4 Derive the recursion relation (2.2.6) from the definition of the universal functions (2.2.5).

2.5 Show that if $g(x)$ is a solution of (2.2.7), then $g_\mu(x) = \mu g(x/\mu)$ is also a solution.

2.6 Following the method used in Sec. 2.6 for the conservative case, derive the universal Feigenbaum numbers for the dissipative case.

2.7 For $B = 0$ (the logistic map) use the RG equations to determine the values of C_k and compare with the following table of exact values [6]:

k	$-C_k$
1	0.618034
2	0.7492808
3	0.7773203
4	0.7833336
5	0.7846217
6	0.7848976
7	0.7849566
8	0.7849693

2.8 Show that every feature of the standard map for a value C of the control parameter is mirrored at another value $C' = 1 + B - C$.

2.9 Show that an area A in the (y_{t-1}, y_t) plane is mapped to an area $A' = BA$, $0 \leq B \leq 1$, in the (y_t, y_{t+1}) plane under the map (2.4.6).

2.10 Show that the standard map (2.4.6) with $B = 0$ can be recast in the form (2.1.2).

2.11 The fractal dimension of the aperiodic attractor is given by [11]

$$\lim_{n \to \infty} \sum_{j=1}^{2^n-1} \frac{1}{|x_j - x_{j+2^{n-1}}|^D} = 1$$

where the x_j are the iterates of $x_0 = 1/2$ under the map (2.1.2) with $C = C_\infty$. Write a simple computer program to generate a series of approximations for d for increasing values of n.

2.12 Find the range of values $C_2 \leq C \leq C_1$ over which the period-2 orbit (2.5.6) is stable.

REFERENCES

1. Arnold, V. I. (1963) *Russian Math. Surveys* **18** 85.
2. Cvitanovic, P. and Jensen, M. H. (1981) Nordita preprint.
3. Eminhizer, C. R. (1980) Brookhaven National Laboratory Report.
4. Feigenbaum, M. J. (1978) *J. Statist. Phys.* **19** 25.
5. Feigenbaum, M. J. (1979) *J. Statist. Phys.* **21** 699.
6. Feigenbaum, M. J. (1979) *Phys. Lett. A* **74** 375.
7. Feigenbaum, M. J. (1980) *Comm. Math. Phys.* **77** 65.
8. Feigenbaum, M. J. (1980) *Los Alamos Science* **4**.
9. Grassberger, P. (1981) *J. Statist. Phys.* **26** 173.
10. Gustavson, F. (1966) *Ap. J.* **71** 670.
11. Hao, B.-L. (1984) *Chaos*. World Scientific Publishing, Singapore.
12. Helleman, R. H. G. (1980) In *Fundamental Problems in Statistical Mechanics* (E. Cohen, ed.) vol. 5. North-Holland, Amsterdam.
13. Helleman, R. H. G. (1984) In *Long-Time Prediction in Hamiltonian Systems* (L. Reichl, ed.). Wiley, New York.
14. Kolmogorov, A. N. (1954) *Akad. Nauk. SSSR Dokl.* **98** 527.
15. Metropolis, N., Stein, M. L., and Stein, P. R. (1973) *J. Comp. Theory* **15** 22.
16. Moser, J. (1962) *Nachr. Akad. Wiss. Göttingen Math. Phys. Kl.* **2**.
17. Moser, J. (1968) *Amer. Math. Soc.* **81** 1.

3

RENORMALIZATION APPROACH TO PERCOLATION

3.0 INTRODUCTION

The flow of oil through fractured rock, dielectric breakdown, the statistical properties of macromolecules, gellation, random resistor networks, and the behavior of disordered media are all problems, to name just a few, which involve the idea of *percolation*. Clear discussion of percolation can be found in references [9, 10] and a good review of real-space renormalization group approaches to this problem is reference [8]. In this chapter we will be concerned with models of percolation on a lattice, in particular the uncorrelated *site* and *bond* problems.

Mathematically, the problem of percolation is most easily formulated on a lattice. We will begin the chapter by treating site percolation, which is concerned with the distribution of occupied and unoccupied lattice sites. Later we will examine bond percolation in which all sites are assumed occupied and a fraction of them have connections or bonds to some of their nearest-neighbor sites. For uncorrelated site percolation, each site is occupied with probability p or empty with probability $1 - p$. Similarly, for bond percolation each bond on the lattice is assigned an independent probability to be present or absent. We begin our study of site percolation by describing the connectivity properties of the lattice: A group of sites, called a cluster, is said to be connected if all sites in the cluster are connected to at least one other site in the cluster.

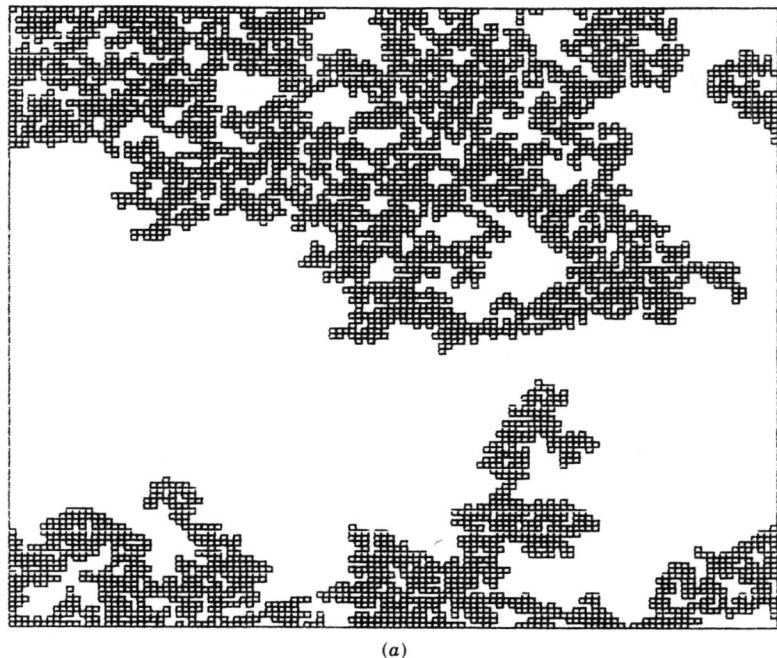

(a)

Fig. 3.0.1 Largest connected cluster on $L \times L$ square lattices with periodic boundary conditions. (a) $p = 0.58$, $L = 120$. (b) $p = p_c = 0.59275$, $L = 100$. (c) $p = 0.60$, $L = 120$.

For small values of p the lattice is essentially empty, and one finds mainly isolated occupied sites and very few clusters. As p increases both the number of clusters and their mean size grow until, at a critical value of the probability called the percolation threshold p_c, many of the isolated clusters finally merge into a highly ramified cluster that spans the infinite lattice. At percolation, the fraction of all sites in the incipient infinite cluster is quite small, and in fact the perimeter of the incipient infinite cluster is fractal. Above percolation the infinite cluster fills out and the number of sites in isolated clusters decreases to 0 as p approaches unity.

The dramatic change in the connectivity of the lattice as one passes through the percolation threshold is illustrated in Figs. 3.0.1a–c, which show the largest connected clusters for p near p_c. In Fig. 3.0.1a, $p = 0.58$, slightly below percolation, and the largest cluster does not span the lattice. At $p_c = 0.59275$ the largest cluster, which contains many more sites than the cluster at $p = 0.58$, spans the

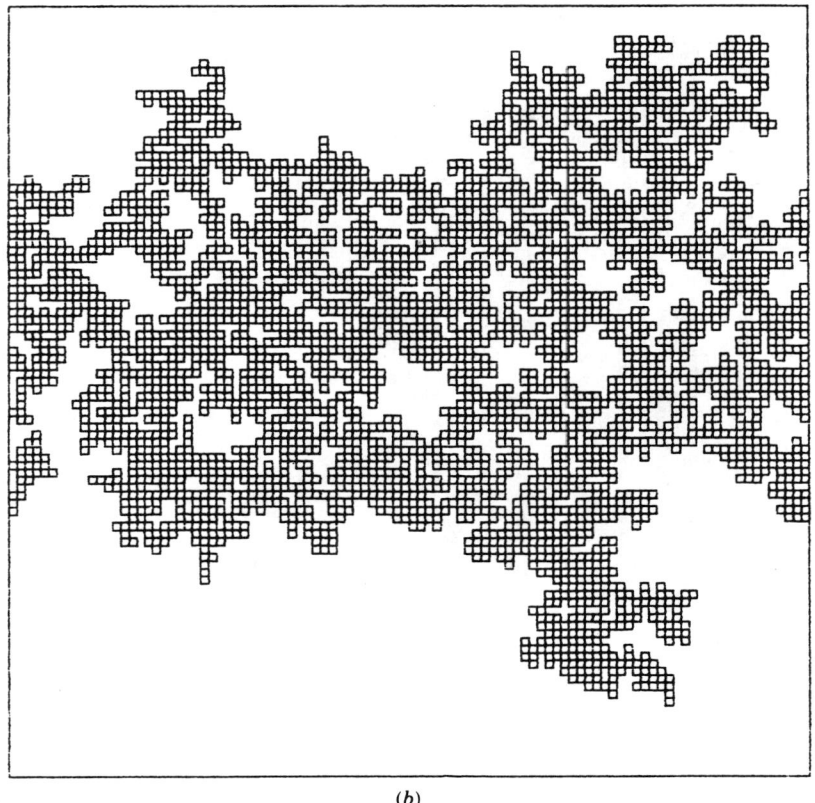

(b)

Fig. 3.0.1 *(Continued)*

lattice, but just barely. Increasing the concentration just 1%, to $p = 0.60$, gives the cluster shown in Fig. 3.0.1c, which fills a large fraction of the lattice.

The behavior of percolating systems is best described in terms of the cluster number $n_s(p)$, which is the number of clusters of s sites per lattice site. To get some idea of how one can calculate the cluster numbers, let us consider $n_s(p)$ for small s on the square lattice. The probability of finding a single isolated occupied site ($s = 1$) is

$$n_1(p) = p(1-p)^4 \qquad (3.0.1)$$

because the probability that the one site is occupied is p and the probability that each of the four neighboring sites are empty is

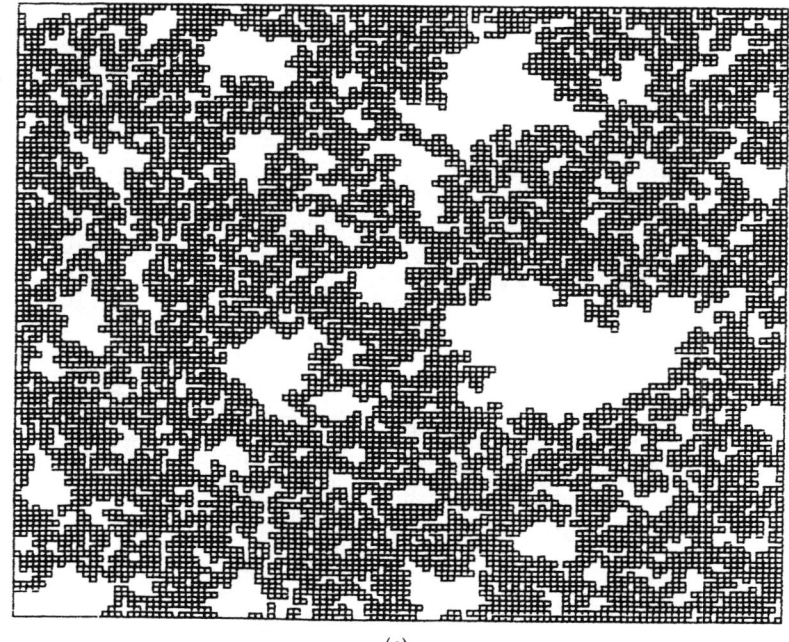

(c)

Fig. 3.0.1 *(Continued)*

$(1 - p)^4$. Similarly, the probability of finding a cluster of two sites is

$$n_2(p) = 4p^2(1 - p)^6 \qquad (3.0.2)$$

The factor of 4 in (3.0.2) arises from the four possible orientations of the 2-cluster.

For larger clusters there are many possible ways of arranging the occupied sites, and each of these different shapes will have its own number of boundary sites b. If we denote the number of distinct clusters of s sites with b boundary sites by g_{sb}, we obtain

$$n_s(p) = \sum_b g_{sb} p^s (1 - p)^b \qquad (3.0.3)$$

The configurations g_{sb} associated with the s, b clusters are called "lattice animals" because they resemble the various conformations of single-celled organisms such as paramecia that are seen under a microscope.

3.0 INTRODUCTION 73

The numbers g_{sb} are themselves independent of the concentration p and depend only on the details of the lattice. One approach to percolation theory is to try and calculate the number of lattice animals g_{sb} for reasonably large s, a task that is not trivial as one may discover by calculating the g_{sb} for s up to 4 in Prob. 3.2. In defining $n_s(p)$ we specifically exclude the infinite cluster, if one exists, and we assume that the lattice is sufficiently large so that clusters in contact with the outer boundary can be neglected.

The total number of clusters per lattice site Γ is then

$$\Gamma = \sum_s n_s(p) \qquad (3.0.4)$$

Γ is small both for $p \ll 1$ and $p \lesssim 1$ and exhibits a cusp singularity at the percolation threshold or critical point p_c,

$$\Gamma \sim |p - p_c|^{2-\alpha} \qquad (3.0.5)$$

which defines the critical exponent α.

The order parameter for percolation is the probability P that a site chosen at random will belong to the infinite cluster. The behavior of P with p is shown in Fig. 3.0.2.

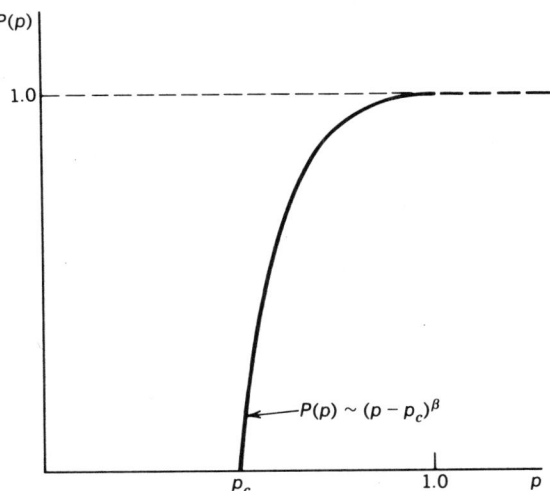

Fig. 3.0.2 Percolation order parameter P as a function of the occupation probability p.

P is given in terms of the cluster numbers by the relation

$$P + \sum_s sn_s(p) = p \qquad (3.0.6)$$

Below p_c there is no infinite cluster and $P = 0$, whereas as $p \to 1$ the fraction of sites *not* in the infinite cluster vanishes, and $P \to 1$ as well. For $0 < p - p_c \ll 1$, P increases as

$$P \sim (p - p_c)^\beta \qquad (3.0.7)$$

which defines the order parameter exponent β.

The probability that a site chosen at random is part of an s cluster is $sn_s(p)$, and so the relative probability that, among all finite clusters, the site belongs to an s cluster is

$$w_s(p) = \frac{sn_s(p)}{\sum_{s'} s' n_{s'}(p)} \qquad (3.0.8)$$

Note that by (3.0.3) if $p < p_c$, then the denominator is simply p, but above p_c the denominator tends to 0 and strongly modifies $w_s(p)$. Given $w_s(p)$, we can calculate the average number of sites in finite clusters

$$S = \sum w_s(p) n_s(p) = \frac{\sum_s s^2 n_s(p)}{\sum_s sn_s(p)} \qquad (3.0.9)$$

As $p \to p_c$, S diverges as shown in Fig. 3.0.3, according to

$$S \sim |p - p_c|^{-\gamma} \qquad (3.0.10)$$

which defines the critical exponent γ.

In the context of the RG the exponent most often encountered is the correlation length exponent ν defined by

$$\xi \sim |p - p_c|^{-\nu} \qquad (3.0.11)$$

where ξ is the average size of a cluster, or correlation length.

The definition of the correlation length requires some care. We begin by defining the "radius of gyration" of an s cluster

$$R_s^2 = \frac{1}{s^2} \sum_{i<j} |\mathbf{r}_i - \mathbf{r}_j|^2 \qquad (3.0.12)$$

where the double sum runs over all pairs of sites in the s cluster.

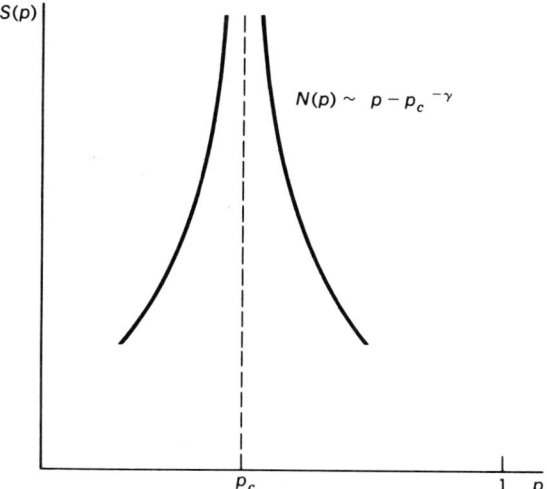

Fig. 3.0.3 Average numbers of sites in a (finite) cluster S as a function of the occupation probability.

The statistical weight attached to the s clusters is the probability $sn_s(p)$ that a given site is a member of an s cluster, times the number of sites s in the cluster, or $s^2 n_s(p)$. Therefore, the average radius of gyration of all the finite clusters is given by

$$\xi^2 = \frac{\sum_s R_s^2 s^2 n_s(p)}{\sum_s s n_s(p)} \quad (3.0.13)$$

Although an exact enumeration of $n_s(p)$ has not yet been achieved, for p close to p_c and large s it is believed that $n_s(p)$ is of the form

$$n_s(p) \sim s^{-\tau} f[(p - p_c) s^{\sigma}] \quad (3.0.14)$$

where the form of the function $f(x)$ is not known a priori. Equation (3.0.14) is called the scaling hypothesis, and it has been found to be of quite universal applicability independent of the dimensions and the details of the system. Its power lies in the fact that all the critical exponents we have defined thus far can be expressed in terms of σ and τ without knowing $f(x)$ explicitly.

To see how this works, we will calculate α and leave the calculation of β and γ as an exercise (Prob. 3.3). We first replace the sum in

(3.0.4) by an integral, which gives

$$\Gamma = \int_0^\infty s^{-\tau} f[(p - p_c)s^\sigma] \, ds \qquad (3.0.15)$$

Making the change of variable $x = |p - p_c|s^\sigma$ gives

$$\Gamma = \frac{1}{\sigma}|p - p_c|^{(\tau-1)/\sigma} \int_0^\infty x^{-(\tau+\sigma-1)/\sigma} f(\text{sgn}(p - p_c)x) \, dx \qquad (3.0.16)$$

Comparison of this expression with (3.0.5) gives

$$\alpha = (2\sigma - \tau + 1)/\sigma \qquad (3.0.17)$$

If all the critical exponents are determined by σ and τ, it follows that there must exist relations between them, and it can be shown that the scaling hypothesis implies the scaling law

$$\alpha + 2\beta + \gamma = 2 \qquad (3.0.18)$$

As we mentioned earlier, the incipient infinite cluster is fractal, and in fact for $p \lesssim p_c$ the largest clusters are also fractal with the same dimension as the incipient infinite cluster. Above p_c, however, the largest cluster occupies a finite fraction of the lattice ($P > 0$), and the dimension is just equal to the Euclidean dimension of the lattice.

For $p \lesssim p_c$ let us assume that the radius of gyration for s clusters, where $s \gg 1$, scales as

$$R_s \sim s^{1/D} \qquad (3.0.19)$$

where D is the fractal dimension of the cluster. Substituting this into (3.0.13) for the correlation length and using the scaling form (3.0.14) for the cluster numbers, we have

$$\xi^2 = \frac{\int_0^\infty s^{2/D} s^{2-\tau} f[(p - p_c)s^\sigma] \, ds}{\int_0^\infty s^{2-\tau} f[(p - p_c)s^\sigma] \, ds} \qquad (3.0.20)$$

Making the same changes of variable as before, we have

$$\xi \sim |p - p_c|^{1/\sigma D} \qquad (3.0.21)$$

and comparing this with (3.0.11), we see that the correlation length exponent ν is given by

$$\nu = \frac{1}{\sigma D} \tag{3.0.22}$$

In order to determine ν and D, we need another relation between them, which we can find by the following argument. For $p \lesssim p_c$, consider a region of the lattice of size $L \sim \xi$. The number of sites in the largest cluster within this region is $S_{max} \sim PL^d$. In addition, we have $S_{max} \sim \xi^D \sim L^D$ from which we see that, by comparing powers of $|p - p_c|$,

$$\beta - \nu d = -\nu D \tag{3.0.23}$$

Putting this together with (3.0.22), we find

$$\nu d = \frac{\tau - 1}{\sigma} \tag{3.0.24}$$

and expressing β in terms of σ and τ (see Prob. 3.3)

$$D = \frac{d}{\tau - 1} \tag{3.0.25}$$

Although neither the bond nor the site problems have been solved completely on any two- or three-dimensional lattice, some exact results for two-dimensional lattices have been found. First, the bond percolation problem on a lattice L is equivalent to site percolation on the "covering lattice" of L, called L^*. The covering lattice is formed by placing a site in L^* at the center of every bond in L and connecting those sites in L^* whose corresponding bonds share a site in L. For example, the covering lattice of the square lattice is shown in Fig. 3.0.4.

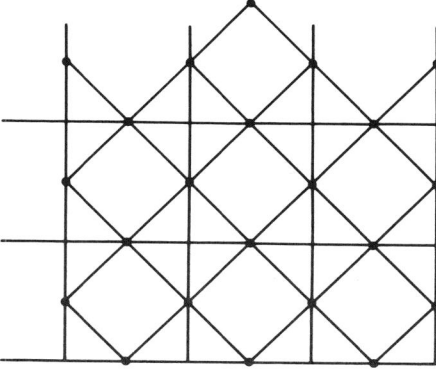

Fig. 3.0.4 Covering lattice for the square lattice.

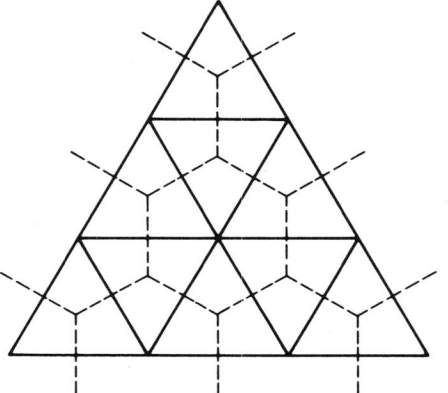

Fig. 3.0.5 Construction of the dual lattice. The dual of the triangular lattice is the honeycomb.

Second, it is easy to see that bond percolation on a lattice L with bond probability p is equivalent to bond percolation on the dual lattice of L, denoted by $\tilde{L}$, with probability $1 - p$. (A similar "duality transformation" exists for the partition function of the Ising model; see Sec. 5.3.) The dual lattice $\tilde{L}$ is constructed by placing a site at the center of each elementary polygon of L and connecting these sites by a bond that bisects one (and only one) bond in L. The square lattice is easily seen to be self-dual, whereas the triangular and hexagonal lattices form a dual pair, as shown in Fig. 3.0.5. A bond in $\tilde{L}$ is open if the bond in L which it crosses is open. Instead of labeling the bonds open and closed, we will call them black and white, respectively. It is easy to see that a connected cluster of white bonds on L gives rise to a connected cluster of black bonds on $\tilde{L}$. Because there is a one-to-one correspondence between bonds on L and $\tilde{L}$, and because by the previous prescription the statistical weight of the two configurations is the same, we see that under the dual transformation $p \to 1 - p$. This is illustrated for the square lattice in Fig. 3.0.6. If we denote the critical bond percolation probability on a lattice L by $p_b(L)$, then it follows that

$$p_{b,c}(L) + p_{b,c}(\tilde{L}) = 1 \qquad (3.0.26)$$

Because the square lattice is self-dual and assuming there is a unique percolation probability, (3.0.26) implies that $p_{b,c}(SQ) = 1/2$ for the square lattice.

A result similar to the duality relations exists for site percolation on a lattice L with site probability p and its "matching lattice" L^+ [2] with site probability $1 - p$. The matching lattice of L is constructed

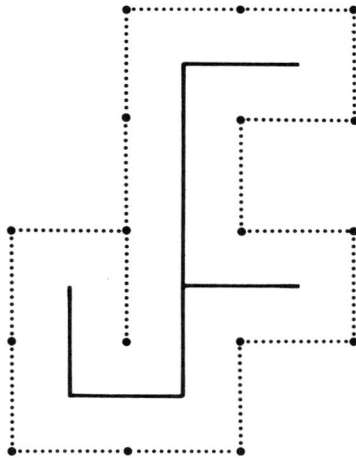

Fig. 3.0.6 A bond configuration on the square lattice and the corresponding configuration on the dual lattice.

by completing each elementary polygon or plaquette of L, that is, connecting each vertex of the polygon by bonds and replacing completed polygons by the corresponding plaquette. For example, the matching lattice of the square lattice is the square lattice with second-neighbor bonds, and the lattice in Fig. 3.0.7 is self-matching. Just as with bond percolation on pairs of dual lattices, we have for site percolation on matching lattices L, L^+:

$$p_{s,c}(L) + p_{s,c}(L^+) = 1 \qquad (3.0.27)$$

The triangular lattice is self-matching because each of its elementary polygons is triangular, and therefore, again assuming a unique critical

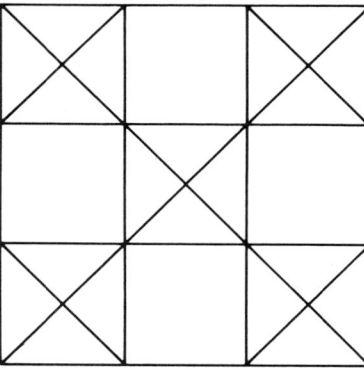

Fig. 3.0.7 A self-matching lattice.

TABLE 3.0.1 Percolation Thresholds for Site and Bond Percolation[a]

Lattice	d	q	p_c (site)	p_c (bond)
Linear chain	1	2	1^b	1^b
Simple square	2	4	0.59275^c	0.50^b
Triangular	2	6	0.500^b	0.3473^b
Honeycomb	2	3	$0.698(3)^d$	0.6527^b
Diamond	3	4	$0.4299(8)^e$	$0.3886(5)^e$
Simple cubic	3	6	$0.3117(3)^e$	$0.2492(2)^f$
BCC	3	8	$0.2464(7)^e$	$0.1795(3)^e$
FCC	3	12	0.1998^e	$0.1198(3)^e$
Cayley tree	∞	q	$\dfrac{1^*}{q-1}$	$\dfrac{1^*}{q-1}$

[a] The dimensions d, number of nearest neighbors q, and critical probability p_c for site and bond cases are given for each lattice. The figure in parentheses is probable error in last figure.
[b] Exact results.
[c] D. Stauffer, *Introduction to Percolation Theory*. Taylor and Francis (1985).
[d] M. F. Sykes and M. Glen, *J. Math. Phys.* **9** 87 (1976).
[e] D. S. Gaunt and M. F. Sykes, *J. Phys. A* **16** 783 (1983).
[f] S. Wilks, *Phys. Lett. A* **96** 344 (1983).

probability p_c,

$$p_{s,c}(\text{tri}) = \tfrac{1}{2} \qquad (3.0.28)$$

In Tables 3.0.1 and 3.0.2 we have compiled data on percolation for lattices in $d = 1, 2$, and 3 dimensions, and for the Cayley tree which in some sense corresponds to $d = \infty$.

TABLE 3.0.2 Critical Exponents for Percolation

d	α	β	γ	ν	σ	τ	D
1	1	0	1	1	1	2	1
2^b	$-2/3$	$5/36$	$43/18$	$4/3$	$36/91$	$187/91$	$91/48$
2	—	$0.138(5)^c$	$2.425(5)^d$	$1.330(5)^e$	—	—	—
3	—	$0.403(8)^f$	$1.73(3)^g$	$0.89(1)^h$	—	—	—
∞^a	-1	1	1	$1/2$	$1/2$	$5/2$	4

[a] (Cayley tree).
[b] These values are presumed to be exact.
[c] P. J. Reynolds, H. E. Stanley, and W. Klein, *J. Phys. A* **8** L198 (1978).
[d] P. J. Reynolds, H. E. Stanley, and W. Klein, *Phys. Rev. B* **21** 1223 (1980).
[e] B. Derrida, *J. Phys. A* **14** L5 (1981).
[f] D. W. Herrmann and D. Stauffer, *Z. Phys. B* **40** 113 (1981).
[g] D. S. Gaunt and M. F. Sykes, *J. Phys. A* **16** 783 (1983).
[h] D. W. Herrmann and D. Stauffer, *Z. Phys. B* **44** 399 (1981).

3.1 PERCOLATION IN ONE DIMENSION

In one dimension it is possible to solve the percolation problem exactly and we will do this for the site problem. Because the "boundary" of any cluster is just two sites, and there is only one way of arranging s sites in a connected cluster, the enumeration of lattice animals in one dimension is trivial, and the cluster numbers $n_s(p)$ are

$$n_s(p) = p^s(1-p)^2 \qquad (3.1.1)$$

All of the sums that we will encounter in the following calculation are trivial and can be performed almost by inspection. In general, we are not so fortunate, and it is often helpful to define a generating function $G(\lambda)$ by

$$G(\lambda, p) = \sum_s \lambda^s n_s(p) \qquad (3.1.2)$$

The usefulness of the generating function is that the kth moment of the distribution $n_s(p)$ can be derived from $G(\lambda, p)$ simply by differentiation:

$$\sum_s s^k n_s(p) = \left(\lambda \frac{\partial}{\partial \lambda}\right)^k G(\lambda = 1, p) \qquad (3.1.3)$$

By (3.1.1) we have, for $\lambda p < 1$,

$$G(\lambda, p) = \sum_{s=1}^{\infty} \lambda^s (1-p)^2 p^s$$

$$= (1-p)^2 \frac{\lambda p}{1 - \lambda p} \qquad (3.1.4)$$

The mean number of clusters per lattice site Γ is just the zeroth-moment of $n_s(p)$, so setting $k = 0$ in (3.1.3) we have $\Gamma(p) = G(1, p)$ and

$$\Gamma(p) = p(1-p) \qquad (3.1.5)$$

In order to calculate P we need the first moment, $\sum_s s n_s(p)$, which, by setting $k = 1$, $\lambda = 1$ in (3.1.3) and (3.1.4), is

$$\sum_s sn_s(p) = p(1 - p) + p^2 = p \tag{3.1.6}$$

By (3.0.6) it is clear that for all $p < 1$, $P = 0$. At $p = 1$ every site of the one-dimensional lattice is occupied and we have an infinite cluster, so the critical value for p in one dimension is $p_c = 1$. Because there are no values of $p > p_c$, the value of β is hard to define, but by the scaling law (3.0.18) we see that β should be taken to be 0.

The average number of sites in a cluster S is given by (3.0.9), and we find on taking a second derivative of $G(\lambda, p)$,

$$S = \frac{2p}{1 - p} \tag{3.1.7}$$

The mean number of sites per cluster diverges as $|p - p_c|^{-1}$, which tells us that $\gamma = 1$ in one dimension. Returning to (3.1.50, we have for $p \lesssim p_c = 1$,

$$\Gamma(p) \sim (p_c - p) \tag{3.1.8}$$

and therefore $2 - \alpha = 1$ or $\alpha = 1$.

For a linear cluster the radius of gyration of an s cluster is just proportional to s. By (3.0.13) then

$$\xi^2 \sim \frac{\sum_s s^4 n_s(p)}{\sum_s s^2 n(p)} \tag{3.1.9}$$

Performing the requisite derivatives, we find

$$\xi^2 \sim \frac{1 + 12p}{(1 - p)^2} \tag{3.1.10}$$

In the limit $p \to 1$ we have

$$\xi \sim \frac{1}{1 - p} \tag{3.1.11}$$

from which see that $\nu = 1$.

The cluster numbers given in (3.1.1) can be cast in the scaling form (3.0.14). First note that $p^s = e^{s \ln p}$, and that near $p_c = 1$, $\ln p \cong$

$-(p - p_c)$. Furthermore, $(1 - p)^s = s^{-2}[s(p - p_c)]^2$, so we have in one dimension

$$n_s = s^{-2} f[(p - p_c)s] \qquad (3.1.12)$$

where in this case the scaling function and exponents are known explicitly: $f(x) = x^2 e^{-x}$, with $\sigma = 1$ and $\tau = 2$. From (3.0.17), $\alpha = 1$, and from Prob. 3.3 we find $\beta = 0$ and $\gamma = 1$.

Some of the wide variety of applications of percolation are covered in Deutscher, Zallen, and Adler [2].

3.2 PERCOLATION ON A CAYLEY TREE

To further illustrate the idea of percolation, we consider another exactly solvable case, percolation on the Cayley tree. The Cayley tree, also known as a Bethe lattice, is a branched structure in which each site has f nearest neighbors and in which no two branches are allowed to cross. The Cayley tree with $f = 3$ is shown in Fig. 3.2.1. The restriction on crossing precludes the possibility of closed paths on the Cayley tree. Also note that the Cayley tree is self-similar, each branch of the tree is itself an entire tree.

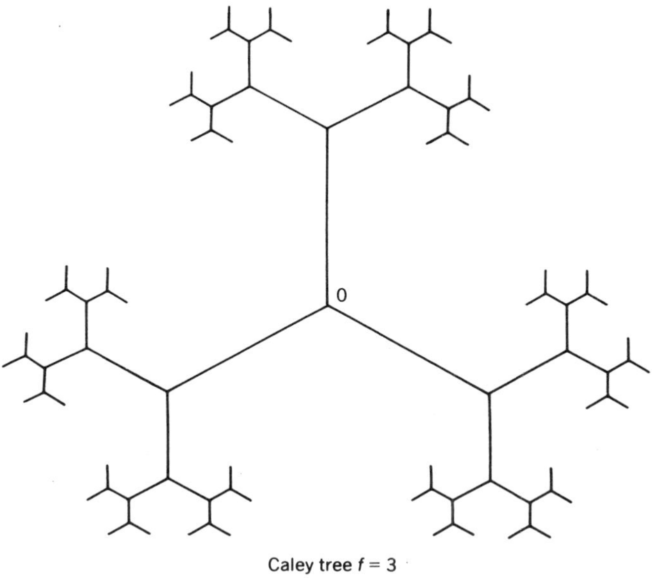

Caley tree f = 3

Fig. 3.2.1 Cayley tree with coordination number $f = 3$.

Site and bond percolation are equivalent on the Cayley tree, so for definiteness we will consider bond percolation. We first wish to establish that there is a critical value of the bond probability p_c at which percolation occurs. To do this, we follow the argument of Flory [4], which can also be found in Stauffer [10].

Imagine moving outward along a connected cluster on the Cayley tree. At each site, $f - 1$ new branches connect us to $f - 1$ new sites. On the average, $p(f - 1)$ of these bonds will be present. Because there are no closed loops on the Cayley tree, each step outward from the root of the tree is statistically independent. An infinite cluster requires an infinite sequence of at least one connected branch leading from each site. If $p(f - 1) < 1$, the probability of an infinite connected cluster must vanish, and the critical value of p is then

$$p_c = \frac{1}{f - 1} \qquad (3.2.1)$$

The "order parameter" for percolation is the probability P that a site chosen at random will belong to an infinite cluster. We will now determine P for the Cayley tree.

Following Essam [3], define $P' = 1 - P$, the probability that all clusters emanating from a given site are finite in size. Again, because each branch is statistically independent, we can write

$$P' = Q^f \qquad (3.2.2)$$

where Q is the probability that the cluster on a particular branch is finite. Because each branch is self-similar to its $(f - 1)$ subbranches, we can write, with $q = 1 - p$,

$$Q = q + pQ^{f-1} \qquad (3.2.3)$$

In (3.2.3) we have taken the probability that the first bond of the branch is absent, q, and added it to the probability that, given the first bond is not closed, the clusters on all $f - 1$ subbranches are finite in size.

For $p < p_c$ we expect $Q = 1$, and indeed this is a solution of (3.2.3), although as $p \to 1$ we expect $Q \to 0$. The behavior of $Q(p)$ is shown in Fig. 3.2.2.

We can verify our result for p_c by differentiating (3.2.3) with respect to p. At p_c, $Q = 1$ and Q' is discontinuous. We have then for $p = p_c$,

$$Q'(p_c)[1 - (f - 1)p_c] = 0 \qquad (3.2.4)$$

3.2 PERCOLATION ON A CAYLEY TREE

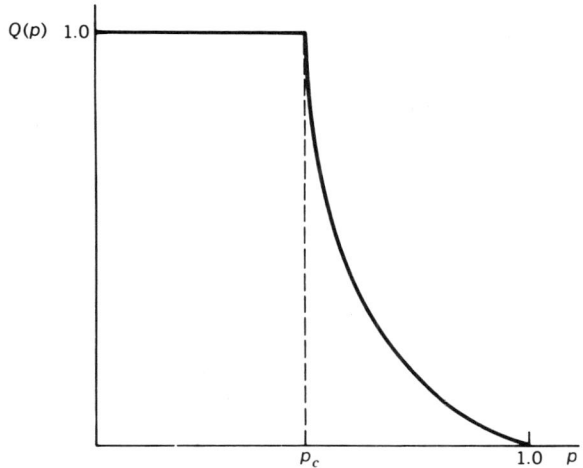

Fig. 3.2.2 The probability Q that all paths on a particular branch of a Cayley tree are finite in length as a function of the bond probability p.

For $p = p_c^-$, $Q'(p_c^-) = 0$, whereas for $p = p_c^+$, $Q'(p_c^+) \neq 0$ and (3.2.4) then leads to (3.2.1). For $p > p_c$, P is an analytic function of p, and vanishes with finite slope as $p \to p_c$. Therefore, the order parameter exponent β defined in (3.0.7) is $\beta = 1$.

The self-similarity of the Cayley tree can also be used to determine the mean cluster size S. The argument we will present here is due to Stauffer [10]. The central site, labeled "0" in Fig. 3.2.1, is connected to f branches. Because each branch is identical, each will contribute an average number B of sites to the cluster. By hypothesis the central site must be part of the cluster, so we may write

$$S = (1 + pfB) \qquad (3.2.5)$$

In the same way, each branch is composed of $f - 1$ subbranches, each of which also contributes B sites to the cluster, and we may write

$$B = [1 + p(f-1)B] \qquad (3.2.6)$$

or

$$B = (1 - (f-1)p)^{-1} \qquad (3.2.7)$$

Substituting (3.2.7) into (3.2.5) gives

$$S = \frac{p(1 + p)}{1 - (f - 1)p} \tag{3.2.8}$$

Note that S diverges as $p \to p_c = 1/(f - 1)$, and near the percolation threshold we can write

$$S \sim (p - p_c)^{-1} \tag{3.2.9}$$

which, by comparison with the definition of γ, (3.0.10), gives $\gamma = 1$.

3.3 RENORMALIZATION GROUP FOR BOND PERCOLATION ON A HIERARCHICAL LATTICE

In addition to the Cayley tree, another useful class of "artificial" lattices consists of the hierarchical lattices. The "diamond" hierarchical lattice (DHL) is illustrated in Fig. 3.3.1. Just as with fractals, a hierarchical lattice is constructed recursively by a generator. In the case of the DHL the generator consists of a cluster of four bonds which, at each iteration, replaces every bond in the lattice.

Hierarchical lattices have several pathological properties which limit their applicability. First, each site has an infinite number of near neighbors, and the ratios of the number of bonds connecting sites on different orders of iteration are different. Nevertheless, like fractals, the hierarchical lattice is self-similar, and this permits the construction of exact RG transformations in many cases. In addition, the RG transformation, which is exact for a hierarchical lattice, is sometimes related to an approximate transformation on a Bravais lattice. For example, we shall see in Section 3.5 that the diamond hierarchical lattice corresponds to an approximate RG transformation on the square lattice.

The decimation phase of the RG transformation essentially reverses the iterative procedure used in constructing the hierarchical lattice. Each elementary cluster of bonds is renormalized into a single bond. The new bond probability is equal to the probability that the "terminal sites" of the clusters, indicated by open circles in Fig. 3.3.2, are connected. Figure 3.3.3 shows each of the states of the cluster that span the cell together with their statistical weights. Summing all these

3.3 BOND PERCOLATION ON HIERARCHICAL LATTICE

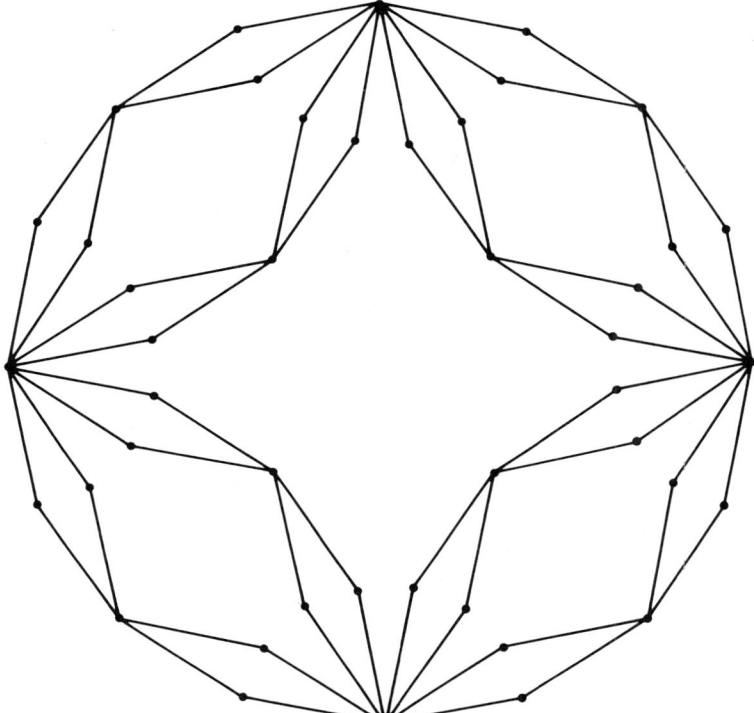

Fig. 3.3.1 The "diamond" hierarchical lattice.

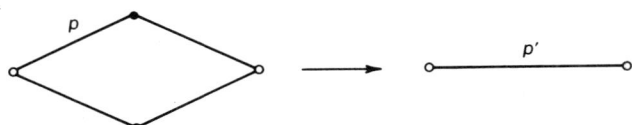

Fig. 3.3.2 Cluster of bonds renormalized to a single bond for the hierarchical lattice.

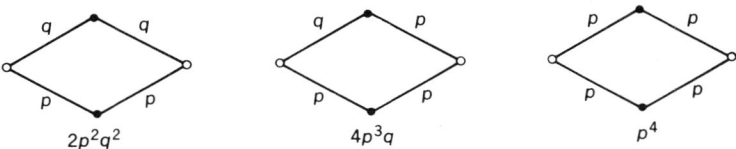

Fig. 3.3.3 States of the bond cluster that contribute to a renormalized bond together with their statistical weights.

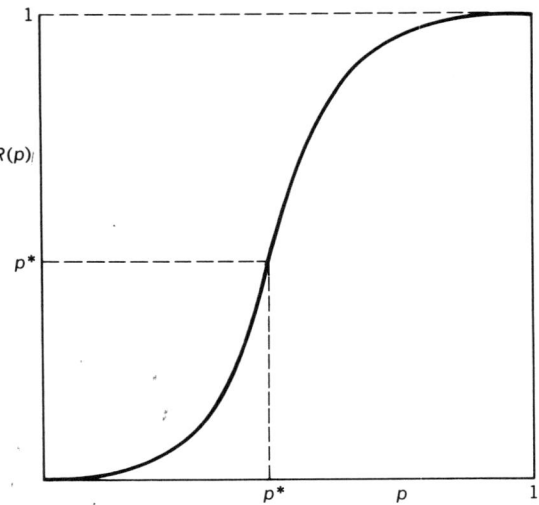

Fig. 3.3.4 The renormalized bond probability $p' = R(p)$ as a function of p.

contributions, we find ($q = 1 - p$)

$$p' = R(p) = 2p^2 - p^4 \qquad (3.3.1)$$

From Fig. 3.3.4 it is clear that for values of $p < p^*$, where $R(p^*) = p^*$, the renormalized bond probability p' is less than p. If the RG is applied many times, which corresponds to looking at the system at larger and larger scales, p tends to 0. Under the RG finite clusters are "renormalized away." Above p^* the renormalized values of p increase and approach unity, which is characteristic of the behavior of the system above percolation. Evidently, p^* corresponds to the percolation threshold p_c.

The decimated and original lattices are geometrically similar. In order to determine the scale factor b for the RG transformation, we must first have some idea about how to define distances on the hierarchical lattice. A simple approach, which is adequate for the present discussion, defines the distance between two sites as proportional to the least number of bonds needed to connect the two sites. Because the decimation transformation replaces a cell that is two bonds in length with a single bond, the scale factor in this case is $b = 2$.

We want to calculate the correlation length $\xi(p)$, which is simply the average size of a connected cluster for a given probability p. If $\xi(p)$ is the correlation length on the original lattice in units of the original lattice spacing with probability p and $\xi(p')$ is the correlation length measured in units of the coarse-grained lattice spacing on the coarse-grained lattice with probability p', then we must have

$$\xi(p) = b\xi(p') \qquad (3.3.2)$$

where the scale factor b is just the ratio of the coarse-grained lattice spacing to the original lattice spacing. Note that $\xi(p')$ must be the same function of p' as $\xi(p)$ is of p because of the self-similarity of the hierarchical lattice.

The critical value for the probability is given by

$$R(p_c) = p_c \qquad (3.3.3)$$

Again, the critical point of the system is associated with a nontrivial fixed point of the RG transformation. This is the central idea of the RG, and embodies the concept that at the critical point the system is self-similar. The renormalization equation (3.3.1) also has two "trivial" fixed points, $p = 0$ and $p = 1$. The first corresponds to an empty lattice and the second to a completely connected lattice. Generally speaking, each distinct "phase" of a system will be characterized by a trivial fixed point, whereas the critical point is associated with a nontrivial fixed point.

Setting $p = p' = p_c$ as in (3.3.3), (3.3.2) then has only two solutions, $\xi = 0$ and $\xi = \infty$; the latter possibility corresponds to critical behavior as indicated in (3.0.11).

Equation (3.3.3) states that the critical probability is a *fixed point* of the RG transformation. The critical exponents, which characterize the behavior of the system near the critical point, are determined by examining the RG equations in the neighborhood of the fixed point. If p is close to p_c, then ξ is given by (3.0.11) and (3.3.2). The RG equation for p, (3.3.1), is analytic, and we may therefore write, for p close to p_c,

$$p' - p_c = \frac{\partial R}{\partial p}(p_c)(p - p_c) \qquad (3.3.4)$$

Substituting (3.3.4) into (3.3.2) and using the scaling form (3.0.11), one finds

$$|p - p_c|^{-\nu} = b[R'(p_c)]^{-\nu}|p - p_c|^{-\nu} \qquad (3.3.5)$$

which gives for the correlation length exponent

$$\nu = \frac{\ln b}{\ln R'(p_c)} \qquad (3.3.6)$$

Note the similarity of this expression for the fractal dimension, (1.0.2) and (1.1.3). In Prob. 3.4 you are asked to calculate p_c and ν from (3.3.3) and (3.3.6). These results should be compared with the results of numerical studies for the square lattice, $p_c = 0.593$ and $\nu = 1.33$.

In Sec. 3.3 we treated the RG for the rather artificial hierarchical lattice. Although these calculation contain all the features of the RG on real lattices, a true Bravais lattice does not have the built in self-similarity which permits an exact solution, and so approximate RG transformations must be constructed. We shall see that one has a great deal of freedom in constructing approximate RG transformations, and it is here that one's physical insight into the problem is essential.

There is strong numerical evidence [5] that the phenomenon of percolation (both site and bond) is universal for all Bravais lattices in a given (Euclidean) dimension. What this means is that, although the percolation thresholds p_c for site percolation and bond percolation on a given lattice are different, and on different lattices they are different, the values of the critical exponents β, γ, ν, and so on are the same for all lattices in a given dimension. Furthermore, the addition of second-neighbor couplings lowers the percolation threshold, but does not change the critical exponents. This is true of the Cayley tree where the critical probability depends on the coordination number, but the exponents do not.

Because an exact solution of the percolation phenomena is lacking even in two dimensions, the universality of percolation has not been proved.

In a more restricted context the equivalence of site and bond percolation on a lattice and its covering lattice, the equivalence of bond percolation on a lattice and its dual, and the equivalence of site percolation on a lattice and its matching lattice all support the conjecture of universality. In the context of critical phenomena, one of the most significant results of the Wilson RG [11] is that within the ε expansion the critical exponents of a system depend on only two parameters; the number of components of the order parameter and the dimensionality of the space.

3.4 RENORMALIZATION GROUP FOR THE TRIANGULAR LATTICE

Physically, the universality hypothesis is quite reasonable. Remember that the critical region near percolation is characterized by a loss of scale; connected clusters of all size appear and in addition average properties, such as the mean cluster size, are dominated by scaleless fluctuations. Local characteristics of the lattice, such as the number of nearest neighbors and second neighbors, determine the value of the percolation threshold and perhaps the width of the critical region, but the critical exponents of the system are determined by the universal, large-scale fluctuations.

3.4 MAJORITY-RULE RENORMALIZATION GROUP FOR THE TRIANGULAR LATTICE

To give some idea of the flexibility one has in defining an RG transformation, we first consider a very simple approximate RG transformation, which gives good results for site percolation on the triangular lattice (see Fig. 3.4.1a). The value of p_c comes out exactly, and ν agrees to within a few percent of the best theoretical estimate.

The RG for this problem is based on breaking up the original lattice into clusters of three sites each, as shown in Fig. 3.4.1b. Note that the clusters themselves form a triangular lattice. The RG transformation maps a cluster on the original lattice to a site on a new lattice with a lattice constant that is larger by the factor $\sqrt{3}$. A cluster maps into an occupied site if a majority of the sites in the cluster are occupied, and the cluster maps into an empty site otherwise.

This is an example of a majority-rule RG. The states of the cluster that contribute to an occupied site on the renormalized lattice are shown in Fig. 3.4.2 together with their statistical weights. The renormalization equation for the probability that a site is occupied is therefore

$$p' = R(p) = 3p^2 - 2p^3 \quad (3.4.1)$$

In addition to the usual trivial fixed points $p^* = 0$ and $p^* = 1$, this transformation has a fixed point at $p^* = 0.5$, which is known to be the exact percolation threshold for the triangular lattice because the triangular lattice is self-matching. The value of the order parameter exponent ν is calculated as in (3.3.6). The scale factor for this

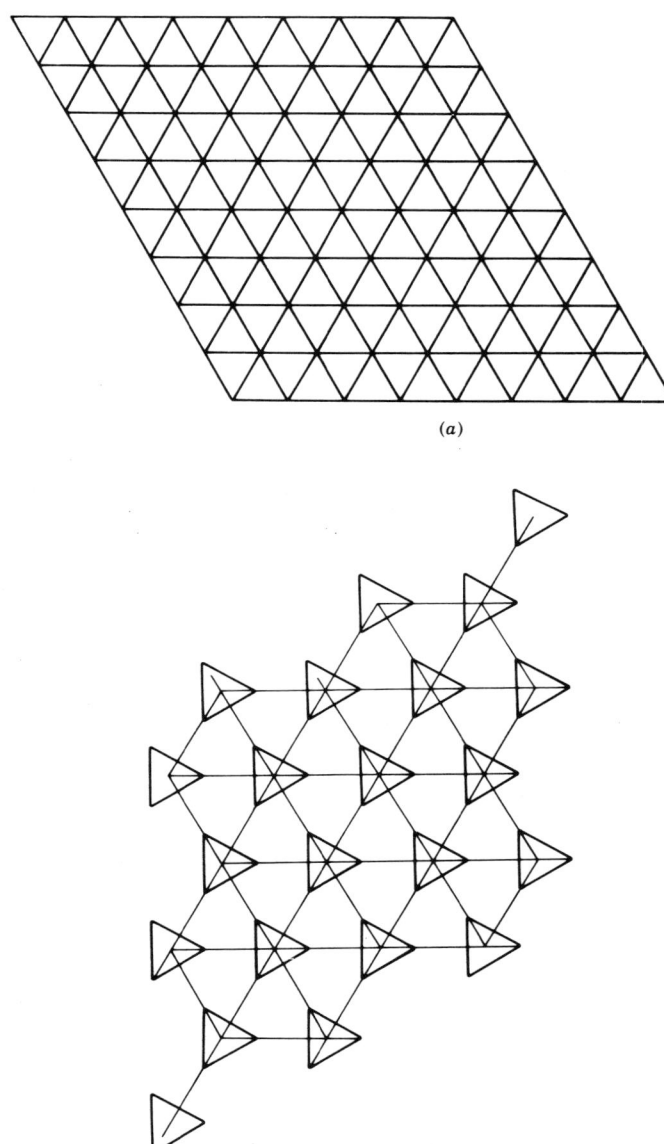

Fig. 3.4.1 (*a*) Original triangular lattice. (*b*) Triangular lattice of three-site cells.

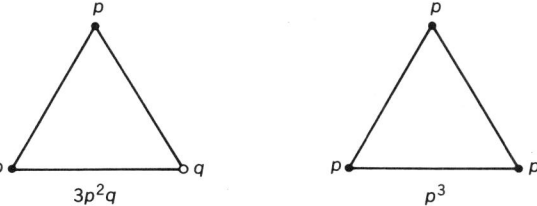

Fig. 3.4.2 Configurations of the three-site cell on the triangular lattice that contribute to the renormalized site probability p'.

transformation is $\sqrt{3}$, which gives for ν,

$$\nu = \frac{1}{2}\frac{\ln 3}{\ln 3/2} = 1.3547\ldots \tag{3.4.2}$$

This value agrees remarkably well with the best estimate from Monte Carlo studies of the square lattice [7] and the conjecture of den Nijs [1] that ν is 4/3. To some extent the agreement with the exact value for ν must be considered fortuitous. To see the nature of the approximation made, consider the two clusters shown in Fig. 3.4.3.

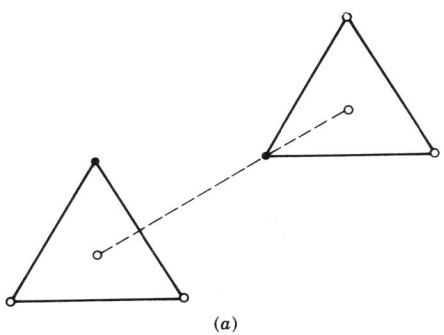

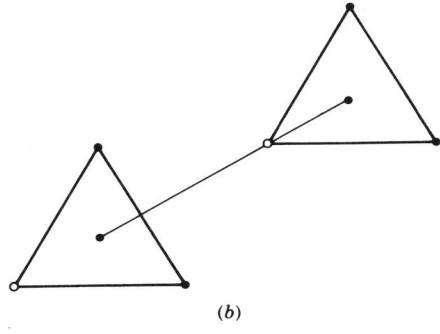

Fig. 3.4.3 (a) Connected cells that map into disconnected sites under the majority rule RG. (b) Disconnected cells that map into disconnected sites under the majority rule RG.

In the first case, the two cells are connected, but the majority rule maps them on to a disconnected pair, whereas in the second case no connection exists between the cells, and yet the renormalized sites are connected. This cancellation of errors makes it difficult to evaluate the accuracy of the approximation a priori.

3.5 SITE PERCOLATION ON THE SQUARE LATTICE: CELL RENORMALIZATION

Despite its apparent success, the majority rule does not incorporate the connectivity properties which are at the heart of percolation phenomena. Here we consider a group of cell or block transformations defined by the following rule: If a connected cluster spans the cell from left to right, then the cell maps to an occupied site; otherwise the cell maps to an empty site. There are other possible

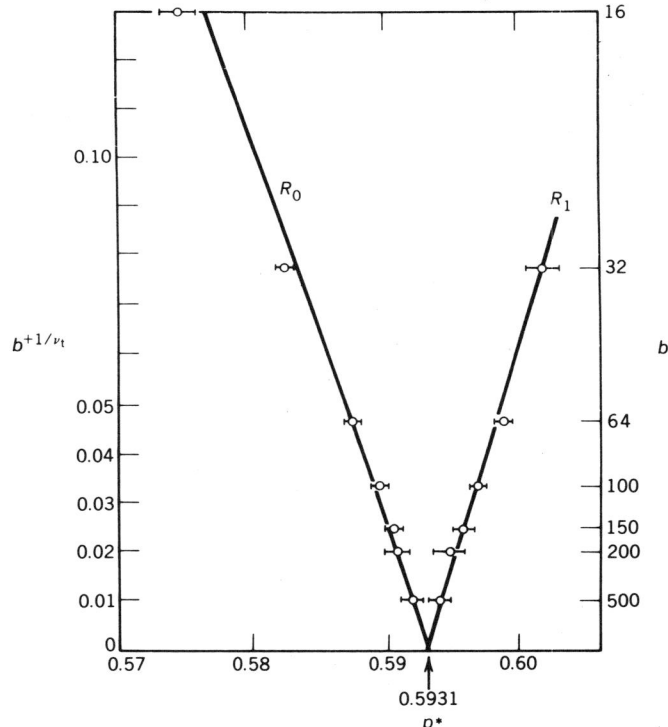

Fig. 3.5.1 Extrapolation of large-cell Monte Carlo calculations on the square lattice [7].

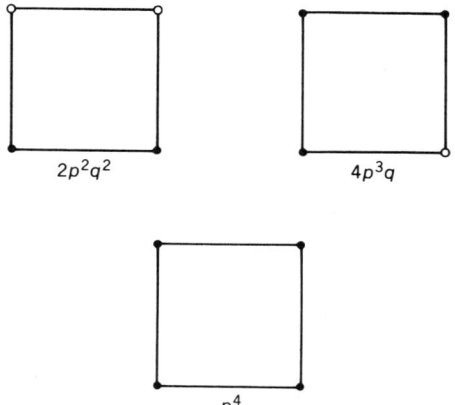

Fig. 3.5.2 Configurations of the four-site cluster on the square lattice that contribute to the renormalized site probability p'.

rules that one might apply. For example, one could count states that span either left to right or top to bottom, or only count states that span both left to right and top to bottom. In the limit of very large cells, these different rules should yield results for p_c and ν that converge to the exact value.

One advantage of the large-cell method [7] is that, although the accuracy of the approximation cannot be determined a priori, it can be improved systematically by considering cells of increasing size. Figure 3.5.1 shows the value of p^* determined by Monte Carlo methods for cells of up to 500 lattice spacings in size [7].

For small cells of up to five lattice spacings in size, it is possible to calculate the RG transformation exactly. Figure 3.5.2 shows the states of a cell of size $b = 2$ that span the cell left to right together with their proper statistical weights. The renormalization transformation is therefore

$$p' = R(p) = 2p^2 - p^4$$

As usual there are the trivial fixed points $p^* = 0$ and $p^* = 1$, and in addition there is the nontrivial fixed point $p^* = 0.618$. Evaluating ν as in (3.3.6), we find $\nu = 1.635$.

Note that this is exactly the same transformation as was derived for the hierarchical lattice, (3.3.1). As we mentioned in Sec. 3.1, some approximate RG transformations on Bravais lattices are exact on hierarchical lattices, which is one reason a good deal of research has gone into understanding systems on hierarchical lattices and their correspondence with more physical lattices.

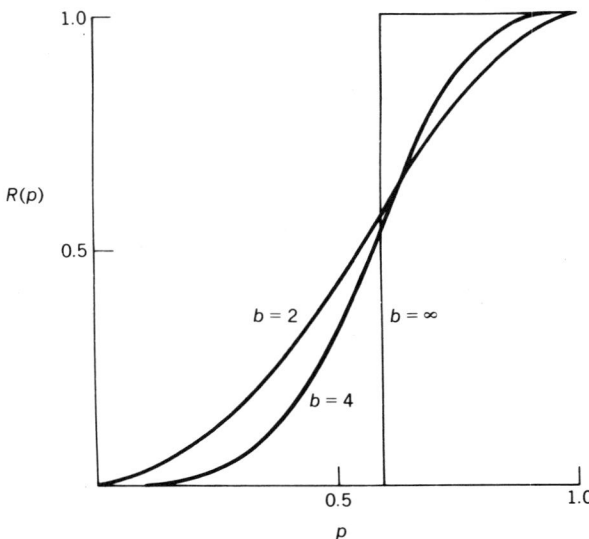

Fig. 3.5.3 Renormalization transformations for cells of increasing size.

The best estimate for the percolation threshold on the square lattice is $p_c = 0.593$, so our simple RG does not do too badly. The value for ν is not as good, and to improve on it one really needs to study fairly large cells. You are invited in Prob. 3.8 to construct the renormalization transformation for a cell of size $b = 3$. The behavior of $R_b(p)$ for increasingly large values of b is shown in Fig. 3.5.3. Note that as b increases the function $R(p)$ tends to a step function.

The problem of percolation remains essentially unsolved, and yet it is deceptively easy to formulate: Analytic approaches have thus far been limited by the necessity of counting nearly every possible state of the cell, which becomes impractical, even on a computer, for cells larger than $b = 5$. Monte Carlo methods can handle larger cells, but convergence requires truly enormous ($b = 500$) cells, which limits the number of configurations that can be sampled. It will be interesting to see how this problem is eventually solved.

3.6 REAL-SPACE RENORMALIZATION GROUP FOR BOND PERCOLATION

A renormalization group treatment for bond percolation can be formulated along lines similar to those of the self-avoiding walk of

3.6 REAL-SPACE RENORMALIZATION GROUP FOR BOND PERCOLATION

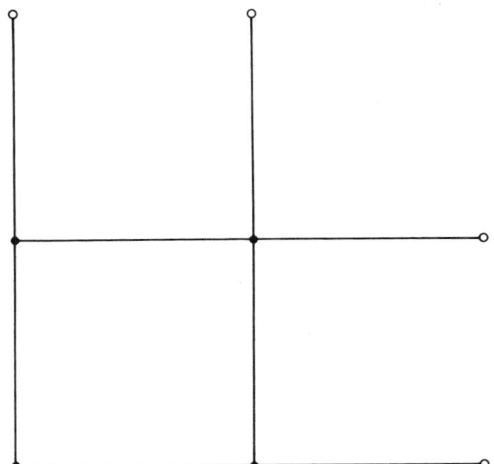

Fig. 3.6.1 Cell used in the RSRG for bond percolation.

Chap. 1. The basic cell, shown in Fig. 3.6.1, for which $b = 2$, is the same as was used for the SAW problem.

The renormalized horizontal bond is open if there is a connected path from the left to the right edge, and the vertical renormalized bond is open if there is a connected path from the bottom to the top edge. Figure 3.6.2 shows the configurations of bonds that contribute to an open renormalized bond. Note that three of the bonds have no effect on the connectivity of the cell and can be ignored.

From this we see that the renormalization transformation is

$$p' = R(p,q) = p^5 + 5p^4 p + 8p^3 q^2 + 2p^2 q^3$$
$$= 2p^5 - 5p^4 + 2p^3 + 2p^2 \qquad (3.6.1)$$

As in the case of site percolation on the triangular lattice, this RG transformation gives p_c exactly; solving for the nontrivial fixed point, one finds $p_c = 0.5$. The scale factor for the transformation is $b = 2$, which gives for ν,

$$\nu = \frac{\ln b}{\ln R'(p_c)} = 1.428 \qquad (3.6.2)$$

98 RENORMALIZATION APPROACH TO PERCOLATION

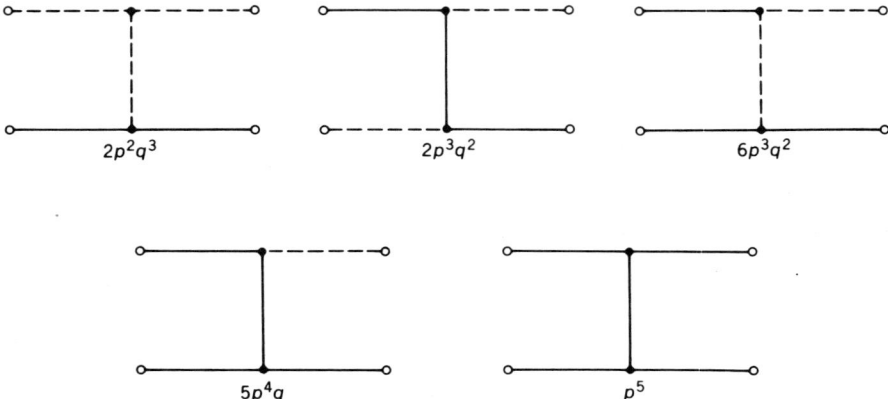

Fig. 3.6.2 Contributions to renormalized horizontal open bond together with their associated statistical weights probability p.

There is an interesting analogy between the bond percolation problem and a random network of resistors (Fig. 3.6.3). If each relevant bond in the cell is replaced by a resistor whose value is r_0 with probability $1 - p$ or 0 with probability p (connected), then the renormalization function $R(p)$ is the sum over all states of the network for which the equivalence resistance is 0, weighted by the probability of finding that configuration. The probability distribution for each individual resistor can be expressed as a sum of two delta functions

$$P(r) = p\delta(r) + (1 - p)\delta(r - r_0) \tag{3.6.3}$$

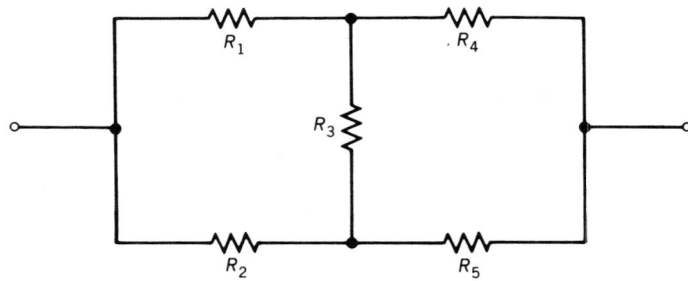

Fig. 3.6.3 Network of resistors equivalent to the cell of bonds, Fig. 3.6.1.

Then $R(p)$ is given by

$$R(p) = \int dr_1 \cdots dr_5 \, \delta(r_{\text{eq}}) p(r_1) p(r_2) \cdots p(r_5) \quad (3.6.4)$$

where r_{eq} is the equivalent resistance of the network.

3.7 SITE–BOND DILUTION: THE SOL–GEL TRANSITION

A beautiful application of the methods that have been developed thus far is the real-space renormalization group treatment of the sol–gel phase transition due to Nakanishi and Reynolds [6]. The gellation transition corresponds to the formation of an infinite connected network of bonds between gel molecules. In the sol phase the static shear modulus and elastic properties of the solution are those of a viscous fluid. In the gel phase the static shear modulus is nonzero and the system behaves as a solid. In the model presented here, each site of the lattice can be occupied by a solvent molecule with probability $1 - s$ or a gel molecule with probability s (s stands for "site"). If two gel molecules happen to be neighbors, then there is also a probability b that a bond will form between them. In a physically realistic model this probability of bond formation would depend on external factors such as pH and temperature. The only approximation in this picture is that we assume the sites are occupied randomly without correlation. In a real sol–gel solution one would expect that molecular forces would introduce some correlation. The general wisdom, however, tells us that these microscopic details are probably "irrelevant," and the critical behavior of the real system is essentially identical to that of the model system. The universality of the results of RG calculations is one of their great strengths.

The cell we will use is the same as we used for bond percolation, but the criterion for the renormalized bond to be present will be modified slightly because we must take into account the possibility that a site may be empty (occupied by a solvent molecule). Following Nakanishi and Reynolds, we will use the "symmetric" rule for site percolation: The cell will renormalize to an occupied site if a connected cluster spans from left to right or top to bottom.

To find the renormalized site probability s', consider first the configuration with all sites occupied (Fig. 3.7.1a). In this case the cluster will span unless all the bonds are absent, which occurs with probability $(1 - b^4)$, so the contribution to the renormalized site

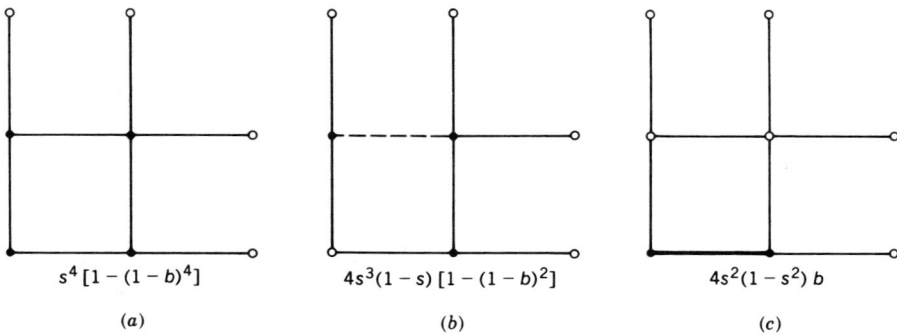

Fig. 3.7.1 Contributions of the renormalized site probability in the site-bond percolation problem. (*a*) All sites occupied. (*b*) Three sites occupied. (*c*) Two sites occupied.

probability from states with four occupied sites is

$$s^4\left[1 - (1-b)^4\right] \tag{3.7.1}$$

Now consider states in which three sites are occupied, as shown in Fig. 3.7.1*b*. In this case the only way the cell is not spanned is if at least two of the bonds are absent. By symmetry the four different states with three occupied sites are equivalent, so the contribution of these states to the renormalized site probability is

$$4s^3(1-s)\left[1 - (1-b)^2\right] \tag{3.7.2}$$

Finally, with just two sites occupied, as in Fig. 3.7.1*c*, the bond connecting them must be present so the contribution of this term is

$$4s^2(1-s)^2 b \tag{3.7.3}$$

Adding these three terms together, the renormalized site probability s' is

$$s' = s^4\left[1 - (1-b)^4\right] + 4s^3(1-s)\left[1 - (1-b)^2\right] + 4s^2(1-s)^2 b \tag{3.7.4}$$

Note that if we set $b = 1$ (all bonds present) we regain the renormalization equation for s using the symmetric rule (Prob. 3.8).

3.7 SITE-BOND DILUTION: THE SOL-GEL TRANSITION

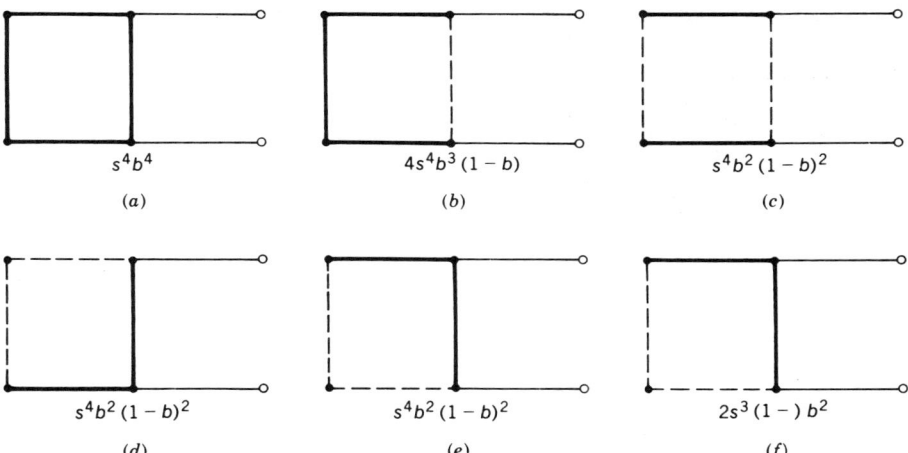

Fig. 3.7.2 Bond configurations with three or four occupied sites that contribute to the renormalized horizontal bond if either terminal bond is open. Bonds shown as dotted lines are closed.

Calculating the renormalized bond probability is a little trickier. The renormalized bond must start on an occupied site and connect with the opposite edge, so what we are really interested in is the renormalized product $s'b'$, rather than b' alone.

To begin, define the bonds that connect sites in the cell to sites in a neighboring cell as "terminal bonds". We consider two classes of states: those that span if either of the two terminal bonds are present and those that span only if a particular terminal bond is present.

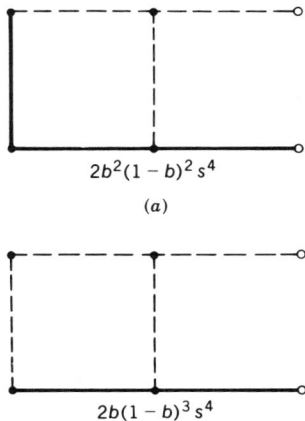

Fig. 3.7.3 Bond configurations with four occupied sites that contribute to the renormalized horizontal bond. Note that one terminal bond must be closed.

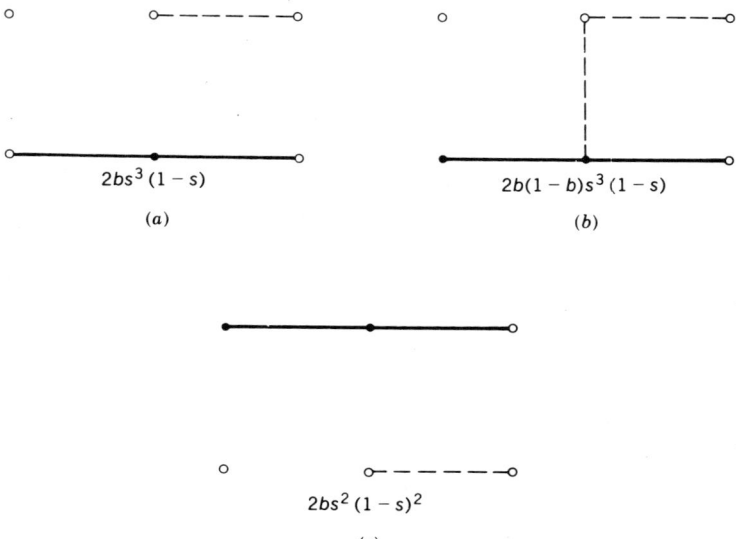

Fig. 3.7.4 Bond configuration with two and three occupied sites and one closed terminal bond that contribute to the renormalized horizontal bond.

Figure 3.7.2 shows all configuration of bonds of order s^4 and stand s^3 that contribute to the renormalized horizontal bond (three or four sites in the cell occupied). Because all of these terms span if either terminal bond is present, they are multiplied by the factor $[1 - (1 - b)^2]$.

The remaining states that contribute to the renormalized bond probability with all four sites occupied are shown in Fig. 3.7.3. Note that these states span only if the indicated terminal bond is present. Bonds that must be absent (to avoid overcounting) are shown as dotted lines.

Contributions from states with two and three sites occupied are shown in Fig. 3.7.4. Adding the contributions from all these terms, we have for the renormalized product $s'b'$,

$$s'b' = \left\{s^4\left[b^4 + 4b^3(1-b) + 3b^2(1-b)^2\right] \right.$$
$$\left. + 2s^3(1-s)b^2\right\}\left\{1 - (1-b)^2\right\}$$
$$+ 2s^4\left[b^3(1-b)^2 + b^2(1-b^3)\right]$$
$$+ 2s^3(1-s)b^2(2-b) + 2s^2(1-s)^2b^2 \qquad (3.7.5)$$

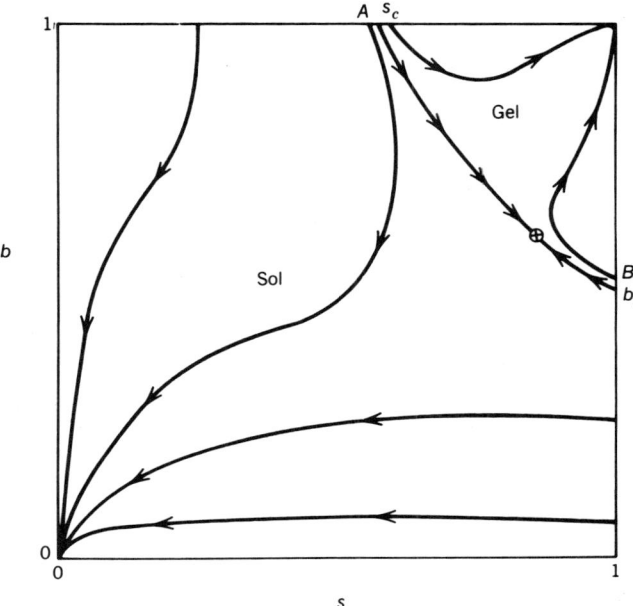

Fig. 3.7.5 Global RG trajectories for the sol–gel problem. The flow on the critical line with the end points A at $(s_c, 1)$ and B at $(1, b_c)$ is toward the nontrivial fixed point x.

The nontrivial fixed point of the coupled RG equations (3.7.4) and (3.7.5) can be found numerically with the result $s^* = 0.879$ and $b^* = 0.586$. These values for s and b have no particular physical significance. To get a feeling for the phase diagram of the sol–gel system, we must examine the RG trajectories shown in Fig. 3.7.5. Note that the critical line AB separates the RG trajectories into two classes; those that flow to the trivial fixed point $s = 0$, $b = 0$ and those that flow to the other trivial fixed point $s = 1$, $b = 1$. For values of s and b on the critical line, the RG trajectory is into the nontrivial fixed point (s^*, b^*). The end points of the critical line correspond to pure site (A) and pure bond percolation (B). Numerically, the intercepts are $s_c = 0.570$ (0.593 exact) and $b_c = 0.5$ (exact). The fact that the critical line connects these two points in the phase diagram, and that the critical behavior is determined by a single fixed point, constitutes a demonstration, within this approximation, of the universality of site and bond percolation.

As in the case of the one-parameter RG transformations we have studied so far, the critical exponents are determined by the properties of the linearized RG in the neighborhood of the fixed point. In this

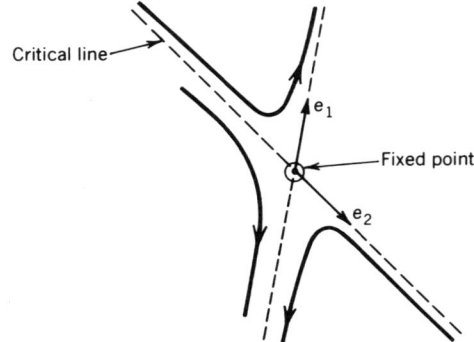

Fig. 3.7.6 RG trajectories in the neighborhood of the sol–gel fixed point.

case the linearized RG equations can be written as

$$\begin{bmatrix} \delta s' \\ \delta b' \end{bmatrix} = \begin{bmatrix} \dfrac{\partial s'}{\partial s} & \dfrac{\partial s'}{\partial b} \\ \dfrac{\partial b'}{\partial s} & \dfrac{\partial b'}{\partial b} \end{bmatrix}_{\substack{s=s^* \\ b=b^*}} \begin{bmatrix} \delta s \\ \delta b \end{bmatrix} \qquad \begin{array}{l} \delta s = s - s^* \\ \delta b = b - b^* \end{array} \qquad (3.7.6)$$

The eigenvalues λ_1 and λ_2 of the matrix of first derivatives and the associated eigenvectors e_1 and e_2 characterize the RG trajectories in the neighborhood of the fixed point (Fig. 3.7.6). The eigenvalues can be determined numerically with the result

$$\lambda_1 = 1.60 \qquad \lambda_2 = 0.536 \qquad (3.7.7)$$

Because λ_1 is greater than unity, repeated applications of the RG will cause small perturbations in the direction of e_1 to grow exponentially, and the RG trajectory will run away from the fixed point. On the other hand, because λ_2 is less than unity, a small displacement from the fixed point in the direction of e_2 will decrease in amplitude as one applies the RG, and the trajectory will flow into the fixed point (along the tangent to the critical line). In the language of the RG, the linear scaling field e_1 is relevant because it gives rise to nonanalytic behavior in the cluster distribution function, and so on. The linear scaling field e_2 is called irrelevant because no matter what the initial amplitude is in the e_2 direction, it tends to zero exponentially under repeated renormalization and does not contribute to the dominant nonanalytic behavior.

PROBLEMS

3.1 Prove the identity

$$P + \sum_{s} n_s(p) = p$$

where $n_s(p)$ is the average number of finite clusters of s sites.

3.2 Calculate the number of lattice animals g_{sb} for s up to 4 on the simple square lattice.

3.3 Near percolation, the cluster distribution function $n_s(p)$ scales according to (3.0.14)

$$n(s) \sim s^{-\tau} f[(p - p_c)s^\sigma]$$

(a) Show that

$$\frac{\sum_s s^2 n_s(p)}{\sum_s s n(p)} \sim |p - p_c|^{-\gamma}$$

and relate γ to the exponents σ and τ.

(b) Using the identity of Prob. 3.1, show that

$$P \sim (p - p_c)^\beta \qquad p < p_c$$

and relate β to σ and τ.

(c) For percolation on the Cayley tree, $\beta = 1$ and $\gamma = 1$. What are the values of σ and τ for the Cayley tree?

3.4 Calculate the critical probability and correlation exponent for bond percolation on the diamond hierarchical lattice (Fig. 3.3.1).

3.5 Construct the dual of the Kagomé lattice, shown in Fig. 3.P.1.

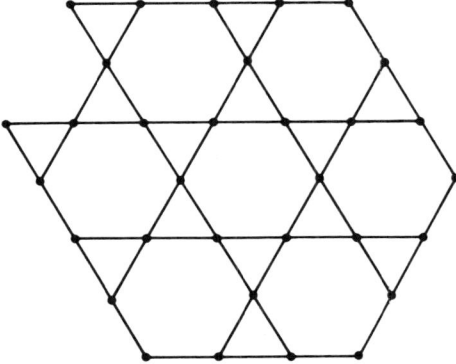

Fig. 3.P.1 The Kagomé lattice (Prob. 3.5).

3.6 Solve (3.2.3) for $Q(p, q)$ explicitly for $d = 3$, and find p_c.

3.7 For site percolation on the square lattice, use the alternate rule that a cell is "occupied" if it can be spanned either left to right or top to bottom. Find the fixed point and calculate the critical exponent.

3.8 Construct the RG transformation for site percolation on the square lattice using a 3×3 cell. Write a simple computer program to find the fixed point and evaluate the correlation exponent. There are $2^q = 512$ possible states of the cell so you really want to be a little clever; use symmetry, and so on, and for six to nine occupied sites it is easier to count states that do not span the cell rather than those that do.

3.9 Show that the percolation threshold for nearest-neighbor site percolation in two dimensions must satisfy $p_c \geq 0.5$. Does a similar restriction hold for bond percolation? *Hint:* If p_c is less than 0.5, then both occupied and empty sites percolate for values of p in the range $p_c < p < 1 - p_c$. Note that for the square lattice ($p_c = 0.593$) there is a range of values of p in which neither occupied nor empty sites percolate.

3.10 Construct the RG transformation for site percolation on the triangular lattice using the cell shown in Fig. 3.P.2. Note that the triangular lattice can be thought of as a square lattice with an extra bond connecting one diagonal.

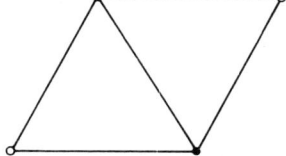

Fig. 3.P.2 Four-site cell for site percolation on the triangular lattice (Prob. 3.10).

3.11 Following the method outlined here, calculate the coupled RG equations for the site–bond problem using the left-to-right rule rather than the symmetric rule.

3.12 For the RG transformation (3.3.1), show that the trivial fixed points $p^* = 0$ and $p^* = 1$ are stable in the sense that small perturbations from these fixed points tend to 0 under repeated renormalizations.

3.13 For the diamond hierarchical lattice show that

$$\lim_{k \to \infty} \frac{N_{k+1}}{N_k} \to 4 = b^{d_e}$$

where N_k is the number of sites in the lattice after k iterations of the generator and d_e is the effective Euclidean dimension of the lattice. Because $b = 2$, $d_e = 2$, and the hierarchical lattice is effectively two dimensional.

3.14 Derive the scaling law (3.0.18).

REFERENCES

1. den Nijs, M. P. M. (1979) *J. Phys. A* **12** 1857.
2. Deutscher, G., Zallen, R., and Adler, J. (1983) *Percolation Structure and Processes*. Adam Higler, Bristol.
3. Essam, J. W. (1972) In *Phase Transitions and Critical Phenomena*. (C. Domb and M. S. Green, eds.) vol. 2. Academic, New York.
4. Flory, P. J. (1941) *J. Amer. Chem. Soc.* **63** 3083.
5. Marro, J. (1976) *Phys. Lett. A* **59** 180.
6. Nakanishi, H. and Reynolds, P. J. (1979) *Phys. Lett. A* **41** 252.
7. Reynolds, P. J., Stanley, H. E., and Klein, W. (1980) *Phys. Rev. B* **21** 1223.
8. Stanley, H. E., Reynolds, P. J., Redner, S., and Family, F. (1982) In *Real Space Renormalization* (T. W. Burkhardt and J. M. J. van Leeuwen, eds.) Springer, Berlin.
9. Stauffer, D. (1979) *Phys. Rep.* **54C** 1.
10. Stauffer, D. (1985) *Introduction to Percolation Theory*. Taylor and Francis, London.
11. Wilson, K. G. and Kogut, J. (1974) *Phys. Rep.* **12C** 75.

4

RENORMALIZATION GROUP AND CRITICAL PHENOMENA

4.0 INTRODUCTION

Figure 4.0.1 shows the phase diagram of a typical chemically pure substance. The pressure–temperature plane is divided into regions in which the system exists in a homogeneous thermodynamic phase. Within each phase the properties of the system are analytic functions of the temperature and pressure, the correlation length is of microscopic dimensions, and in principle one can compute the free energy of the system.

On the other hand, at the boundaries between phases the properties of the system change suddenly, and nonanalytic behavior in the free energy is manifest. Clearly, analytic approximations will be of little use in describing the transition from one phase to another, and it is here that RG methods have proved most effective.

In discussing a phase transition, it is useful to consider an operator whose thermodynamic average vanishes identically in one phase, which we call the "disordered" phase, and is nonzero in the "ordered phase." The ensemble average of this operator is called the order parameter. If the order parameter changes discontinuously across the phase boundary, as, for example, in the familiar case of melting, then the phase transition is said to be "first order." If the order parameter is a continuous function, rising from 0 as the boundary is crossed, then the transition is called "second order," and the thermodynamic func-

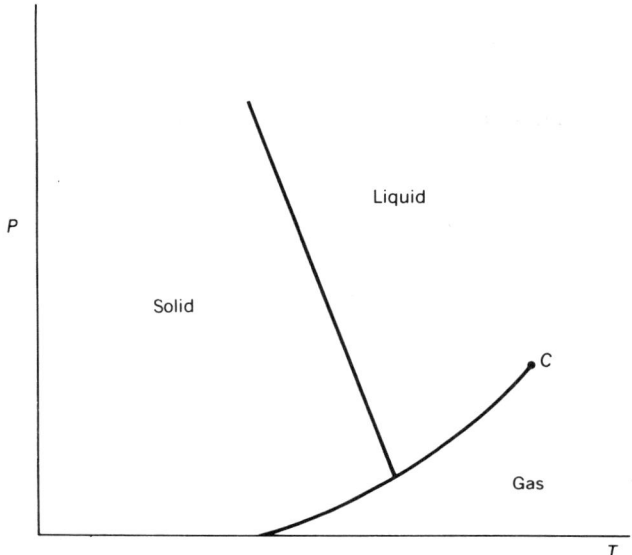

Fig. 4.0.1 Phase diagram for a typical pure substance. Point C is the liquid–gas critical point.

tions exhibit "critical," or scaling, behavior. The end point of the liquid–gas phase boundary, labeled C in Fig. 4.0.1, is the critical point and the liquid–gas transition at this point is second order.

The essential ingredient in critical behavior is the divergence of the correlation length ξ at the critical point. The correlation length can be though of as the distance over which fluctuations in the order parameter are correlated, and it sets a natural scale in the system. At the critical point the correlation length is infinite and fluctuations in the order parameter are strong on all length scales. At the liquid–gas critical point, where the order parameter is essentially the difference in density between the liquid and gaseous phases, light is so strongly scattered by the critical density fluctuations that the fluid becomes opaque, a phenomenon called "critical opalescence." In magnetic systems these critical fluctuations lead to a divergence in the magnetic susceptibility.

From the point of view of the RG, it is again the loss of scale as ξ tends to ∞ that leads to scaling behavior. Taking the temperature to be the parameter that controls how far the system is from the critical point, the critical behavior of the basic thermodynamic functions is summarized in Table 4.0.1, which also defines the critical exponents α, β, γ, and ν.

TABLE 4.0.1 Critical Exponents for the Basic Thermodynamic Functions

Order parameter	$m \sim (T_c - T)^\beta$	$T < T_c$
Specific heat	$C \sim |T - T_c|^{-\alpha}$	
Susceptibility	$\chi \sim |T - T_c|^{-\gamma}$	
Correlation length	$\xi \sim |T - T_c|^{-\nu}$	

TABLE 4.0.2 Scaling Relations for the Order Parameter m and Correlation Function

Order parameter	$m \sim h^{1/\delta}$	$T = T_c$
Correlation function	$G(r) \sim r^{-d+2-\eta}$	$T = T_c$

In addition, at $T = T_c$ the correlation function $G(r)$ decays as a power of r and the order parameter scales with an external field h, which couples directly to the order parameter, as shown in Table 4.0.2.

One of the central goals of the RG approach is to calculate the critical exponents α, β, γ, δ, ν, and η. As we will see this task is considerably simplified by the existence of relations, or scaling laws, between the exponents which reduce the number of independent exponents to just 2.

4.1 CONSTRUCTION OF THE RENORMALIZATION GROUP TRANSFORMATION

In this section we will consider some particular ways in which the RG transformation can be constructed and investigate the general behavior of the RG transformation in the neighborhood of a fixed point.

We begin with the Hamiltonian H for the system we are interested in, which is a function of a set of dynamical variables, or fields, s_i. The Hamiltonian is usually composed of simple, local, functions of the fields O_k, which we will refer to as "operators." Each such operator in the Hamiltonian appears multiplied by a coefficient, or parameter h_k, which we say is conjugate to that operator. In general, we have

$$H = \sum_k h_k O_k \tag{4.1.1}$$

One can think of the set of parameters $\{h_k\}$ as comprising a space, and each point in this space represents a particular Hamiltonian. Generally, this space is infinite in dimension and it is usually necessary to identify those operators that are physically most relevant and truncate parameter space to include just these.

The RG transformation is formulated in terms of the partition function Z, which is given as a trace over the fields

$$Z = \operatorname*{Tr}_{\{s_i\}} e^{-H[\{s_i\}]} \qquad (4.1.2)$$

Note that the usual factor of $\beta = 1/(k_B T)$ has been absorbed into the definition of H. Coarse-grained variables $\{S_i\}$ are introduced through the projection operator $T(\{S_i\}, \{s_i\})$, which contains all the information connecting the coarse-grained state with the original state. The Hamiltonian for the coarse-grained variables is given by

$$e^{-H'[\{S_i\}]} = \operatorname*{Tr}_{\{s_i\}} T(\{S_i\}, \{s_i\}) e^{-H[\{s_i\}]} \qquad (4.1.3)$$

Not every choice of the projection operator T will lead to a meaningful RG transformation, and in constructing T we must let physical insight guide us. The transformation must have the following properties:

1. The fundamental length scale (e.g., the lattice spacing a or the inverse of the momentum cutoff Λ^{-1}) in the two Hamiltonians are related by the scale factor $b > 1$,

$$\frac{a'}{a} = b \qquad (4.1.4)$$

2. The coarse-grained, or renormalized, variables must be of the same kind as the original. The RG transformation should not change the dimension or symmetry of the internal space of the dynamical variables.
3. The states of the original system corresponding to homogeneous phases should transform into states of the same kind under the RG. In particular, the RG must preserve the character and symmetry of the ordered phases of the system. This ensures that the thermodynamic phases will correspond to stable fixed points of the RG transformation.

4. The coarse-grained partition function, and therefore the coarse-grained free energy, should be numerically equal to the original free energy. This invariance of the free energy resembles the more familiar symmetry properties of Hamiltonian systems and lends some support to the "group" aspect of the RG. This constraint can be satisfied by requiring that

$$\operatorname*{Tr}_{\{s_i\}} T(\{S_i\}, \{s_i\}) = 1 \tag{4.1.5}$$

The first restriction ensures that the transformation does what we want it to do; that is, it averages out the microscopic details of the problem. This reflects the idea that the critical behavior of the system is dominated by fluctuations on the largest length scales. The other three restrictions are necessary if we are to locate fixed points of the RG transformation. We have seen in the examples of the previous chapters that scaling behavior goes hand in hand with the existence of fixed points of the RG.

In order to make the preceding discussion more concrete, we will consider several examples of the operator T, which we will also use in later chapters. The first two examples are formulated in terms of variables defined on a lattice and give rise to what are called "real space" RG transformations. The third example makes use of the Fourier transform variables and leads to the "momentum space" RG.

The first of the real-space transformations is due to Kadanoff, and in the context of spin models has come to be called the "block spin transformation." Figure 4.1.1 shows a triangular lattice, and at each lattice site a "spin" variable, $\{s_i\} = \pm 1$, is defined. Such a model is called an "Ising" model, and we will have more to say about Ising models in the next chapter. For now, however, we are simply interested in the idea of coarse graining and the construction of the operator T.

Again referring to Fig. 4.1.1, the lattice sites are grouped into three spin clusters. The coarse-grained, or block, spin is defined by

$$\{S_i\} = \operatorname{sgn}\left(\sum_{k=1}^{3} s_k^{(i)}\right) \tag{4.1.6}$$

where $\operatorname{sgn}(x) = +1$ if $x > 0$ and $\operatorname{sgn}(x) = -1$ if $x < 0$. This type of rule relating the original and coarse-grained variables is called a "majority rule" for obvious reasons. Note that the block spin also takes on just the values ± 1, and therefore the coarse-grained system

4.1 THE RENORMALIZATION GROUP TRANSFORMATION

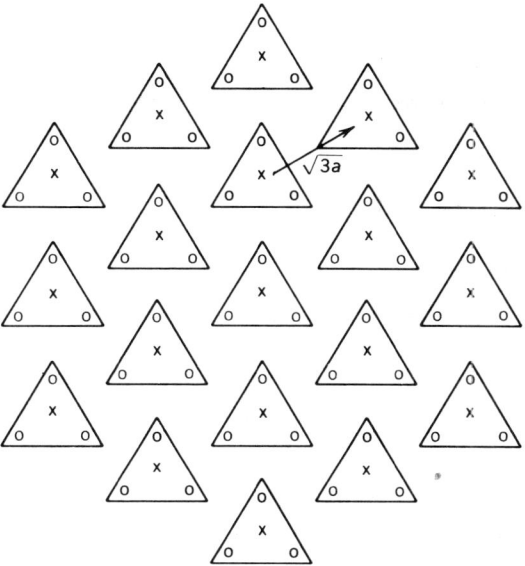

Fig. 4.1.1 Block spins for the triangular lattice. Note that the blocks themselves form a triangular lattice of size $a' = \sqrt{3}\, a$.

is also an Ising model. The coarse-grained lattice has a lattice constant $a' = \sqrt{3}\, a$, so the first two of our criteria are satisfied. If the Ising model undergoes a ferromagnetic transition so that in the ordered state all the spins are parallel, the corresponding coarse-grained state will also have all spins parallel. This satisfies condition 3. Finally, we need an explicit form for the operator $T(\{S_i\}, \{s_i\})$, which implements the rule (4.1.6),

$$T(\{S_i\}, \{s_i\}) = \prod_i \delta\left(S_i - \mathrm{sgn} \sum_{k=1}^{3} s_k^{(i)}\right) \quad (4.1.7)$$

The delta function in (4.1.7) is, in this case, to be taken as a Kronecker delta. If we were dealing with dynamical variables defined over a continuous range, then the delta function would be a Dirac delta. In either case the normalization condition of the delta function

$$\operatorname*{Tr}_{\{s_i\}} \delta\left(S_i - \mathrm{sgn} \sum_{k=1}^{3} s_k^{(i)}\right) = 1 \quad (4.1.8)$$

ensures that condition 4 and (4.1.5) are satisfied.

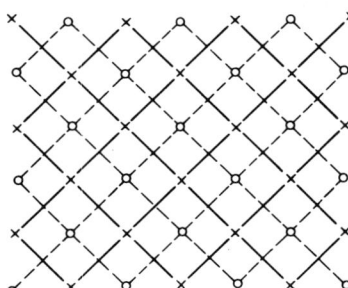

Fig. 4.1.2 Two sublattice structures of the square lattice. Sites on sublattice A are marked by circles; those on B by crosses.

Another realization of the operator T is a decimation transformation in which a subset of the dynamical variables is traced over. Figure 4.1.2 shows a square lattice that can be decomposed into two simple square sublattices with lattice constant $a' = \sqrt{2}\,a$. The sites on sublattice A are marked by open circles and those on sublattice B are marked by crosses. The Hamiltonian can be regarded as a function of the two sets of dynamical variables $\{s^A\}$ and $\{s^B\}$, and the trace (4.1.2) can be performed in two steps

$$Z = \operatorname*{Tr}_{\{s_i^A\}} \operatorname*{Tr}_{\{s_i^B\}} e^{-H[\{s_i^A\},\{s_i^B\}]} \qquad (4.1.9)$$

The coarse-grained Hamiltonian is defined just as in (4.1.3)

$$e^{-H'[\{s_i^A\}]} = \operatorname*{Tr}_{\{s_i^B\}} e^{-H[\{s_i^A\},\{s_i^B\}]} \qquad (4.1.10)$$

Obviously, the coarse-grained variables are of the same kind as the original, and (4.1.10) ensures the invariance of the partition function.

In many problems the system is translationally invariant, and it is convenient to work with the Fourier transform or momentum-space variables

$$s(k) = \frac{1}{\sqrt{N}} \sum_i e^{ikR_i} s_i \qquad (4.1.11)$$

where R_i are the lattice vectors. If the system is defined on a lattice, as in the previous example, then there is a natural limit to the values of the wave vectors k, namely the inverse of the lattice spacing. If the system is defined on a continuous space, as in a liquid or a gas, then the upper limit to the wave vector, called the momentum cutoff Λ, is of the order of the inverse of the interparticle spacing. The important point is that the cutoff is of microscopic dimensions, whereas the

4.1 THE RENORMALIZATION GROUP TRANSFORMATION

fluctuations near the critical point are macroscopic. This suggests that the fields with wave vectors near the cutoff should not contribute in any essential way to the critical behavior of the system. Coarse graining is accomplished by decimating those degrees of freedom near the cutoff.

The partition function is given as an integral over the Fourier coefficients $s(k)$, which we write in the compact form

$$Z = \int D[s(k)] e^{-H[s(k)]} \qquad (4.1.12)$$

In the case of a lattice, the momentum-space variables are also defined on a lattice (the reciprocal lattice), and the integration symbol in (4.1.12) simply stands for an ordinary multiple integral. In the continuum case the symbol in (4.1.12) stands for a "functional integral." For the purposes of this discussion, we can simply think in terms of the discrete case.

Coarse graining can be accomplished by dividing the integration variables into two sets, which we label $s(p)$ and $s(q)$. The wave vectors q satisfy $0 < |q| < \Lambda/b$, whereas the wave vectors p satisfy $\Lambda/b < |p| < \Lambda$. This division of momentum space is illustrated in Fig. 4.1.3.

Dividing the integration variables into the two groups, the partition function becomes

$$Z = \int D[s(q)] \int D[s(p)] e^{-H[\{s(p)\}, \{s(q)\}]} \qquad (4.1.13)$$

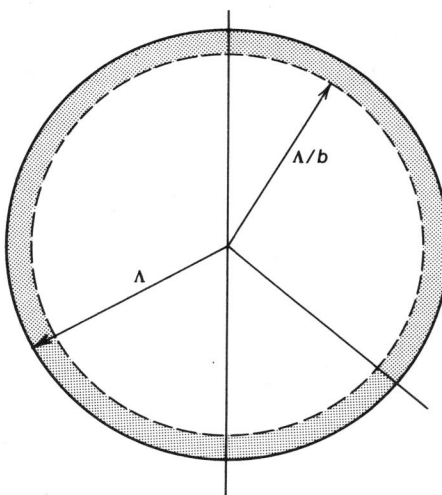

Fig. 4.1.3 Partitioning of momentum space variables $s(k)$. The variables with wave vectors in the shell near the cutoff Λ (shaded region) are integrated out.

where the dependence of H on both sets of degrees of freedom is made explicit. The coarse-grained Hamiltonian is defined in a manner similar to (4.1.10)

$$e^{-H'[s(q)]} = \int D[s(p)] e^{-H[\{s(p)\},\{s(q)\}]} \qquad (4.1.14)$$

From the definition of the procedure as simply a partial evaluation of the trace, it is clear that the free energy is kept invariant. The coarse-grained variables are just the remaining, or undecimated, degrees of freedom, and the cutoff for the coarse-grained Hamiltonian is reduced from Λ to Λ/b. Thus the momentum-space procedure satisfies all the criteria for a good RG transformation.

The second step of the RG transformation is to rescale the unit of length and the dynamical variables so as to restore the lattice spacing or cutoff to its original value. For a system on a lattice, the coarse-grained lattice constant is related to the original lattice constant by the scale factor b. In order to compare the coarse-grained Hamiltonian with the original one, we must redefine the unit of length, and the scaling equation for the coordinates is

$$r' = b^{-1} r \qquad (4.1.15)$$

All quantities with the dimension of length will scale as in (4.1.15), and, for example, the volume, which has dimension L^d, will scale as

$$V' = b^{-d} V \qquad (4.1.16)$$

Because the free energy is invariant under the RG, when lengths are rescaled the renormalized free-energy density will be given by

$$f' = b^d f \qquad (4.1.17)$$

In the momentum-space picture of (4.1.13) and (4.1.14), the fundamental length is the inverse of the cutoff Λ^{-1}. After decimation, that is, integrating out the fields $s(p)$, the new cutoff is $\Lambda' = \Lambda/b$. Scaling all momenta by b then restores the cutoff to its original value.

In most real-space RG transformations, there is no need to explicitly rescale the unit of length. For example, the decimated square lattice with lattice constant a of Fig. 4.1.2 is automatically a square lattice of lattice constant $\sqrt{2}\,a$. On the other hand, rescaling is an essential part of the momentum-space RG.

Typically, there will be a term in the Hamiltonian of the form

$$H_2 = \tfrac{1}{2} \sum_k k^2 s(k) s(-k) \qquad (4.1.18)$$

which tends to suppress rapid spatial changes in the order parameter. In general, decimation of the fields $s(p)$ will generate further terms of this form. In order to keep the coefficient of H_2 fixed at $1/2$, it is necessary to rescale the fields as well as the momenta, and conventionally the scaling of the fields is

$$s'(k) = b^{(2-\eta)/2} s(k) \qquad (4.1.19a)$$

or, for real-space fields,

$$s'(x) = b^{(d-2+\eta)/2} s(x) \qquad (4.1.19b)$$

The decimation part of the RG transformation can be regarded as a partial evaluation of the trace (4.1.2). In each successive application of the RG, which we can label with an index n, more of the original variables are eliminated, and part of the renormalized Hamiltonian, which is independent of the fields $f_0(n)$, is generated. This term represents the contribution of the decimated degrees of freedom to the free-energy density.

Because at each iteration of the RG transformation the volume scales as b^{-d}, we can write

$$f = \sum_{n=1}^{\infty} b^{-nd} f_0(n) \qquad (4.1.20)$$

which expresses the free-energy density as the sum of a function over the trajectory, or orbit, of the Hamiltonian through parameter space. We will find (4.1.20) useful in discussing the behavior of the free energy in the critical region.

4.2 RENORMALIZATION GROUP IN THE NEIGHBORHOOD OF A FIXED POINT

As we have seen in the simple examples already considered, scaling behavior is closely associated with the existence of fixed points of the RG. It is the basic assumption of RG theory that the critical point at which a second-order phase transition occurs maps to a fixed point of the RG, and that the critical region in which scaling holds maps into a neighborhood of the fixed point under the RG. In this section we consider the behavior of the RG in the neighborhood of a simple fixed point and show how this does in fact lead to scaling behavior.

We denote the action of the RG transformation on H by $R_b(H)$, where the scale factor b is shown explicitly. The fixed-point Hamiltonian H^* satisfies

$$R_b(H^*) = H^* \qquad (4.2.1)$$

Now consider a perturbation of the form $H = H^* + \varepsilon O$, where $\varepsilon \ll 1$ and O is any operator. A linear RG transformation can be defined by

$$L_b O = \lim_{\varepsilon \to 0} \frac{1}{\varepsilon} [R_b(H^* + \varepsilon O) - H^*] \qquad (4.2.2)$$

As with any linear operator, we are interested in its eigenvalues λ_k and eigenvectors ψ_k, which in the context of the RG are called linear scaling fields and which satisfy the eigenvalue equation

$$L_b \psi_k = \lambda_k(b) \psi_k \qquad (4.2.3)$$

We should emphasize that L_b is an operator in parameter space, whereas the linear scaling fields are operators in the sense of being functions of the dynamical variables.

If the RG is applied twice, first with scale factor b and then with scale factor b', the net result should be identical to a single renormalization with scale factor bb'. That is, we require

$$R_b(R_{b'}(H)) = R_{bb'}(H) \qquad (4.2.4)$$

Applying the composition rule to (4.2.3) gives

$$\lambda(bb') = \lambda(b)\lambda(b') \qquad (4.2.5)$$

from which we see that the eigenvalues are generally of the form

$$\lambda_k(b) = b^{y_k} \qquad (4.2.6)$$

Just as in quantum mechanics, the eigenvectors of the linear operator b form a useful basis on which to describe the action of the RG

4.2 RENORMALIZATION GROUP IN A FIXED POINT

transformation. In the neighborhood of the fixed point the Hamiltonian can be written as

$$H[\{h_k\}] = H^* + \sum_k h_k \psi_k \qquad (4.2.7)$$

where the parameters $\{h_k\}$ are "small." Applying the RG once, we have

$$R_b(H[\{h_k\}]) = H^* + \sum_k b^{y_k} h_k \psi_k \qquad (4.2.8)$$

from which we see that

$$R_b(H[\{h_k\}]) = H[\{b^{y_k} h_k\}] \qquad (4.2.9)$$

If $y_k < 0$, then under repeated applications of the RG the amplitude h_k of the scaling field ψ_k tends to 0 and that operator is said to be "irrelevant." On the other hand, if $y_k > 0$, the amplitude of the scaling field increases and the operator is "relevant." In the special case $y_k = 0$, the operator is called "marginal."

There is always at least one relevant scaling field of the linearized RG, but in a sense it is trivial and does not lead to critical behavior. A constant of the form cV, where V is the volume, can always be added to the Hamiltonian without changing the properties of the system. As we have seen, the volume scales as b^{-d}, and because the free energy is invariant, c must scale as $c' = b^d c$. The additive constant is relevant but innocuous because it is always possible to redefine the Hamiltonian so that $c = 0$.

In the typical case the Hamiltonian itself is a relevant operator and the conjugate parameter is the reduced temperature

$$t = \frac{T - T_c}{T_c} \qquad (4.2.10)$$

Note that in (4.2.7) the parameters h_k are defined in such a way that at the fixed point all the $h_k = 0$. The reduced temperature is defined so that $t = 0$ also corresponds to the fixed point.

The critical surface (it can be a single point, a line, or a set of higher dimension) in parameter space is the set of points whose trajectories or orbits under the RG tend to the fixed point. In the language of the theory of maps, the critical surface associated with a fixed point is its basin of attraction. The critical region is the set of points whose orbits come sufficiently close to the fixed point that the linear approximation discussed previously is applicable. The critical point for a system is the intersection of the critical surface with the subspace spanned by the physical parameters, which we will refer to simply as the "physical subspace." In simple RGs there may be just a single parameter, in which case the fixed point and the critical point are identical. More typically, the physical subspace is of higher dimension, in which case the intersection of the critical surface with the physical subspace can be a line of critical points or even a manifold of higher dimension.

4.3 CORRELATION LENGTH AT A FIXED POINT

The concept of the "correlation length" plays a central role in the theory of phase transitions. The correlation length is defined in terms of the order parameter correlation function

$$G(r_{i,j}) = \langle s_i s_j \rangle - \langle s_i \rangle \langle s_j \rangle \tag{4.3.1}$$

by

$$\xi^{-1} = \lim_{r \to \infty} \left[-\frac{\partial}{\partial r} \ln G(r) \right] \tag{4.3.2}$$

Close to the fixed point the correlation length in the original system is related to that in the renormalized system by

$$\xi(\{h_k\}) = b\xi(\{b^{y_k} h_k\}) \tag{4.3.3}$$

which states that the correlation length is a generalized homogeneous function of the parameters $\{h_k\}$. If we now set all the parameters to 0 in the sense of (4.2.7) and take for H the fixed-point Hamiltonian, we have

$$\xi(\{0\}) = b\xi(\{0\}) \tag{4.3.4}$$

which has two solutions, $\xi(\{0\}) = 0$ and $\xi(\{0\}) = \infty$.

In the first case, which is sometimes called "trivial," the fixed point corresponds to a homogeneous phase either at infinite or zero temperature. Although it is clear that the correlation length should vanish as $T \to \infty$, it is perhaps less obvious that it also vanishes in the ordered state as $T \to 0$. To see this, note that as the temperature tends to 0 the fluctuations of the order parameter about its average value vanish and so the second term in (4.3.1) tends to cancel the first. The remainder decays exponentially at large separations, and the characteristic length ξ tends to 0 with T.

Because the RG transformation itself should be nice and analytic, and all thermodynamic quantities in the pure phases are analytic functions of the parameters, we can assume that the correlation length is an analytic function of the parameters as well. This implies that all the scaling fields in the neighborhood of a trivial fixed point are irrelevant (Prob. 4.1) and the fixed point is stable. The dimension of the "critical surface" in this case is just the dimension of the parameter space itself (Prob. 4.2).

The other solution of (4.3.4), $\xi = \infty$, corresponds to critical behavior. The correlation length sets the natural scale for the system, and the divergence of ξ at the critical point signals the loss of scale. In contrast to the trivial fixed point considered previously, a "nontrivial" fixed point is characterized by a least one relevant scaling field (see Prob. 4.3).

It is clear that a necessary condition for a point to lie on the critical surface is that the amplitudes of all the relevant scaling fields must vanish. In the usual situation one has control over only a few external parameters such as the temperature, magnetic field, and so on, and so the fixed points that are responsible for phase transitions are characterized by just a few, and often only one, relevant scaling field. The dimension of the critical surface in this case is less than that of the whole parameter space.

4.4 HOMOGENEITY PROPERTIES OF THE FREE ENERGY

In most situations the temperature is a relevant parameter, and one usually expresses thermodynamics averages in terms of the reduced temperature, (4.2.10).

Let us assume for simplicity that there are no external fields and the temperature is the only relevant parameter. Very close to the fixed point, we can ignore the other (irrelevant) scaling fields, and we have

$$\xi = b\xi(tb^{y_T}) \qquad (4.4.1)$$

where we have labeled the eigenvalue of the RG associated with the temperature parameter as y_T. From this it follows that the correlation length diverges as

$$\xi \to |t|^{-1/y_T} \tag{4.4.2}$$

and we can identify the eigenvalue y_T with the reciprocal of the correlation length exponent ν.

Similarly, by (4.1.17) the free-energy density can be written as a homogeneous function of t of the form

$$f(t) = b^{-d} f(tb^{y_T}) \tag{4.4.3}$$

In (4.4.3) we have rearranged things so that all the dependence on the scale factor b is on the right-hand side. The scale parameter b in (4.4.3) is arbitrary, and in particular we can choose b such that $b|t|^{1/\nu} = 1$. Then the free energy as a function of t in the critical region varies as

$$f(t) = |t|^{\nu d} f(\text{sgn}(t)) \tag{4.4.4}$$

Differentiating the free energy twice with respect to t yields the specific heat, which varies with temperature by (4.4.4) as

$$C(t) = |t|^{-(2-\nu d)} f_{tt}(\text{sgn}(t)) \tag{4.4.5}$$

and therefore the specific heat exponent α is related to the correlation length exponent and the dimension of space by the "scaling law"

$$\alpha = 2 - \nu d \tag{4.4.6}$$

Now let us consider a slightly more complex situation where, in addition to the temperature parameter, there is also an external symmetry-breaking field h. The exponent associated with the external field is y_h, and the free-energy density becomes a generalized homogeneous function of both t and h:

$$f(t, h) = b^{-d} f(tb^{y_T}, hb^{y_h}) \tag{4.4.7}$$

The order parameter $m = \langle s \rangle$ can be calculated from the free-energy density by differentiation with respect to the external field

$$m(t, h) = -\frac{\partial f(t, h)}{\partial h} \tag{4.4.8}$$

4.4 HOMOGENEITY PROPERTIES OF THE FREE ENERGY

Differentiating both sides of (4.4.7), we have

$$m(t, h) = b^{d-y_h} m(tb^{y_T}, hb^{y_h}) \tag{4.4.9}$$

Comparing the order parameter on the left-hand side of (4.4.9) with the renormalized order parameter on the right-hand side, we see that the order parameter scales as

$$m' = b^{d-y_h} m \tag{4.4.10}$$

If we again choose b such that $|t|b^{y_T} = 1$, we arrive at the "equation of state," which gives the order parameter as a function of the external field and the temperature

$$m(t, h) = -|t|^{\nu(d-y_h)} f_h(\text{sgn}(t), h|t|^{-\nu y_h}) \tag{4.4.11}$$

If we now take $h \to 0$, the order parameter vanishes when t goes to 0 as t^β, and β is related to ν and y_h by

$$\beta = (d - y_h)\nu \tag{4.4.12}$$

The susceptibility, which is the derivative of the order parameter with respect to the external field, can be calculated from (4.4.12) and we have

$$\chi(h, t) = -|t|^{\nu(d-2y_h)} f_{hh}(\text{sgn}(t), h|t|^{\nu y_h}) \tag{4.4.13}$$

The zero-field susceptibility is singular at the critical temperature, varying as $|t|^{-\gamma}$, and we see that the critical exponent γ is given in terms of ν and y_h by

$$\gamma = -(d - 2y_h)\nu \tag{4.4.14}$$

Finally, if we return to (4.4.9) and set $T = T_c$, that is, $t = 0$, and choose b such that $|h|b^{y_h} = 1$, we find

$$m(h) = -|h|^{(d-y_h)/y_h} f_h(0, \text{sgn}(h)) \tag{4.4.15}$$

Comparing this with the definition of the critical isotherm exponent δ, we see that

$$\delta = \frac{y_h}{d - y_h} \tag{4.4.16}$$

We can use the homogeneous form of the free energy to derive an important inequality for the scaling exponents h_k. The natural extension of (4.4.7) to the case of an arbitrary number of scaling fields is

$$f(\{h_k\}) = b^{-d} f(\{b^{y_k} h_k\}) \qquad (4.4.17)$$

The derivative of the free-energy density with respect to the parameter h_k yields the ensemble average of the density m_k conjugate to h_k. For example, the energy density is conjugate to the reduced temperature and the magnetization per spin is conjugate to the external magnetic field. Each of these densities is an intensive quantity bounded at all temperatures, and they cannot diverge. If we assume that h_k scales as b^{y_k}, then it follows that

$$m_k = b^{-d+y_k} f_{h_k}(\{b^{y_k} h_k\}) \qquad (4.4.18)$$

If we now take $b^{y_k}|h_k| = 1$, we have

$$m_k = |h_k|^{(d-y_k)/y_k} f_{h_k}\!\left(\{|h_k|^{-y_j/y_k} h_j\}, \operatorname{sgn}(h_k)\right) \qquad (4.4.19)$$

In order for m_k to remain finite as h_k tends to 0, we must have

$$y_k \leq d \qquad (4.4.20)$$

which is the inequality we were seeking.

4.5 SCALING FORM OF THE CORRELATION FUNCTION

Consider the generalization of the partition function to an ensemble in which the expectation value of the field is fixed at each site. This is accomplished by the introduction of "source fields" η_i at each site, and the "generating function" for the system is given by

$$\Xi(\{\eta_i\}) = \operatorname*{Tr}_{\{s_i\}} \exp\!\left(-\Big(H[\{s_i\}] - \sum_i \eta_i s_i\Big)\right) \qquad (4.5.1)$$

The average of any function of the fields in the original ensemble can be found by evaluating the appropriate derivatives of the generating function with respect to the sources and setting all $\eta_i = 0$.

4.5 SCALING FORM OF THE CORRELATION FUNCTION

We can also define the generating potential Φ in analogy with the free energy

$$\Phi[\{\eta_i\}] = -\ln \Xi[\{\eta_i\}] \qquad (4.5.2)$$

The generating potential has the advantage that its derivatives yield the connected, or cumulant, part of a correlation function. For example, taking derivatives of $\Xi(\{\eta_i\})$ with respect to η_j and η_k will yield $\langle s_j s_k \rangle$, whereas the same derivatives of $\Phi[\{\eta_i\}]$ yield

$$G(r_{i,j}) = \frac{\partial^2 \Phi}{\partial \eta_i \, \partial \eta_j}(\{\eta_i = 0\}) \qquad (4.5.3)$$

which is how the order parameter correlation function is defined in (4.3.1). In general, if a correlation function depends on several coordinates, $r_1, r_2, \ldots, r_k$, the connected, or cumulant, part vanishes as all the coordinate differences $|r_i - r_j|$ tend to ∞. The generating potential, just like the free energy, is invariant under the RG. Although it would appear that the source term in the Hamiltonian is like an external field term, there is a subtle difference. Each term in the Hamiltonian must be extensive and therefore scale invariant. This applies to the term $h \Sigma_i s_i$ as a whole and to each term in the sum $\Sigma_i \eta_i s_i$ individually. For this reason the sources scale differently from an external field: The sources must scale as

$$\eta_i' = b^{-y_s} \eta_i \qquad (4.5.4)$$

Now that we have established the scaling properties of the sources, the generating potential can be expressed as a generalized homogeneous function of the temperature and the sources

$$\Phi(t, \{\eta_i\}) = \Phi(tb^{y_T}, \{\eta_i b^{-y_s}\}) \qquad (4.5.5)$$

Note that, like the free energy, $\Phi[t, \{\eta_i\}]$ is required to be scale invariant. Calculating the correlation function as in (4.5.3) gives

$$G(r) = b^{-2y_s} \Phi_{\eta_i \eta_j}(tb^{y_T}, rb^{-1}, \{\eta_i = 0\}) \qquad (4.5.6)$$

In (4.5.6) we have displayed explicitly the dependence of the correlation function on $r = |r_i - r_j|$, the distance between sites i and j. Because r has dimensions of length, it scales as b^{-1}. In this case it is convenient to choose b such that $b^{-1}r = 1$, which yields the scaling form for the correlation function,

$$G(r) = r^{-2y_s}\Phi_{\eta_i\eta_j}(tr^{1/\nu})$$
$$= r^{-2y_s}g\left(\frac{r}{\xi}\right) \qquad (4.5.7)$$

In the second form we have eliminated the temperature variable in favor of the correlation length by (4.4.2) to emphasize the dimensionless character of the argument of g. Typically, $g(x)$ is an exponentially decreasing function for large x.

As one approaches the critical point, the correlation length diverges and the correlation function decays algebraically as

$$\lim_{t \to 0} G(r, t) \to \frac{g(0)}{r^{-2y_s}} \qquad (4.5.8)$$

Comparing the result (4.5.7) with the scaling form of $G(r)$ in Table 4.0.2, we see that y_s is given by

$$y_s = \tfrac{1}{2}(d - 2 + \eta) \qquad (4.5.9)$$

4.6 SCALING LAWS, HYPERSCALING, AND UNIVERSALITY

We have seen that the consequence of the existence of an isolated nontrivial fixed point of the RG is that all the usual critical exponents for a system can be expressed in terms of the eigenvalues of the linearized RG, y_T and y_h. This implies that only two of the critical exponents are independent, and therefore there must exist relations between the various critical exponents. Why should the critical exponents depend upon just y_T and y_h? We can divide the set of all operators into "even" and "odd" depending on whether they change sign under the operation $S_i \to -S_i$. The Hamiltonian in the absence of any symmetry-breaking field is even, and therefore the temperature parameter will control the amplitude of the most relevant even scaling field. The odd-parity operators enter the picture when we switch on a symmetry-breaking field such as an external magnetic field. The most relevant odd-parity field scales with exponent y_h, and so the two most relevant exponents y_T and y_h are sufficient to describe the scaling behavior of the system very close to the critical point. Here we collect

4.6 SCALING LAWS, HYPERSCALING, AND UNIVERSALITY

the scaling laws, which follow from (4.4.3), (4.4.7), (4.4.13), (4.4.16), (4.4.18), and (4.5.9),

$$\alpha = 2 - \nu d \tag{4.6.1a}$$

$$\beta = \frac{\nu}{2}(d - 2 + \eta) \tag{4.6.1b}$$

$$\gamma = \nu(2 - \eta) \tag{4.6.1c}$$

$$\alpha + 2\beta + \gamma = 2 \tag{4.6.1d}$$

$$\delta = \frac{d + 2 - \eta}{d - 2 + \eta} \tag{4.6.1e}$$

A distinction is sometimes made between scaling laws that explicitly contain the dimension d and those that do not. Those that do, for example, (4.6.1a), (4.6.1b), and (4.6.1e), are called "hyperscaling" relations because they depend directly on the assumption that the free-energy density is a homogeneous function of the scaling fields. This should be contrasted with (4.6.1c), which can also be deduced from the fluctuation–dissipation theorem that relates the fluctuations in a thermodynamic average, in this case the order parameter, to the corresponding susceptibility (see Probs. 4.6 and 4.14).

In addition to the scaling laws, which reduce the number of independent exponents to just 2, it is widely believed that the values of the exponents are determined by the dimension of space and the nature of the order parameter. This concept is called "universality," and in a weak form implies that, for example, the critical exponents for the Ising model are independent of the particular lattice that one studies. More significantly, it asserts that if one has two real crystals that undergo Ising-like phase transitions, although the details of the interactions and crystal structures may be quite different, the critical exponents are the same. This simplifies the "botany" of critical phenomena by placing all similar systems into a rather small set of "universality classes."

Although there is strong numerical and experimental support for the universality hypothesis, it has not yet been proved except in certain restricted cases where transformations can be found which map one problem on to another, for example the equivalence of bond percolation on a lattice and its dual.

4.7 FIXED POINTS AND FIRST-ORDER PHASE TRANSITIONS

In the preceding sections we have been describing the structure of the RG in the neighborhood of the fixed point associated with a second-order phase transition. In such a phase transition the thermodynamic densities are continuous through the transition. However, many systems exhibit first-order transitions in which the order parameter or internal energy is discontinuous. It is natural to inquire as to whether this important class of transitions can also be described within the framework of the RG.

Following Fisher and Berker [2], we can write the RG equations that relate the renormalized parameters to the old parameters in the form

$$h'_i = R_i(\{h_j\}) \tag{4.7.1}$$

where the $R_i(\{h_j\})$ are in general nonlinear functions of the old parameters. Consider again the thermodynamic density m_k, which can be found by differentiating the free-energy density with respect to the parameter h_k,

$$m_k = -\frac{\partial f}{\partial h_k} \tag{4.7.2}$$

If we apply the RG once, we have, because f scales as b^d,

$$m_k = -b^{-d} \sum_j \frac{\partial h'_j}{\partial h_k} \frac{\partial f}{\partial h'_j} = b^{-d} \sum_j m'_j T_{jk} \tag{4.7.3}$$

In (4.7.3) the matrix of derivatives T is just the linear RG transformation. If we now evaluate the densities at a fixed point, we have

$$m^*_k = b^{-d} \sum_j m^*_j T_{jk} \tag{4.7.4}$$

Because the densities must all be finite and cannot all be 0, consistency requires that m^* be an eigenvector of the linearized RG with eigenvalue b^d. This eigenvector includes as a component the additive constant discussed previously, and the associated density in this case is simply the ensemble average of the identity.

The usual second-order transition corresponds to the case where the identity is the only such eigenvector. If, on the other hand, we wish to describe the coexistence of two phases, which we label A and B, each with its own set of distinct thermodynamic densities, then there must exist two eigenvectors ψ^A and ψ^B with eigenvalue b^d. Because these two eigenvectors are degenerate, any linear combination $\phi = \alpha \psi^A + \beta \psi^B$ will also be an eigenvector which scales as b^d and which corresponds to a state that is a mixture of the A and B phases.

We can still set the amplitude of one scaling field to 0 by eliminating the constant term in the Hamiltonian, but now there is a second relevant scaling field whose exponent assumes the maximum thermodynamically allowed value $y = d$. Here we shall consider the case where this scaling field couples to the reduced temperature. The case where the relevant parameter couples to the order parameter is left as an exercise (Prob. 4.14).

The argument follows exactly as in (4.4.3) and (4.4.4), except that now y_T assumes the maximal value $y_T = d$. Therefore, (4.3.4) becomes

$$f(t) = |t| f(\mathrm{sgn}(t)) \qquad (4.7.5)$$

Taking the derivative with respect to t, we find that the discontinuity in the internal energy, or latent heat, is given by

$$L = \lim_{\varepsilon \to \infty} \left[\frac{\partial f(\varepsilon)}{\partial t} - \frac{\partial f(-\varepsilon)}{\partial t} \right] = f(1) + f(-1) \qquad (4.7.6)$$

We therefore come to the conclusion that a first-order transition also corresponds to a simple fixed point of the RG, but one in which the dominant relevant scaling field scales with the maximal exponent $y_T = d$.

4.8 CROSSOVER PHENOMENA

Consider a one-parameter set of Hamiltonians of the form

$$H(\Delta) = H_s + \Delta H_a \qquad (4.8.1)$$

where H_s is invariant with respect to a given symmetry operation and H_a is not. For $\Delta = 0$ the system belongs to the universality class of H_s, but for any value of $\Delta > 0$ the system will belong to the universality class of the asymmetric Hamiltonian and the critical exponents will be different. Evidently, the parameter Δ is a special example of a "relevant parameter."

If Δ is very small there will be a region near the critical temperature in which the thermodynamic functions will scale as if $\Delta = 0$, but as the critical temperature is more closely approached the true scaling behavior characteristic of the asymmetric fixed point will emerge. In the language of the renormalization group, we say that there is a crossover from one universality class to another.

In a sense, we have already encountered this situation in the discussion of the scaling of the free energy in the presence of a field that couples to the order parameter. For any nonzero value of this field, the nontrivial fixed point is unstable because the external field is a relevant variable. In that case the renormalization trajectories tend to one of the trivial fixed points and there is no critical behavior. Here we wish to consider a different situation in which there are two nontrivial fixed points, one of which yields the critical behavior if $\Delta = 0$, and the other yields the critical behavior if $\Delta > 0$.

The free-energy density will take the form

$$f(t, \Delta) = b^{-d} f(tb^{1/\nu}, \Delta b^{1/\nu\phi}) \tag{4.8.2}$$

where we have introduced the crossover exponent ϕ, and ν is the correlation length exponent for the symmetric universality class. Note that the temperature parameter $t = (T - T_c(0))/T_c(0)$ refers to the symmetric critical point. For $\Delta = 0$ the scaling arguments proceed as before, but for $\Delta > 0$ we choose

$$\Delta b^{1/\nu\phi} = 1 \tag{4.8.3}$$

which gives for the free-energy density

$$f(t, \Delta) = \Delta^{\nu d \phi} f(t\Delta^{-\phi}, 1) \tag{4.8.4}$$

In order for (4.8.4) to yield scaling behavior, the function $f(x, 1)$ must be singular at some value x_c, varying as

$$f(x, 1) \sim |x - x_c|^{\nu' d} \tag{4.8.5}$$

where ν' is the correlation length exponent for the asymmetric universality class. Substituting (4.8.5) into (4.8.4) then yields

$$f(t, \Delta) \to \Delta^{(\nu - \nu')\phi d} |t(\Delta)|^{\nu' d} \tag{4.8.6}$$

where

$$t(\Delta) = \frac{T - T_c(\Delta)}{T_c(\Delta)} \qquad (4.8.7)$$

and

$$T_c(\Delta) = T_c(0)(1 + x_c \Delta^\phi) \qquad (4.8.8)$$

Note that by (4.8.8) the shift in the transition temperature $T_c(\Delta) - T_c(0)$ scales with the symmetry-breaking field as Δ^ϕ.

4.9 FINITE-SIZE SCALING

A system will exhibit true critical behavior only in the thermodynamic limit. For any finite system it is easy to see that the free energy must be analytic. In a sense then, the characteristic size of the system L can be regarded as a relevant parameter. For finite L the specific heat will show a maximum, which increases in magnitude and sharpness as $1/L \to 0$. One can estimate how close the system comes to critical behavior by realizing that so long as the correlation length ξ is much less than L, the finite size of the system will play no role. Rounding-off of the sharp singularities will occur when

$$L \sim \xi \qquad (4.9.1)$$

Because $\xi \sim |t|^{-\nu}$, the rounding temperature T_m will be shifted from the true critical temperature by

$$T_m - T_c \sim L^{-1/\nu} \qquad (4.9.2)$$

These ideas can be put on a firm foundation by incorporating $1/L$ as a parameter in the free energy. It is clearly relevant because it scales as

$$\frac{1}{L'} = b \frac{1}{L} \qquad (4.9.3)$$

The free-energy density then takes on the scaling form

$$f\left(t, \frac{1}{L}\right) = b^{-d} f\left(tb^{1/\nu}, b\frac{1}{L}\right) \qquad (4.9.4)$$

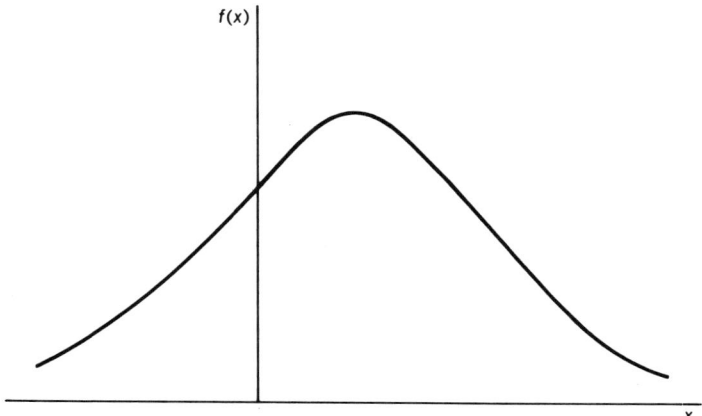

Fig. 4.9.1 Typical form for the scaling function of the specific heat.

Taking $b = L$, (4.9.4) can be written as

$$f\left(t, \frac{1}{L}\right) = L^{-d} f(tL^{1/\nu}, 1) \qquad (4.9.5)$$

The specific heat is then given by

$$C\left(t, \frac{1}{L}\right) = L^{-d+2/\nu} f_{tt}(tL^{1/\nu}, 1) \qquad (4.9.6)$$

If we now take the limit $L \to \infty$, $t \to 0$ and hold $|t|L^{1/\nu} = x_0$ fixed, we find that the specific heat varies as

$$\lim_{L \to \infty} \lim_{t \to 0} C\left(t, \frac{1}{L}\right) = x_0^{\nu-2d} |t|^{2-\nu d} f_{tt}(x_0, 1) \qquad (4.9.7)$$

Typically, the function $f_{tt}(x, 1)$ will be smooth and exhibit a maximum for $x = x_m$ as shown in Fig. 4.9.1. Taking $x_0 = x_m$ in the preceding argument, the temperature at which the specific heat is maximum for a system of size L is given by

$$\frac{T_m - T_c}{T_c} L^{1/\nu} = x_m \qquad (4.9.8)$$

which leads directly to

$$T_m = T_c(1 + x_m L^{-1/\nu}) \qquad (4.9.9)$$

and should be compared to (4.9.2).

When numerical methods such as Monte Carlo simulation are used to evaluate the thermodynamic properties of a system, one is always forced to consider a system of finite size. Therefore, the singularities will be rounded and the determination of the critical temperature and exponents becomes ambiguous. However, if we return to (4.9.6), we can rewrite this in the form

$$L^{-\alpha/\nu} C\left(t, \frac{1}{L}\right) = f_{tt}(tL^{1/\nu}, 1) \qquad (4.9.10)$$

where we have used the scaling law (4.6.1a), $\alpha = 2 - \nu d$. Equation (4.9.10) states that the scaled specific heat

$$\overline{C}\left(t, \frac{1}{L}\right) = L^{-\alpha/\nu} C\left(t, \frac{1}{L}\right) \qquad (4.9.11)$$

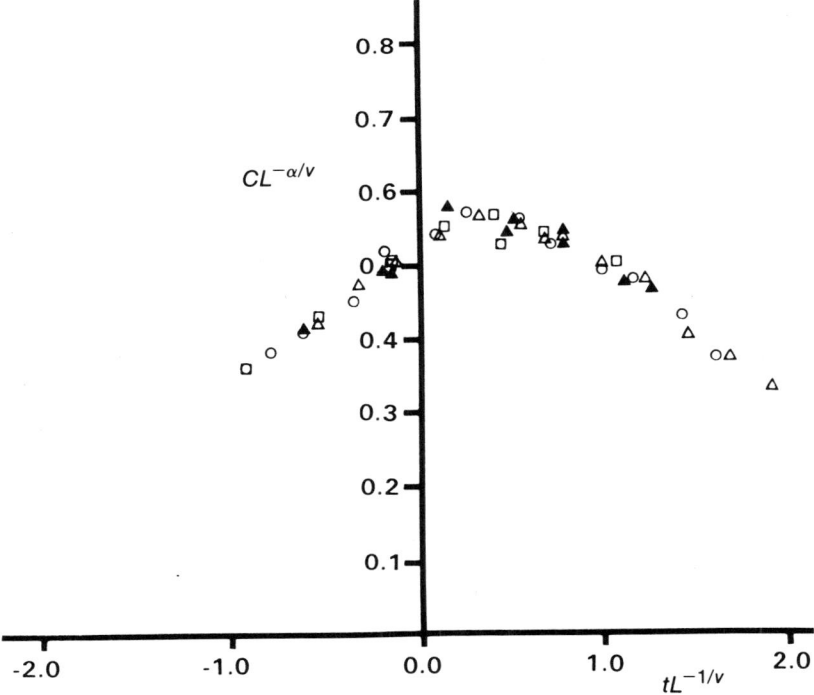

Fig. 4.9.2 Universal behavior of the specific heat for $\Delta = 1.0$ (Ising Model), and $L = 8(\bigcirc)$, $L = 10(\triangle)$, $L = 12(\square)$, $L = 14(\blacktriangle)$, $T_c = 2.269$, $\nu = 1.0$, and $\alpha = 0.0$.

is a universal function of the scaled temperature

$$\bar{t} = tL^{1/\nu} \tag{4.9.12}$$

Therefore, if we calculate the specific heat as a function of temperature for systems of various size and plot the scaled specific heat against the scaled temperature, all the data should fall on a single curve. Because T_c, α, and ν are generally unknown, these parameters are adjusted until the best fit of the data to a single curve is achieved.

The method of finite-size scaling allows one to extract in a systematic way the behavior of an infinite system from that of a sequence of finite systems. In practice, reasonable estimates of the critical temperature and exponents can be found from systems of rather modest size, and if one has access to sufficient computing power, finite-size scaling can yield results that rival other theoretical techniques such as high-temperature expansions.

Figure 4.9.2 shows scaled Monte Carlo data for the two-dimensional Ising model calculated by the authors.

4.10 INVARIANCE OF CRITICAL EXPONENTS WITH RESPECT TO CHANGE OF VARIABLE

In analyzing the behavior of the RG in the neighborhood of a fixed point, it is often useful to introduce new parameters that are, in general, nonlinear functions of the original parameters. Such a change of variables is justified only if the values of the critical exponents, or eigenvalues of the linearized RG transformation, remain unchanged. In this section we consider the conditions under which such a transformation is possible, and then consider some situations where the values of the critical exponents appear to depend on the choice of parameters.

Let us denote the original set of parameters by $\{J_k,\ k = 1, 2, \ldots, n\}$, defined for convenience such that $J_k^* = 0$. In the neighborhood of the fixed point, the RG equations are

$$J_i' = \sum_j \left(\frac{\partial J_i'}{\partial J_j}\right)_0 J_j \tag{4.10.1}$$

and let ϕ_j be a right eigenvector of the linear RG

$$\sum_j \frac{\partial J_i'}{\partial J_j} \phi_j = \lambda \phi_i \tag{4.10.2}$$

4.10 INVARIANCE OF CRITICAL EXPONENTS

Now suppose we take the original parameters to be general functions of a new set of parameters $\{K_i\}$,

$$J_i = J_i(\{K_j\}) \tag{4.10.3}$$

Then by the chain rule

$$\frac{\partial J_i'}{\partial J_j} = \sum_{m,n} \frac{\partial J_i'}{\partial K_m'} \frac{\partial K_m'}{\partial K_n} \frac{\partial K_n}{\partial J_j} \tag{4.10.4}$$

Inserting this expression into (4.10.2), we have

$$\sum_{m,n} \sum_j \frac{\partial J_i'}{\partial K_m'} \frac{\partial K_m'}{\partial K_n} \frac{\partial K_n}{\partial J_j} \phi_j = \lambda \phi_i \tag{4.10.5}$$

We now assume that the transformation between the old and new parameters is non singular at the fixed point so that

$$\det\left[\frac{\partial J}{\partial K}\right] \neq 0 \tag{4.10.6}$$

We can then multiply on the left by $\partial K_p'/\partial J_i$ and sum on i, and because

$$\sum_i \frac{\partial K_p'}{\partial J_i'} \frac{\partial J_i'}{\partial K_m'} = \delta_{p,m} \tag{4.10.7}$$

we have

$$\sum_n \frac{\partial K_p'}{\partial K_n} \psi_n = \lambda \psi_p \tag{4.10.8}$$

where we have defined

$$\psi_p = \sum_i \frac{\partial K_p}{\partial J_i} \phi_i \tag{4.10.9}$$

We see that there is a one-to-one relationship between eigenvalues of the linearized RG in terms of the old and new variables. We only require that the change of variables be nonsingular at the fixed point.

There are situations that arise in which things are not so clear-cut. For example, in the one-dimensional Ising model (see Chap. 5), we will derive the RG equation

$$v' = v^2 \qquad (4.10.10)$$

where $v = \tanh \beta J$. This equation has fixed points at $v^* = 0$ and $v^* = 1$.

At the former fixed point we have

$$\left.\frac{\partial v'}{\partial v}\right|_{v^*=0} = 0 \qquad (4.10.11)$$

Now suppose we define a new parameter w so that

$$v = e^w \qquad (4.10.12)$$

The RG equation (4.10.10) then becomes

$$w' = 2w \qquad (4.10.13)$$

Equation (4.10.13) has fixed points at $w^* = 0$, $w^* = \infty$, and $w^* = -\infty$. The fixed point $w^* = -\infty$ corresponds to $v^* = 0$, and by (4.10.13)

$$\left.\frac{\partial w^*}{\partial w}\right|_{w^*=-\infty} = 2 \qquad (4.10.14)$$

How do we choose between RG equations (4.10.10) and (4.10.13)? Comparison of (4.10.11) and (4.10.14) illustrates the point that the eigenvalues of the RG are invariant only under nonsingular transformations in parameter space.

The question still remains: Which answer for the eigenvalue is correct? The answer must come from our physical insight into the problem, because a priori either RG equation is valid.

In the preceding example the appropriate temperature parameter is $v \sim J/T$ for large T and not $w \sim \ln T$, and therefore we should choose (4.10.10) over (4.10.13).

4.11 GOLDSTONE MODES AND THE LOWER CRITICAL DIMENSION

If a system orders by breaking a continuous symmetry, then there exist excitations above the ground state with arbitrarily low energy. These excitations are called Goldstone modes, or, as is more common in quantum field theory, "Goldstone bosons." Rather than trying to present a very general derivation of this result, we will illustrate the theorem with the simplest possible example, the X–Y model. In this case the continuous symmetry is rotations about the z axis. These rotations form a one-parameter abelian group, $U(1)$. This group is also a symmetry of electrodynamics and the quantum Bose gas.

The Goldstone modes tend to disturb the long-range order in a system, and if the density of states for these modes is sufficiently large, the order parameter fluctuations can wipe out the phase transition completely. Because the density of long-wavelength modes increases as the dimensionality of the system decreases, it follows that there is a lower critical dimension d_c, below which there is no spontaneously broken-symmetry state at finite temperature. For the Landau–Ginzburg type model, Mermin and Wagner [6, 7] and Hohenberg [3] showed that the lower critical dimension is 2.

The Landau–Ginzburg free energy in the absence of a magnetic field is

$$L = \int d^d \left[\tfrac{1}{2} x \, \nabla \phi_\alpha \cdot \nabla \phi_\alpha + U(\phi_\alpha \phi_\alpha) \right] \quad (4.11.1)$$

where U is a local function of the scalar product $\phi_\alpha \phi_\alpha$ and $\alpha = 1, 2$. Obviously, (4.11.1) is symmetric under rotations about the $\alpha = 3$, or z, axis.

We now assume that below some temperature T_c the symmetry is broken and the average of a single field assumes a nonzero value

$$\langle \phi_\alpha(x) \rangle = -m_\alpha \quad (4.11.2)$$

It follows from the original symmetry of the free energy that there exists a one-parameter family of states degenerate with this state where the order parameter is

$$m_\alpha(\theta) = \sum_\beta R(\theta)_{\alpha\beta} m_\beta \quad (4.11.3)$$

where $R_{\alpha\beta}(\theta)$ is the usual rotation operator

$$R_{\alpha\beta}(\theta) = \begin{bmatrix} \cos\theta & -\sin\theta \\ \sin\theta & \cos\theta \end{bmatrix} \quad (4.11.4)$$

In what follows it will be useful to express $R(\theta)$ in exponential form

$$R(\theta) = e^{i\theta L_z} \quad (4.11.5)$$

where the generator of rotations about the z axis is, in this representation of the group, the matrix

$$L_z = \begin{bmatrix} 0 & -i \\ i & 0 \end{bmatrix} \quad (4.11.6)$$

If a global rotation by θ produces a state degenerate with the ground state, then it is reasonable to expect that if we allow the rotation angle θ to vary slowly from point to point we will achieve a state with an energy just slightly above the ground state. Therefore, let us define

$$\bar{\phi}_\alpha(x) = R_{\alpha\beta}(\theta(x))m_\beta \quad (4.11.7)$$

The potential U is still invariant, but the gradient-squared term is no longer 0.

Using the expression (4.11.5), we have

$$\nabla\bar{\phi}_\alpha = i\nabla\theta(L_z)_{\alpha\beta}R_{\beta\gamma}m_\gamma \quad (4.11.8)$$

which gives for the difference in free energies of the two states

$$L[\bar{\phi}] - L[m] = -\tfrac{1}{2}\int d^dx |\nabla\theta|^2 (L_z)_{\alpha\beta} R_{\beta\gamma} m_\gamma (L_z)_{\alpha\mu} R_{\mu\nu} m_\nu$$

$$= -\tfrac{1}{2}\int d^dx |\nabla\theta|^2 m_\gamma \tilde{R}_{\gamma\beta}(\tilde{L}_z)_{\beta\alpha}(L_z)_{\alpha\mu} R_{\mu\nu} m_\nu \quad (4.11.9)$$

Because $\tilde{L}_z(L_z) = -1$ and $\tilde{R} \cdot R = 1$, we have

$$H[\bar{\phi}] - H[m] = \frac{|m|^2}{2}\int d^dx |\nabla\theta|^2 \quad (4.11.10)$$

4.11 GOLDSTONE MODES AND THE LOWER CRITICAL DIMENSION

If we take $\theta = k \cdot x$, the change in the energy per unit volume is

$$\varepsilon(k) = \frac{m^2}{2} k^2 \tag{4.11.11}$$

In the limit $k \to 0$ this excitation energy, which in this example corresponds to a "spin wave," is vanishingly small, which bears out our expectation that there is no gap in the energy spectrum above the ground state.

These modes are important in determining the lower critical dimension of a system because if, as $T \to 0$, there is a sufficient density of these spin waves, the order parameter vanishes.

To see this, consider the correlation function

$$G_{\alpha\alpha}(x - y) = \langle \overline{\phi}_\alpha(x) \overline{\phi}_\alpha(y) \rangle \tag{4.11.12}$$

In (4.11.12) we are interested in averaging over all possible $\theta(x)$, so we have

$$G_{\alpha\alpha}(x - y) = \frac{\int D[\theta] \exp\left(-\frac{m^2}{2} \int d^d x |\nabla \theta|^2\right) m \tilde{R}(\theta(x)) \cdot R(\theta(y)) m}{\int D[\theta] \exp\left(-\frac{m^2}{2} \int d^d x |\nabla \theta|^2\right)}$$

$$\tag{4.11.13}$$

Because we have an abelian group and $\tilde{R}(\theta) = R(-\theta)$, (4.11.13) becomes

$$G_{\alpha\alpha}(x - y) = \frac{\int D[\theta] \exp\left(-\frac{m^2}{2} \int d^d x |\nabla \theta|^2\right) m R(\theta(y) - \theta(x)) m}{\int D[\theta] \exp\left(-\frac{m^2}{2} \int d^d x |\nabla \theta|^2\right)}$$

$$= m^2 \frac{\int D[\theta] \cos(\theta(x) - \theta(y)) \exp\left(-\frac{m^2}{2} \int d^d x |\nabla \theta|^2\right)}{\int D[\theta] \exp\left(-\frac{m^2}{2} \int d^d x |\nabla \theta|^2\right)}$$

$$\tag{4.11.14}$$

We can write this in the form

$$G_{\alpha\alpha}(x-y)$$

$$= m^2 \frac{\int D[\theta]\exp\left(i\int d^dx'[\delta(x'-x) - \delta(x'-y)]\theta(x')\right)\exp\left(-\frac{m^2}{2}\int d^dx|\nabla\theta|^2\right)}{\int D[\theta]\exp\left(-\frac{m^2}{2}\int d^dx|\nabla\theta|^2\right)}$$

(4.11.15)

The integration in (4.11.15) is now Gaussian and can be performed by completing the square. Let θ_0 satisfy

$$m^2\nabla^2\theta_0 + i[\delta(x-x_0) - \delta(x-y_0)] = 0 \quad (4.11.16)$$

If we define the d-dimensional Green function $g(x-x')$ such that

$$\nabla^2 g(x-x') = \delta(x-x')$$

and[†]

$$g(0) = 0 \quad (4.11.17)$$

then we have

$$\theta_0 = -\frac{i}{m^2}[g(x-x_0) - g(x-y_0)] \quad (4.11.18)$$

Writing $\theta = \theta' + \theta_0$, the result is

$$G_{\alpha\alpha}(x-y) = m^2 e^{-(1/m^2)g(x-y)} \quad (4.11.19)$$

Now let us examine the behavior of the Green function $g(x-y)$ in various dimensions. Fourier transforming (4.4.17), we have

$$g(x) = \int \frac{d^dk}{(2\pi)^d} \frac{e^{ik\cdot x}}{k^2} \quad (4.11.20)$$

If we let $x = \lambda x'$ and make the change of variable $k = \lambda^{-1}k'$, we have

$$g(\lambda x) = \lambda^{2-d} g(x) \quad (4.11.21)$$

[†]This condition is a reflection of the lattice on which the system was originally defined. In a true continuum theory $g(0)$ diverges. Ultimately we only need the behavior of g for large arguments.

4.11 GOLDSTONE MODES AND THE LOWER CRITICAL DIMENSION

which tells us that $g(x)$ is a homogeneous function of x of order $2 - d$. For $d > 2$, $g(x)$ falls off as

$$\lim_{x \to \infty} g(x) \to \frac{1}{x^{d-2}} \qquad (4.11.22)$$

so that the spin wave contribution to the correlation function in the limit $r \to \infty$ behaves as

$$\lim_{r \to \infty} G_{\alpha\alpha}(r) \to m^2 \exp\left(-\frac{1}{m^2 r^{d-2}}\right) \cong m^2 \qquad d > 2 \quad (4.11.23)$$

However, for $d < 2$ the correlation function decays as

$$\lim_{r \to \infty} G_{\alpha\alpha}(r) \to m^2 e^{-r^{2-d}/m^2} \qquad d < 2 \qquad (4.11.24)$$

and there is no long-range order for $d < 2$. Right at $d = 2$ we have

$$\lim_{x \to \infty} g(x) \to \ln x \qquad (4.11.25)$$

and so

$$\lim_{r \to \infty} G_{\alpha\alpha}(r) \to m^2 |r|^{-1/m^2} \qquad (4.11.26)$$

In this case the correlations decay algebraically, a situation sometimes called "short" long-range order, and there is a "zero temperature" phase transition.

Evidently, $d = 2$ is the boundary between systems that can order by spontaneously broken continuous symmetries and those that cannot. Note that it is essential that the symmetry is continuous. The Ising model, which has a discrete symmetry, orders in two dimensions quite nicely. This result is originally due to Mermin and Wagner [6, 7] and Hohenberg [3]. It states that the spontaneous breaking of a continuous symmetry is thwarted in $d \leq 2$ by the strength of the fluctuations in the order parameter.

The Mermin–Wagner–Hohenberg theorem does not, however, preclude phase transition in systems with a continuous symmetry. The most famous example of such a phase transition is the Kosterlitz–Thouless transition in the two-dimensional $X-Y$ model [4, 5, 8], which is the subject of Chap. 11.

PROBLEMS

4.1 Show that if the correlation length is an analytic function of the parameters, then all the scaling fields are irrelevant. Note, however, that this does not imply that the scaling exponents themselves are integers.

4.2 Show that for a trivial fixed point the "critical surface" has the same dimension as parameter space itself.

4.3 Show that the existence of at least one relevant scaling field at a fixed point implies that the correlation length at the fixed point must be infinite.

4.4 Prove that the eigenvalues of the linearized RG are as given in (4.2.6). What is y_k in terms of $\lambda_k(b)$ and b? Is y_k independent of b?

4.5 Show that (4.4.2) follows from (4.4.1).

4.6 The susceptibility is related to the correlation function by

$$\chi = \sum_{\mathbf{r}} G(\mathbf{r})$$

Use this to derive the scaling law (4.6.1c).

4.7 Using the generating potential, rederive the equation of state (4.4.11).

4.8 Show that the magnetization varies as $h^{1/\delta}$ at $t = 0$. What is δ in terms of ν and y_h?

4.9 Derive the scaling laws (4.6.1b) and (4.6.1c)

4.10 Derive Rushbrooke's scaling law (4.6.1d)

$$\alpha + 2\beta + \gamma = 2$$

4.11 Derive the scaling law (4.6.1e)

$$\delta = \frac{d + 2 - \eta}{d - 2 + \eta}$$

4.12 An alternative form for the equation of state can be derived by choosing $|h|b^{y_h} = 1$ in (4.4.9). Show that the magnetization is given by

$$m = h^\delta f_h(t|h|^{-1/\Delta\nu}, 1)$$

and determine δ and Δ in terms of y_T and y_h.

4.13 Given that y_T and y_h must obey the inequality $y < d$, what bounds can be placed on the critical exponents α, β, and γ?

4.14 Show that if y_h assumes the maximal value $y_h = d$, then the magnetization is discontinuous at $t = 0$.

4.15 Show that the specific heat is related to the thermodynamic fluctuations in the energy

$$C = \frac{1}{T^2}(\langle H^2 \rangle - \langle H \rangle^2)$$

This is another example of the fluctuation–dissipation theorem.

REFERENCES

1. Farach, H. A., Creswick, R. J., and Poole, C. P., Jr., (1990), *Mod. Phys. Lett. B* **4** 1029.
2. Fisher, M. E. and Berker, A. N. (1982) *Phys. Rev. B* **26** 2507.
3. Hohenberg, P. C. (1967) *Phys. Rev.* **158** 383.
4. Kosterlitz, J. M. (1974) *J. Phys. C* **7** 1046.
5. Kosterlitz, J. M. and Thouless, D. J. (1973) *J. Phys. C* **6** 1181.
6. Mermin, N. D. (1966) *Phys. Rev.* **176** 250.
7. Mermin, N. D. and Wagner, H. (1966) *Phys. Rev. Lett.* **22** 1133.
8. Nelson, D. R. and Kosterlitz, J. M. (1977) *Phys. Rev. Lett.* **39** 1201.

5

THE ISING MODEL

5.0 INTRODUCTION

Of all the models in statistical physics, perhaps none is as simple or as widely studied as the Ising model. This model was originally put forward as a model of magnetism, and indeed the Ising model is still used with great success for this purpose. In addition, it has been used to model binary alloys, the imperfect lattice gas, chemical reactions, and ferroelectric behavior. The reason the Ising model is so popular is that it is quite amenable to both theoretical and numerical techniques, and at the same time it exhibits all the nontrivial critical behavior of more complicated systems.

In this chapter we introduce the Ising model and study its properties in one and two dimensions. For more details the reader is referred to the excellent monographs of Baxter [1] and McCoy and Wu [5].

The Ising, or Ising–Lenz, model was proposed to Ising by his thesis advisor (Lenz) in the early 1920s as a model of ferromagnetism. At that time the only theory of magnetism was the Weiss molecular field theory, which predicts a second-order phase transition in any number of dimensions. Ising was able to solve the problem in one dimension [4] and was disappointed to find that there was no phase transition in that case. A decade or so later, Peierls [8] proved that the Ising model in two dimensions must order and in 1944 Onsager [6] published his celebrated exact solution of the Ising model in two dimensions. Since

Onsager's solution of the Ising model, several other two-dimensional lattice models have been solved exactly, but as yet there is no exact solution for a nontrivial three-dimensional system.

Normally, the Ising model is formulated on a Bravais lattice whose sites we denote by lowercase Latin indices i, j, and so on. The dynamical variables in the Ising model are "spins" S_i, which take the values ± 1. Note that although we use the jargon of quantum mechanics, the Ising model is purely classical and predates the discovery of the spin of the electron. The spins are coupled through "exchange constants" J_{ij} and the Hamiltonian takes the form

$$H = -\frac{1}{2}\sum_{i,j} J_{ij} S_i S_j - h \sum_i S_i \qquad (5.0.1)$$

The second term in (5.0.1) includes the effect of an external magnetic field h. In general, one must consider terms in the Hamiltonian that involve products of three spins, four spins, and so on, and in fact we will find that the RG generates such terms even if the original Hamiltonian is free of them. Because the Hamiltonian (5.0.1) embodies all of the basic properties of critical systems, we will concentrate on this problem for the present.

An important special case is when the exchange constants are 0 unless the lattice sites i and j are nearest neighbors. We denote such pairs by $\langle ij \rangle$, and the Hamiltonian for the nearest-neighbor Ising model is

$$H = -J \sum_{\langle ij \rangle} S_i S_j - h \sum_i S_i \qquad (5.0.2)$$

If the coupling constant J is positive, then the ground state of the system is ferromagnetic; that is, all the spins are aligned. On the other hand, if J is negative, nearest-neighbor spins are antiparallel in the lowest-energy configuration and the system is said to be antiferromagnetic. All lattices admit ferromagnetic order, but an important class of lattices, the so-called close-packed lattices, are incommensurate with antiferromagnetic order. The simplest example of this is afforded by the triangular lattice in two dimensions. Figure 5.0.1 shows a cluster of three spins that are all nearest neighbors of each other. As one can readily see, it is impossible to arrange the spins so that every neighbor of each spin is antiparallel to that spin. This is the phenomenon of frustration, and it plays an essential role in the theory of spin glasses.

146 THE ISING MODEL

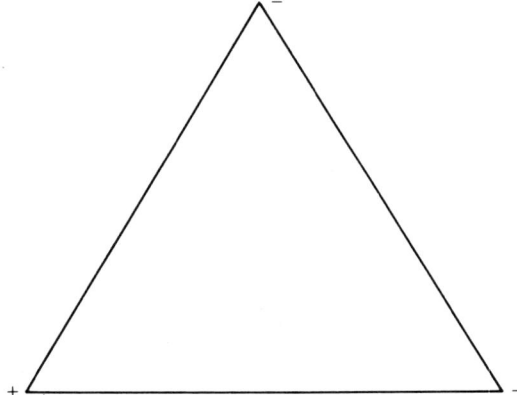

Fig. 5.0.1 Three-spin cluster of nearest-neighbor spins on the antiferromagnetic triangular lattice, There is no way to align all three spins so that each is antiparallel to the other two.

The central object of interest is the partition function Z for the system given by

$$Z = \text{Tr}\, e^{-\beta H[S]} \tag{5.0.3}$$

The trace symbol "Tr" in (5.0.3) indicates that we are to sum over all possible states of the system, that is, all possible combinations of $S_i = \pm 1$. The parameter $\beta = 1/T$ (Boltzmann's constant k_B is set to unity throughout this text).

One possible representation of the spin operator S_i is the Pauli matrix σ_z. In this case the Hamiltonian is a square matrix of dimension $2^n \times 2^n$, and the trace operation is just the familiar matrix trace.

Once the partition function is known, all the interesting thermodynamic quantities, such as the internal energy, entropy, specific heat, magnetization, and susceptibility, can be calculated by differentiation. First, the Gibbs free energy F is given by

$$F = -T \ln Z \tag{5.0.4}$$

and the internal energy E and entropy S are given by

$$E = -\frac{\partial \ln Z}{\partial \beta}$$

$$S = \frac{\partial F}{\partial T} \tag{5.0.5}$$

The specific heat is the derivative of the internal energy with respect to the temperature, which yields

$$C = \frac{\partial E}{\partial T} = \frac{1}{T^2}(\langle H^2 \rangle - \langle H \rangle^2) \qquad (5.0.6)$$

If we include a magnetic field as in (5.0.1), then the magnetization M is given by

$$M = T \frac{\partial}{\partial h} \ln Z \qquad (5.0.7)$$

and the susceptibility χ is

$$\chi = \frac{\partial M}{\partial h} = \frac{1}{T}(\langle M^2 \rangle - \langle M \rangle^2) \qquad (5.0.8)$$

In the language of phase transitions, the magnetization is the order parameter for the Ising model. Note that, in the absence of an external field, the Hamiltonian (5.0.1) is invariant under the transformation $S_i \to -S_i$, and that on the basis of symmetry arguments we would expect the magnetization to be identically 0. However, in two and higher dimensions one finds that this symmetry is spontaneously broken and the system develops a nonzero magnetization even in the absence of an external field.

In addition to these thermodynamic quantities, we will also be interested in the spin–spin correlation function G_{ij}, which is defined as

$$G_{ij} = \langle S_i S_j \rangle - \langle S_i \rangle \langle S_j \rangle \qquad (5.0.9)$$

and the correlation length, which is given in terms of the correlation function as

$$\xi^{-1} = -\lim_{r \to \infty} \frac{\partial}{\partial r} \ln G(r) \qquad (5.0.10)$$

5.1 THE ONE-DIMENSIONAL ISING MODEL

In this section we consider the problem first solved by Ising. We will consider the nearest-neighbor Ising model and impose periodic boundary conditions so that spin S_i and spin S_{N+i} are identified. First,

Fig. 5.1.1 The one-dimensional Ising chain. Spins can be divided into two sublattices denoted by $\bigcirc$, the even spins, and $\times$, the odd spins. The decimation transformation traces over the even spins only.

let $h = 0$, so the Hamiltonian is simply

$$H = -J \sum_i S_i S_{i+1} \tag{5.1.1}$$

The partition function can be easily evaluated in this case, but let us consider the problem from the point of view of the RG. We can divide the spins into those on the even- and odd-numbered sites and calculate the trace over, say, the even spins only. In the language of the RG, this is the decimation part of the RG transformation. This process is illustrated in Fig. 5.1.1.

The trace over one of these even-numbered spins, say S_2, is of the form

$$\mathrm{Tr}\, e^{\beta J S_2 (S_1 + S_3)} = 2 \cosh \beta J (S_1 + S_2)$$
$$= \Delta e^{\beta J' S_1 S_3} \tag{5.1.2}$$

where Δ is a constant independent of the S's. The partition function can therefore be rewritten as

$$Z = \Delta^{N/2} \mathop{\mathrm{Tr}}_{\{S\ \mathrm{odd}\}} \exp\left(\beta J' \sum_i S_i S_{i+2} \right) \tag{5.1.3}$$

Unlike Ising models in two and higher dimensions, the simple decimation of a nearest-neighbor Ising model in one dimension produces another nearest-neighbor Ising model with a renormalized coupling constant. The new lattice spacing is just twice the original, so the scale factor for this RG transformation is

$$b = \frac{a'}{a} = 2 \tag{5.1.4}$$

The renormalized exchange coupling J' and the normalization factor Δ can be calculated by first setting $S_1 = 1$, $S_2 = 1$ and then reversing

S_2. The result is, in terms of $v = \tanh \beta J$,

$$\Delta = 2\sqrt{\cosh 2\beta J} \tag{5.1.5}$$

$$v' = v^2 \tag{5.1.6}$$

Note that because $0 < v < 1$, $v^2 < v$, and the exchange constant tends to 0 as the RG is repeated. The RG equation (5.1.6) therefore has two fixed points, $v^* = 1$ and $v^* = 0$. For any initial temperature different from 0, the parameter v tends to the second of these fixed points, which we may call the high-temperature, or paramagnetic, fixed point. The zero-temperature, or ferromagnetic, fixed point, $v^* = 1$, is unstable and physically inaccessible. Therefore, we come to the conclusion, which disappointed Ising, that the Ising model in one dimension fails to exhibit critical behavior.

We can see this in another way by considering the RG equation for the correlation length. Because the original and renormalized systems are of the same kind but with all lengths scaled by a factor of 2, it follows that the correlation lengths calculated in the two systems must be related by (4.3.3), and because $b = 2$ here we have

$$\xi(v) = 2\xi(v^2) \tag{5.1.7}$$

from which it follows that ξ must have the form

$$\xi = -\frac{K}{\ln v} \tag{5.1.8}$$

where K is a constant. The absence of a phase transition is again reflected in the fact that the correlation length is finite for all temperatures and diverges only as $T \to 0$ as

$$\lim_{T \to 0} \xi(T) \to \frac{K}{2} e^{2J/T} \tag{5.1.9}$$

Physically, the reason the one-dimensional Ising model fails to order at nonzero temperatures can be ascribed to excitations called "kinks." A kink is a point in the chain where the spins flip, and it is the one-dimensional equivalent of a domain wall in higher-dimensional systems. Figure 5.1.2 shows two kinks in the one-dimensional chain.

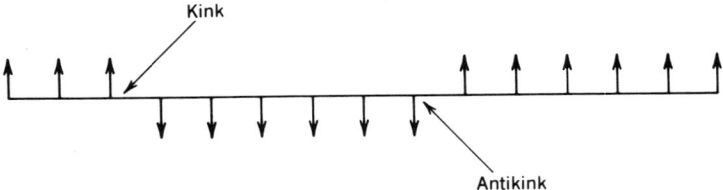

Fig. 5.1.2 A kink–antikink pair on the one-dimensional Ising chain.

Actually, the figure depicts a kink and an antikink. At a kink the spins flip from $+1$ to -1, whereas at an antikink they flip from -1 to $+1$.

The energy to create a single kink from the ground state is $2J$, so if there are n kinks in the chain the energy of excitation is

$$E(n) = 2nJ \qquad (5.1.10)$$

The entropy can be evaluated by enumerating the number of states of n_1 kinks and n_2 antikinks. Assuming that the kinks and antikinks are dilute, that is, the temperature is very low, we can assume that $1 \ll n \ll N$, where N is the number of sites. If we employ periodic boundary conditions, then kinks and antikinks must be created in pairs. Starting with the ground state in which all spins $S = +1$, then each kink–antikink pair defines a domain of spins in which $S = -1$. In order to count the number of states available to n kink–antikink pairs, we need to know how many ways it is possible to divide N sites into $2n$ segments of alternating magnetization, which is equal to the number of ways of distributing the $2n$ kinks and antikinks among the N sites. Therefore, the number of states available to n kink–antikink pairs is just the binomial coefficient

$$\Omega(2n) = \binom{N}{2n} \qquad (5.1.11)$$

Using Stirling's approximation for the factorial, the free energy of the kink gas is

$$F = 4nJ - T[N \ln N - (N - 2n)\ln(N - 2n) - 2n \ln 2n] \qquad (5.1.12)$$

The number of kinks in equilibrium is determined by minimizing the free energy, which gives

$$\frac{n}{N} = e^{-2J/T} \qquad (5.1.13)$$

Note that even at very low temperatures there are still a few kinks here and there, and the magnetization is, on the average, 0. Roughly speaking, the correlation length is equal to the average length of one of these domains, which can be estimated from their density,

$$\xi \sim \left(\frac{n}{N}\right)^{-1} \sim e^{2J/T} \tag{5.1.14}$$

The low-temperature behavior found in (5.1.14) agrees with that of (5.1.9).

The destruction of long-range order by "topological excitations," in this case the kinks, is the subject of intense theoretical interest in condensed matter physics and field theory. The nature of the topological excitations depends on the order parameter and the dimensionality of space. For example, consider the two-dimensional X–Y model where the Hamiltonian is given by

$$H_{xy} = -J \sum_{\langle ij \rangle} \left(S_i^x S_j^x + S_i^y S_j^y \right) \tag{5.1.15}$$

S is now a two-component vector and the model exhibits an exotic phase transition, the so-called Kosterlitz–Thouless (KT) transition. In the KT theory the topological excitations, which in this case are vortices in the spin vector field, make a transition from the low-temperature state in which vortices and antivortices form closely bound neutral pairs to a vortex plasma state. In Chap. 11 we will study the RG approach to the KT theory in detail.

The one-dimensional Ising model is simple enough that we can actually evaluate the correlation function exactly and compare the result with our simple RG calculation. The correlation function $G(r)$ is given by

$$G(r) = \langle S_{i+r} S_i \rangle \tag{5.1.16}$$

which by translational invariance is independent of the site i. In order to evaluate (5.1.16), we first make the observation that because each term in the Hamiltonian commutes, we can write

$$\exp\left(\beta J \sum_{\langle ij \rangle} S_i S_j \right) = \prod_{\langle ij \rangle} e^{\beta J S_i S_j} \tag{5.1.17}$$

Furthermore, because each spin variable takes on only the values ± 1, the exponential in (5.1.17) can be written as

$$e^{\beta J S_i S_j} = \cosh \beta J + S_i S_j \sinh \beta J \qquad (5.1.18)$$

Factoring out the hyperbolic cosine, we have

$$e^{\beta J S_i S_j} = \cosh \beta J (1 + v S_i S_j) \qquad (5.1.19)$$

and $v = \tanh \beta J$. The denominator in (5.1.17) can be evaluated by noting that

$$\operatorname*{Tr}_{S_2} (1 + v S_1 S_2)(1 + v S_2 S_3) = 2(1 + v^2 S_1 S_3) \qquad (5.1.20)$$

so that we find for the correlation function

$$G(r) = v^r \qquad (5.1.21)$$

By the definition of the correlation length, (5.0.10), we have

$$\xi^{-1} = -\ln v \qquad (5.1.22)$$

which is the result obtained by our RG argument (5.1.8) and where the constant K is found to be equal to unity.

5.2 THE ONE-DIMENSIONAL ISING MODEL IN A MAGNETIC FIELD

Now that we have solved the Ising model in one dimension in the absence of a magnetic field, let us consider the problem in the presence of a magnetic field. One might think that an external field would have the effect of suppressing the kinks and encouraging critical behavior, but in fact this is not true. We will see that in two dimensions, where there is spontaneous symmetry breaking in the Ising model, the presence of an external field actually suppresses critical behavior. The Hamiltonian now has the form

$$H = -J \sum_i S_i S_{i+1} - h \sum_i S_i \qquad (5.2.1)$$

We again want to apply the simple decimation transformation, which proved so successful in the absence of a field. The decimation is

5.2 THE ONE-DIMENSIONAL ISING MODEL IN A MAGNETIC FIELD

carried out as before and we write

$$\operatorname*{Tr}_{S_2} e^{\beta J S_2(S_1+S_3) + \beta h S_2} = \Delta e^{\beta J' S_1 S_3 + \beta h''(S_1+S_3)} \qquad (5.2.2)$$

where J' is the renormalized exchange coupling, and h'' is a piece of the renormalized field. As before, if we evaluate the left and right sides for various values of the spins, we can solve for J' and h'', with the result

$$h'' = \frac{1}{4\beta} \ln\left[\frac{\cosh \beta(2J+h)}{\cosh \beta(2J-h)}\right] \qquad (5.2.3)$$

$$\Delta^2 = 4\cosh\beta[\cosh\beta(2J+h)\cosh\beta(2J-h)]^{1/2} \qquad (5.2.4)$$

$$J' = \frac{1}{4\beta} \ln\left[\frac{\cosh \beta(2J+h)\cos \beta(2J-h)}{\cosh^2 \beta h}\right] \qquad (5.2.5)$$

The total renormalized field acting on one of the undecimated sites is $h + 2h''$, because a contribution of h'' comes from the decimation of each of the sites to the left and right of it. Therefore, the renormalization equation for the field is

$$h' = h + \frac{1}{2\beta} \ln\left[\frac{\cosh \beta(2J+h)}{\cosh \beta(2J-h)}\right] \qquad (5.2.6)$$

For small h, (5.2.5) and (5.2.6) become

$$\tanh \beta J' = (\tanh \beta J)^2 \qquad (5.2.7)$$

$$h' = h(1 + \tanh 2\beta J) \qquad (5.2.8)$$

The RG equations (5.2.5) and (5.2.6) have a line of fixed points $J^* = 0$, h^* arbitrary, corresponding to the paramagnetic phase and an isolated ferromagnetic fixed point $J^* = \infty$, $h^* = 0$. The RG equations can be cast in a somewhat simpler form by introducing new temperature and magnetic field parameters $v = \tanh \beta J$ and $w = \tanh \beta h$,

$$\frac{1+w'}{1-w'} = \frac{1+w}{1-w} \frac{1 + 2wv + v^2}{1 + 2wv + v^2} \qquad (5.2.9)$$

$$(1+v')^2 = \frac{(1+v^2) - 4v^2 w^2}{(1-v^2)} \qquad (5.2.10)$$

THE ISING MODEL

At the paramagentic fixed point, $v^* = 0$, these become, to lowest order in v,

$$w' = w \tag{5.2.11}$$

$$v = v^2(1 - w^2) \tag{5.2.12}$$

The temperature and magnetic exponents are given by

$$y_T = \frac{\ln\left[\frac{\partial v'}{\partial v}(v^*, w^*)\right]}{\ln 2} = -\infty \tag{5.2.13}$$

and

$$y_h = \frac{\ln\left[\frac{\partial w'}{\partial w}(v^*, w^*)\right]}{\ln 2} = 0 \tag{5.2.14}$$

The temperature parameter v is irrelevant, whereas, by (5.2.14), the magnetic parameter w is marginal and leads to a line of fixed points rather than a single isolated fixed point.

The ferromagnetic fixed point is somewhat more interesting. Linearizing the RG equations about $w^* = 0$, $v^* = 1$ and writing $v = 1 - t$ with $t \ll 1$, we have

$$\begin{aligned} w' &= 2w \\ t' &= 2t \end{aligned} \tag{5.2.15}$$

which yields $y_T = 1$, $y_h = 1$. We could have found y_T by observing from (5.1.14) or (5.1.23) that

$$\chi \sim t^{-1} \tag{5.2.16}$$

which gives $\nu = 1/y_T = 1$.

In addition, by (5.1.22) we have for $T = T_c = 0$, $G(r) = 1$, and by the definition of the correlation function exponent η, (4.0.6), we see that $\eta = 1$. Note that both y_T and y_h assume their maximal values $y = d = 1$ at the ferromagnetic fixed point. A typical RG trajectory is shown in Fig. 5.2.1.

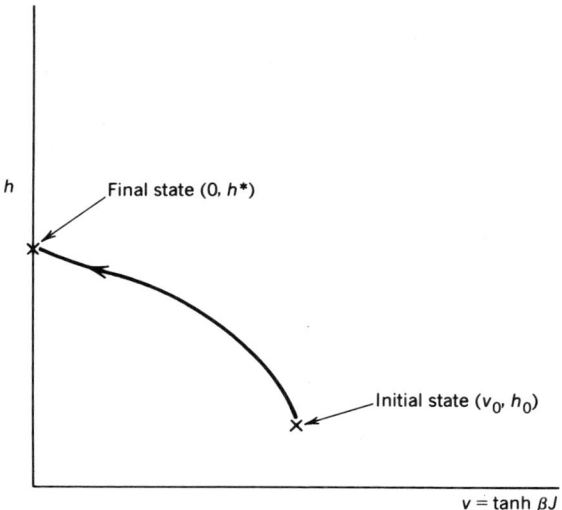

Fig. 5.2.1 Typical RG trajectory for the one-dimensional Ising model in a magnetic field.

5.3 THE FERROMAGNETIC ISING MODEL IN TWO DIMENSIONS

The Ising model in higher dimensions remained unsolved until, in 1944, Onsager published his celebrated solution of the Ising model on the square lattice [6]. Before entering into the mechanics of solving the two-dimensional Ising model, several interesting and useful results can be established which lead one to believe that the Ising model does in fact order at temperatures different from 0.

As in the one-dimensional lattice, all the spins are aligned parallel in the ground state, but the lowest-energy excitations are single flipped spins rather than kinks. The energy needed to excite a state in which n spins are flipped is $2nqJ$, where q is the number of nearest neighbors, or coordination number, for the lattice under consideration. Assuming these flipped spins are distributed randomly around the lattice and $1 \ll n \ll N$ where N is the number of spins in the lattice, the free energy of such a state is, as in (5.1.12),

$$F = 2nqJ - T[N \ln N - (N-n)\ln(N-n) - n \ln n] \quad (5.3.1)$$

Again we have applied Stirling's approximation for the factorial.

THE ISING MODEL

Minimizing the free energy with respect to n gives for the mean density of flipped spins

$$\frac{n}{N} = e^{-2qJ/T} \tag{5.3.2}$$

Now unlike the one-dimensional chain, the low-energy excitations have little effect on the order parameter, or magnetization, which in this approximation is

$$\frac{m}{N} = 1 - 2e^{-2qJ/T} \tag{5.3.3}$$

Therefore, we can conclude that the broken-symmetry, or ordered, phase of the Ising model is stable at finite temperatures in two and higher dimensions.

For simple two-dimensional lattices in the absence of a magnetic field, we can in fact determine the transition temperature without actually solving the problem completely. We first employ the expansion of the exponential equation (5.1.19) and write the partition function as

$$Z = \text{Tr} \prod_{\langle ij \rangle} (\cosh \beta J + S_i S_j \sinh \beta J) \tag{5.3.4}$$

Factoring out the hyperbolic cosine factor from each term, we have

$$Z = (\cosh \beta J)^{N_b} \text{Tr} \prod_{\langle ij \rangle} (1 + v S_i S_j) \tag{5.3.5}$$

where N_b is the number of "bonds" or nearest-neighbor pairs, and again $v = \tanh \beta J$.

We first observe that because $S^2 = 1$, any function of one of the spins can be expressed as a multiple of the identity plus a multiple of the spin variable itself. The trace of the identity is equal to 2, whereas the trace of the spin itself is 0. Therefore, the only terms that contribute to the trace (5.3.5) must involve an even number (including 0) of factors of each spin. There is a simple representation of these terms in the partition function as closed graphs on the lattice. By a graph we mean a figure composed of bonds, which connect the nearest-neighbor sites of the lattice, and vertices, which are lattice sites at which two or more bonds intersect.

5.3 THE FERROMAGNETIC ISING MODEL IN TWO DIMENSIONS

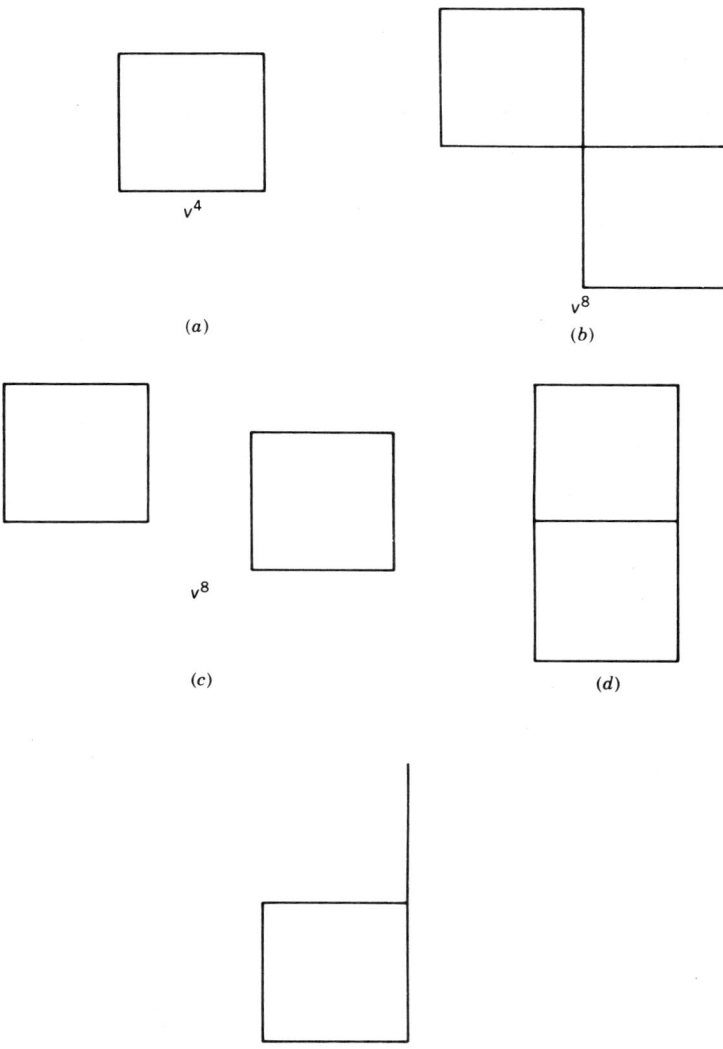

Fig. 5.3.1 Graphs in the expansion of the partition function. (*a*), (*b*), and (*c*) are allowed graphs, (*d*) and (*e*) are forbidden.

The condition that only even powers of the spin variables contribute to the trace restricts our attention to graphs in which an even number of lines emerge from each vertex. The first few allowed graphs for the square lattice are shown in Figs. 5.3.1*a–c*, whereas Figs. 5.3.1*d* and *e* show forbidden graphs. The graph in Fig. 5.3.1*d* is forbidden because there are two vertices at which an odd number of bonds (3)

158   THE ISING MODEL

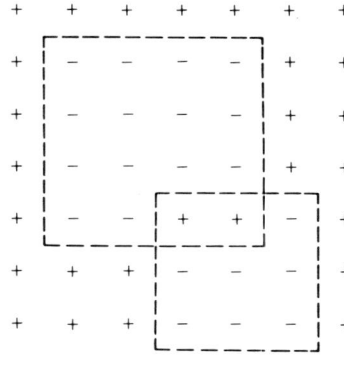

Fig. 5.3.2 Domains of like spins for the FM Ising model on a square lattice. Note that the domain boundaries form closed loops that intersect the broken bonds at right angles.

intersect; Fig. 5.3.1e is forbidden for the same reason and also because the graph is not closed.

Just as the kinks divided the one-dimensional Ising model into domains of + and − spins, the states of the two-dimensional Ising model can also be thought of as composed of domains of parallel spins. Several such domains are shown in Fig. 5.3.2. Note that it is possible for domains to be nested inside one another. The boundary between domains, shown in Fig. 5.3.2 by the dotted line, bisects the bonds connecting spins of opposite sign.

The points at the centers of the bonds also form a lattice, which is called the dual of the original lattice (see Sec. 3.0). Figure 5.3.3 shows

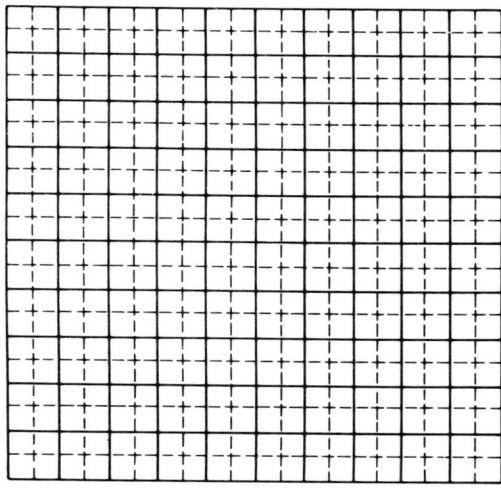

Fig. 5.3.3 The square lattice (solid lines) and the dual of the square lattice (dotted lines), which is also square.

5.3 THE FERROMAGNETIC ISING MODEL IN TWO DIMENSIONS   159

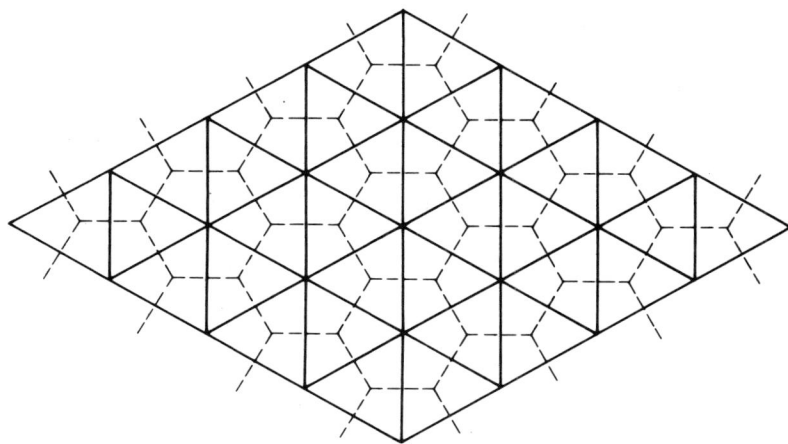

Fig. 5.3.4 The triangular lattice (solid lines) and its dual, the honeycomb lattice (dotted lines).

the square lattice and the dual of the square lattice, which also turns out to be a square lattice. When a lattice and its dual are identical, that lattice is said to be self-dual. On the other hand, as shown in Fig. 5.3.4, the dual of the triangular lattice is the honeycomb lattice. The dual of the dual of a lattice is the original lattice, so the dual of the honeycomb lattice is the triangular lattice.

The boundaries between domains form closed graphs on the dual lattice, and therefore it is possible to establish a two-to-one correspondence between the states of the Ising model on a lattice and closed graphs on the dual of the lattice. The reason the correspondence is two to one is that one can obtain a second state of the Ising model for the same graph simply by reversing all the spins.

If we now return to the representation of the partition function in terms of closed graphs in which each bond of the graph is assigned a weight $v = \tanh \beta J$, a relationship between the partition function on a lattice and the partition function on its dual can be established. If L is the length of the boundary of a domain, the energy required to create such a domain is $E = 2KL$, where K is the nearest-neighbor coupling on the dual lattice. The Boltzmann factor for a given domain with perimeter L is therefore

$$e^{-\beta E} = e^{-2\beta KL} \qquad (5.3.6)$$

The corresponding graph on the original lattice enters with a weight v^L, and therefore the partition function of the Ising model with

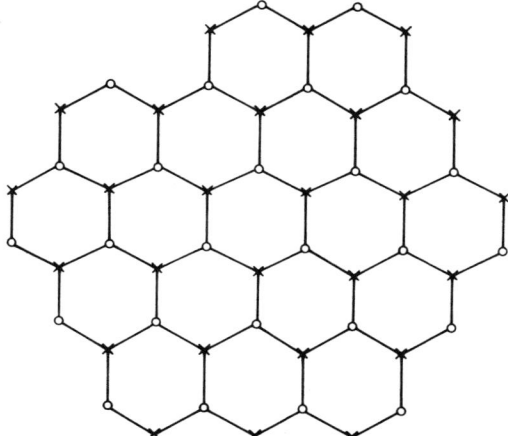

Fig. 5.3.5 The honeycomb lattice consists of two triangular sublattices indicated by open circles and crosses. In the star-triangle transformation the trace is carried out over spins on one sublattice.

coupling J is related to the partition function on the dual lattice with coupling K by

$$e^{-2\beta K} = \tanh \beta J \qquad (5.3.7)$$

It follows that if the partition function is singular at a unique critical temperature $(\beta J)_c$, the partition function on the dual lattice will exhibit an identical singularity at $(\beta K)_c$ given by (5.3.7). Because the square lattice is self-dual and assuming there is only one critical point, the critical temperature of the Ising model on a square lattice is given by

$$e^{-2\beta J_c} = \tanh \beta J_c \qquad (5.3.8)$$

There exists another useful transformation that relates the partition function of the Ising model on the triangular and honeycomb lattices, the so-called star–triangle transformation. The honeycomb lattice is composed of two triangular sublattices, which in Fig. 5.3.5 are indicated by the open circles and crosses. The trace for the partition function can be done in two steps, first tracing over spins on the sublattice marked by crosses and then tracing over the remaining spins

5.3 THE FERROMAGNETIC ISING MODEL IN TWO DIMENSIONS

marked by open circles,

$$Z_h = \operatorname*{Tr}_{\{S^A\}} \operatorname*{Tr}_{\{S^B\}} \exp\left(\beta J \sum_{\langle ij \rangle} S_i^A S_j^B\right) \tag{5.3.9}$$

If we decimate the spins on the sublattice marked by crosses, the remaining sites form a triangular lattice. What is more, because each of the decimated spins is coupled to only its three nearest neighbors, the highest-order spin terms generated by the decimation are just two-spin nearest-neighbor couplings. The decimations are of the form

$$\operatorname*{Tr}_{S_0^A} = e^{\beta J S_0^A (S_1^B + S_2^B + S_3^B)} = \Delta e^{\beta K (S_1^B S_2^B + S_1^B S_3^B + S_2^B S_3^B)} \tag{5.3.10}$$

and therefore the partition function for the honeycomb lattice is

$$Z_h = \Delta^{N/2} \operatorname*{Tr}_{\{S^B\}} \exp\left(\beta K \sum_{\langle ij \rangle} S_i^B S_j^B\right) \tag{5.3.11}$$

Because the remaining trace is just the partition function for the triangular lattice with exchange coupling K, where K is related to J by (5.3.10), the star–triangle transformation relates the partition function of the Ising model on the honeycomb lattice to that of the Ising model on the triangular lattice by

$$Z_h^N(\beta J) = \Delta^{N/2} Z_t^{N/2}(\beta K) \tag{5.3.12}$$

This relation can now be used in conjunction with the dual transformation to determine the critical temperatures for both the honeycomb and triangular lattices. By (5.3.7) the critical temperature for the honeycomb lattice must satisfy

$$2 \cosh 2\beta J - 1 = e^{-4\beta K} \tag{5.3.13}$$

In Table 5.3.1 the transition temperatures for the three two-dimensional lattices considered here are shown. Note that as one would

TABLE 5.3.1 Critical Temperatures for Several Two-Dimensional Ising Models

	q	T_c/J	v_c
Honeycomb	3	1.5187	0.57735
Square	4	2.2692	0.41412
Triangular	6	3.6410	0.26795

162 THE ISING MODEL

expect, the transition temperature increases with the coordination number q of the lattice.

5.4 EXACT SOLUTION OF THE ISING MODEL ON THE SQUARE LATTICE: THE PARTITION FUNCTION

In (5.3.5) we have a representation of the partition function in terms of the parameter $v = \tanh \beta J$. As we have seen, the partition function is given as a sum over closed graphs on the lattice with weights v^L, where L is the number of bonds in the graph. If we let $g(L)$ be the number of graphs of L bonds, then the partition function is

$$Z = (\cosh \beta J)^{N_b} 2^N \sum_L g(L) v^L \qquad (5.4.1)$$

A direct calculation of $g(L)$ is a rather laborious task for even modest values of L, but we can employ a rather clever trick which makes the counting easy. This method was developed by Burgoyne [2] and Vdovichenko [10, 11], and here we follow the notation of Stanley [9]. The idea is to establish a many-to-one relationship between paths on the lattice, which we will define momentarily, and graphs.

A "path" associated with a given graph is defined as a particular directed sequence of steps around the graph. Note that the path is not allowed to double-back on itself, but is otherwise unrestricted. There are many possible paths for a given graph, and to each path we assign a sign which is computed as follows. At each vertex of the graph, we assign a phase factor $e^{i\theta/2}$, where θ is the angle through which the path turns as it passes through the vertex, and with each subloop, or disconnected part, we assign a factor of -1. As shown in Fig. 5.4.1, there may be several ways of traversing a single graph, but the trick is that if we include all these possibilities, the sum of their individual contributions is just the weight of the original graph. In addition to the multiplicity of choices for paths that arise from a self-intersection, paths are also allowed to traverse the same bond more than once, as shown in Fig. 5.4.2. Although it is relatively easy to show that the sum over paths without multiple bonds produces the sum over graphs, it is trickier to prove that the sum over all paths with multiple bonds vanishes, and the interested reader should refer to App. C.

Let us denote the total contribution from paths with L bonds and containing n subloops by $G(L, n)$. Then

$$g(L) = \sum_n (-1)^n G(L, n) \qquad (5.4.2)$$

5.4 SQUARE LATTICE ISING MODEL: PARTITION FUNCTION

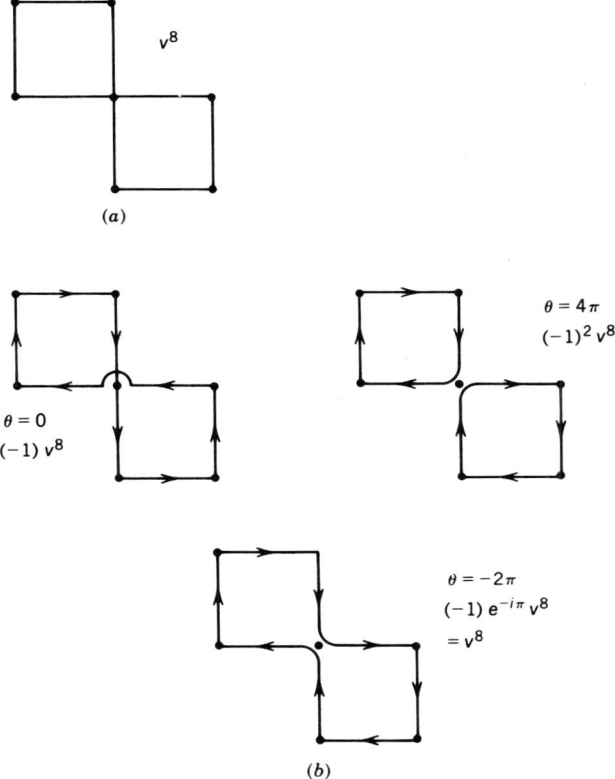

Fig. 5.4.1 (a) A graph representing a contribution v^8 to the partition function. (b) Different paths associated with the original graph together with the appropriate weights.

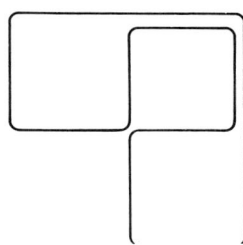

Fig. 5.4.2 Paths that traverse the same bond more than once.

Fig. 5.4.3 The weight associated with a path containing two simple subloops is the product of the weights for each simple loop.

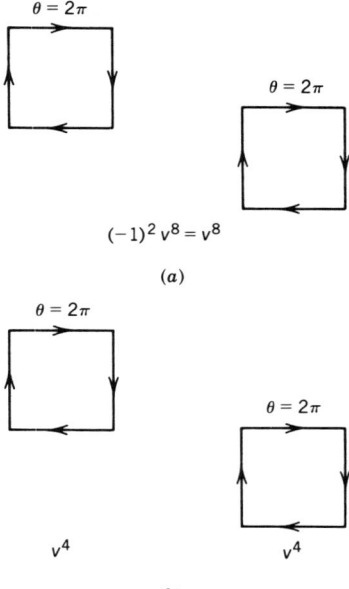

Figure 5.4.3 shows a graph consisting of two simple subloops (a simple subloop is one that is connected). The weight from such a path with disconnected parts is simply the product of weights for each subloop. If we let $D(l)$ be the contribution of a simple loop of l bonds, then the function $G(L, n)$ can be written as

$$G(L, n) = \frac{1}{2^n} \frac{1}{n!} \sum_{l_1} \cdots \sum_{l_n} D(l_1) \cdots D(l_k) v^L \delta_{l_1 + l_2 + \cdots + l_n, L} \quad (5.4.3)$$

The factor of $1/n!$ takes into account the fact that a simple permutation of the indices $l_1, l_2, \ldots$ does not yield a new term, and the factor of $1/2^n$ comes from the fact that each loop can be traversed in one of two possible directions. The Kronecker delta assures us that the total number of bonds in each term is L.

We now have for the sum in (5.4.1)

$$Z = (\cosh \beta J)^{N_b} 2^N \sum_L \sum_n \frac{1}{2^n} \frac{1}{n!} \sum_{l_1}$$
$$\times \cdots \sum_{l_n} D(l_1) \cdots D(l_k) v^L \delta_{l_1 + l_2 + \cdots + l_n, L} \quad (5.4.4)$$

5.4 SQUARE LATTICE ISING MODEL: PARTITION FUNCTION

The sum over L takes care of the Kronecker delta, and if we define $D(0) = 1$, we have

$$Z = (\cosh \beta J)^{N_b} \sum_n \frac{1}{2^n} \frac{(-1)^n}{n!} \left[\sum_l v^l D(l) \right]^n \tag{5.4.5}$$

The partition function is therefore given by

$$Z = 2^N (\cosh \beta J)^{N_b} \exp\left(-\frac{1}{2} \sum_{l=1}^{\infty} v^l D(l) \right) \tag{5.4.6}$$

In order to evaluate the $D(l)$, we introduce a new function $M(p, p')$, which is the sum over all paths from the site $p = (i, \alpha)$ to the site $p' = (j, \beta)$. The index α takes on the values 1, 2, 3, or 4, corresponding to the four possible directions in which the path can leave the site, whereas i denotes the site. This notation is illustrated in Fig. 5.4.4. Therefore, $D(l)$ can be written as

$$D(l) = \frac{1}{l} \sum_p M_l(p, p) \tag{5.4.7}$$

The $1/l$ comes from the fact that each simple loop of l bonds can be started from any of the l vertices.

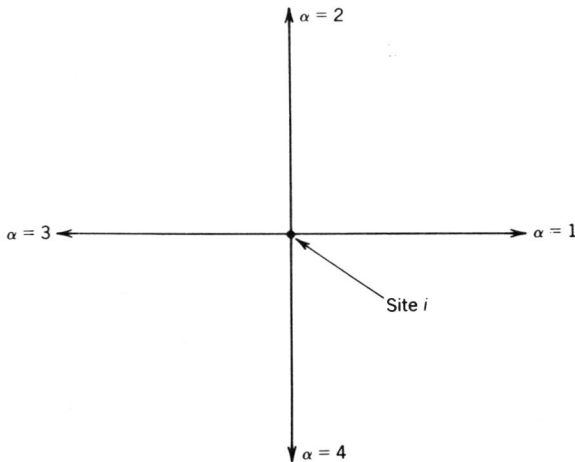

Fig. 5.4.4 The four possible directions α leading away from the lattice site i are labeled $\alpha = 1, 2, 3,$ and 4.

166   THE ISING MODEL

Each of the paths contributing to $M_l(p, p')$ are made up of individual single-bond steps, and we may therefore write

$$M_l(p, p') = \sum_{p_1 \cdots p_{l-1}} M_1(p, p_1) M_1(p_1, p_2) \cdots M_1(p_{l-1}, p') \quad (5.48)$$

Regarding this as a matrix equation, we have simply

$$M_l(p, p') = M_1^l(p, p') \quad (5.4.9)$$

and the function $D(l)$ now becomes

$$D(l) = \frac{1}{l} \operatorname{Tr} M_1^l \quad (5.4.10)$$

The matrix $M_1(p, p')$ is a $4N \times 4N$ matrix, where N is the number of sites on the square lattice. Denoting the eigenvalues of this matrix by m_k, we can write for the partition function

$$Z = (\cosh \beta J)^{N_b} 2^N \exp\left(-\frac{1}{2} \sum_{j=1}^{N} \sum_{l=1}^{\infty} \frac{1}{l} v^l m_j^l\right) \quad (5.4.11)$$

The sum over l is just the logarithm, and after a little algebra we finally have

$$Z = (\cosh \beta J)^{N_b} 2^N \left[\prod_{j=1}^{4N} (1 - v m_j)\right]^{1/2} \quad (5.4.12)$$

The problem now is to find the eigenvalues of the $4N \times 4N$ matrix M_1. M_1 has a simple structure, the only nonzero elements being

$$M_1(x, y, 1; x', y', \alpha') = \delta_{x, x'+1} \delta_{y, y'} \delta_{1, \alpha'} + e^{-i\pi/4} \delta_{x, x'} \delta_{y, y'+1} \delta_{2, \alpha'}$$
$$+ e^{+i\pi/4} \delta_{x, x} \delta_{y, y'-1} \delta_{4, \alpha'} \quad (5.4.13)$$

$$M_1(x, y, 2; x', y', \alpha') = e^{i\pi/4} \delta_{x, x'+1} \delta_{y, y'} \delta_{1, \alpha'} + \delta_{x, x'} \delta_{y, y'+1} \delta_{2, \alpha'}$$
$$+ e^{-i\pi/4} \delta_{x, x'-1} \delta_{y, y'} \delta_{3, \alpha'} \quad (5.4.14)$$

$$M_1(x, y, 3; x', y', \alpha') = e^{i\pi/4} \delta_{x, x'} \delta_{y, y'+1} \delta_{2, \alpha'} + \delta_{x, x'-1} \delta_{y, y'} \delta_{3, \alpha'}$$
$$+ e^{-i\pi/4} \delta_{x, x} \delta_{y, y'-1} \delta_{4, \alpha'} \quad (5.4.15)$$

$$M_1(x, y, 4; x', y', \alpha') = e^{-i\pi/4} \delta_{x, x'+1} \delta_{y, y'} \delta_{1, \alpha'} + e^{i\pi/4} \delta_{x, x'-1} \delta_{y, y'} \delta_{3, \alpha'}$$
$$+ \delta_{x, x'} \delta_{y, y'-1} \delta_{4, \alpha'} \quad (5.4.16)$$

5.4 SQUARE LATTICE ISING MODEL: PARTITION FUNCTION

This huge matrix can be put into a manageable block-diagonal form by taking the Fourier transform. By translational symmetry $M_1(r, r')$ is a function of $r - r'$ only and we can write

$$M_1 = \frac{1}{N} \sum_q e^{iq(r-r')} M_1(q) \tag{5.4.17}$$

where $M_1(q)$ now assumes the block-diagonal form

$$\begin{bmatrix} Q_1 & e^{-i\pi/4}Q_2 & 0 & e^{i\pi/4}Q_2^* \\ e^{i\pi/4}Q_1 & Q_2 & e^{-i\pi/4}Q_1^* & 0 \\ 0 & e^{i\pi/4}Q_2 & Q_1^* & e^{-i\pi/4}Q_2^* \\ e^{-i\pi/4}Q_1 & 0 & e^{i\pi/4}Q_1^* & Q_2^* \end{bmatrix} \tag{5.4.18}$$

where $Q_i = e^{-iq_i}$. For each of the 4×4 block matrices, we have

$$\prod_{j=1}^{4}(1 - vm_j) = \det|1 - vM(q)|$$

$$= (1 + v^2)^2 - 2v(1 - v^2)\mathcal{R}(Q_1 + Q_2) \tag{5.4.19}$$

which yields for the partition function

$$Z = 2^N(1-v^2)^{-N_b} \prod_q \left\{(1-v^2)^2 - 2v(1-v^2)(\cos q_1 + \cos q_2)\right\}^{1/2} \tag{5.4.20}$$

In the limit of an infinite lattice, the sum over lattice vectors can be replaced by an integral to give for the free energy per spin

$$\beta F = \ln 2 - \ln(1 - v^2)$$
$$+ \frac{1}{8\pi^2} \int dq_1 \int dq_2 \ln\left\{(1+v^2)^2 - 2v(1-v^2)(\cos q_1 + \cos q_2)\right\} \tag{5.4.21}$$

The integrand will diverge for small q when

$$A(v) = (1 + v^2)^2 - 4v(1 - v^2) = 0 \tag{5.4.22}$$

which has the solution $v_c = \sqrt{2} - 1$ and is the critical value found by the duality transformation (5.3.8).

168 THE ISING MODEL

The singular part of the free energy can be evaluated by observing that it is the small q part of the integral that will diverge. This is a reflection of the fact that as the system approaches its critical temperature the free energy is dominated by very long wavelength fluctuations.

For small q and taking $v = v_c + t$ where $|t| \ll 1$, the integrand in (5.4.22) becomes

$$\frac{1}{8\pi^2} \int dq_1 \int dq_2 \ln\left[A''(v_c)t^2 + B(v_c)q^2 \right] \qquad (5.4.23)$$

$B(v) = v(1 - v^2)$. Evaluating the integral in the neighborhood of the origin in q space gives for the singular part

$$F_s \sim t^2 \ln t \qquad (5.4.24)$$

The specific heat is given by the second derivative of the free energy with respect to the temperature, and therefore by (5.4.24) the specific heat diverges as $T \to T_c$ as

$$C \sim \ln(T - T_c) \qquad (5.4.25)$$

Because the logarithmic singularity is weaker than any power, the value of the critical exponent α in this case is taken to be 0.

5.5 EXACT SOLUTION FOR THE ISING MODEL ON THE SQUARE LATTICE: CORRELATION FUNCTION

The method of the previous section can be extended to evaluate the correlation function [11]

$$G_{ij} = \frac{\text{Tr} \prod_{\langle pq \rangle} (1 + v S_p S_q) S_i S_j}{\text{Tr} \prod_{\langle pq \rangle} (1 + v S_p S_q)} \qquad (5.5.1)$$

The numerator in (5.5.1) can again be represented as a sum over graphs, but now these graphs are rooted in the sites i and j. By "rooted" we mean that there must be an odd number of bonds (1 or

5.5 SQUARE LATTICE ISING MODEL: CORRELATION FUNCTION

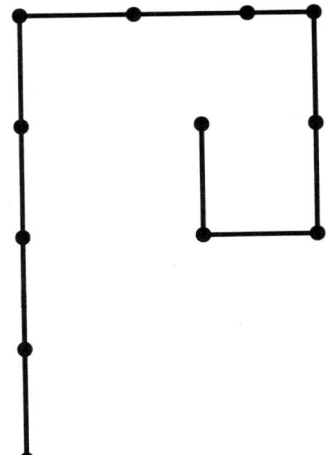

Fig. 5.5.1 Graph for G_{ij} which has a nonzero winding number.

3) emanating from these two sites and an even number of bonds (0, 2, or 4) emanating from every other site.

We can again make the many-to-one correspondence between rooted paths and rooted graphs by assigning a factor of (-1) for each self-intersection of the path. Unfortunately, the simple device of counting self-intersections by weighting each path by the product of $e^{i\theta/2}$, where θ is the angle through which the path turns at each vertex, does not work for rooted graphs. For example, consider the graph shown in Fig. 5.5.1. This graph has no self-intersections and so there is only one path associated with it. However, if we calculate the winding number of the path we get 2π, and so the phase is, incorrectly, (-1).

This difficulty is not, however, insurmountable. Define a sequence of sites $1, 2, 3, \ldots, l+1$, which form a connected chain from site i to site j, so that $S_1 = S_i$ and $S_{l+1} = S_j$. Because, for any k, $S_k^2 = 1$, we have

$$(S_1 S_2)(S_2 S_3) \cdots (S_l S_{l+1}) = S_1 S_{l+1} \qquad (5.5.2)$$

the numerator in (5.5.1), G_{ij}, can be written as

$$\Gamma_{ij} = \text{Tr} \prod_{\langle pq \rangle} (1 + v S_p S_q) \prod_{k=1}^{l} S_k S_{k+1} \qquad (5.5.3)$$

Because sites k and $k+1$ are nearest neighbors, a term of this type occurs in the first factor of (5.5.3), and we have

$$(1 + vS_kS_{k+1})S_kS_{k+1} = v\left(1 + \frac{1}{v}S_kS_{k+1}\right) \quad (5.5.4)$$

Defining

$$\tilde{v}_{pq} = \begin{cases} v & \text{if } (pq) \text{ is not in the chain} \\ \dfrac{1}{v} & \text{if } (pq) \text{ is in the chain} \end{cases} \quad (5.5.5)$$

we can then write

$$\Gamma_{ij} = v^l \, \text{Tr} \prod_{\langle pq \rangle} (1 - \tilde{v}_{pq} S_p S_q) \quad (5.5.6)$$

The trace in (5.5.6) is now just like the trace for the partition function, the only difference being that the matrix of bonds is no longer translationally invariant. We can, however, apply all the tricks we learned for the partition function to evaluate Γ. By arguments completely parallel to those leading to (5.4.19), we have

$$G_{ij} = v^l \sqrt{\frac{\det|1 - \tilde{v}M|}{\det|1 - vM_1|}} \quad (5.5.7)$$

where $\tilde{v}$ is given by (5.5.5) and the nonzero matrix elements of M are given by (5.4.13) to (5.4.16).

If we define

$$\tilde{v}_{pq} = v + g_{pq} \quad (5.5.8)$$

we can write

$$\det|1 - \tilde{v}M| = \det|1 - vM_1|\big(1 - (1 - vM)^{-1}gM_1\big)$$
$$= \det|1 - vM|\det|1 - (1 - vM)^{-1}gM_1| \quad (5.5.9)$$

The first factor in (5.5.9) cancels the denominator in (5.5.7), and we have

$$G_{ij} = v^l \det|1 - (1 - vM_1)^{-1}gM_1| \quad (5.5.10)$$

5.5 SQUARE LATTICE ISING MODEL: CORRELATION FUNCTION

The matrix gM_1 has nonzero elements only for those bonds that make up the chain introduced in (5.5.2). Because, from the point of view of critical phenomena, we are only interested in the behavior of the correlation function when the sites are very far apart, for the sake of simplicity let us take the sites i and j to both lie in the first row, site i at $(1, 1)$ and site j at $(l + 1, 1)$. The chain of sites then simply consists of the set $\{(x, 1), 1 \leq x \leq l + 1\}$.

The only nonzero matrix elements of $\Delta = gM_1$ are for $1 \leq x' \leq l$,

$$\Delta(x, 1, \alpha; x - 1, 1, 1) = g(\delta_{\alpha,1} + e^{i\pi/4}\delta_{\alpha,2} - e^{-ix/4}\delta_{\alpha,4}) \quad (5.5.11)$$

and for $1 \leq x \leq l$,

$$\Delta(x, 1, \alpha; x + 1, 1, 3) = g(\delta_{\alpha,3} + e^{i\pi/4}\delta_{\alpha,2} - e^{-i\pi/4}\delta_{\alpha,4}) \quad (5.5.12)$$

Because the matrix $(1 - vM_1)$ is translationally invariant, its inverse can be written as

$$(1 - vM_1)^{-1}(p, p') = \frac{1}{(2\pi)^2} \int dq^2\, e^{iq(r-r')}[1 - vM_1(q)]^{-1}_{\alpha,\beta} \quad (5.5.13)$$

Substituting these results into (5.5.10), we have

$$1 - g(1 - vM_1)^{-1}\Delta = 1 - g\int \frac{d^2q}{(2\pi^2)} \sum_{r''\alpha''} e^{iq(r-r'')}$$
$$\times \left\{[1 - vM_1(q)]^{-1}_{\alpha,\alpha''} \times \delta_{x'',x'+1}\delta_{y'',y}\delta_{y',1}\delta_{\alpha',1}\theta(x'-1)\theta(l-x')\right.$$
$$\times [\delta_{\alpha'',1} + e^{i\pi/4}\delta_{\alpha'',2} + e^{-i\pi/4}\delta_{\alpha'',4}]$$
$$+ [1 - vM_1(q)]^{-1}_{\alpha,\alpha''}\delta_{x'',x'-1}\delta_{y'',y'}\delta_{y',1}\delta_{\alpha',3}\theta(x''-1)\theta(l-x'')$$
$$\left.\times [\delta_{\alpha'',3} + e^{-i\pi/4}\delta_{\alpha'',2} + e^{i\pi/4}\delta_{\alpha'',4}]\right\} \quad (5.5.14)$$

where $\theta(x)$ is the usual step function.

Evidently, this matrix has the block structure

$$1 - g(1 - vM_1)^{-1}\Delta = \begin{bmatrix} P_{11} & 0 & P_{13} & 0 \\ P_{21} & 1 & P_{23} & 0 \\ P_{31} & 0 & P_{33} & 0 \\ P_{41} & 0 & P_{43} & 1 \end{bmatrix} \quad (5.5.15)$$

Each block in (5.5.15) is itself an $N \times N$ matrix. The determinant of (5.5.15) is just

$$\det|1 - g(1 - vM_1)^{-1}\Delta| = \det|P_{11}|\det|P_{33}| - \det|P_{13}|\det|P_{31}| \quad (5.5.16)$$

172   THE ISING MODEL

it turns out that $\det|P_{13}| = \det|P_{31}| = 0$ by symmetry,

$$P_{11} = 1 - g \int \frac{d^2q}{(2\pi)^2} e^{-iq_1} e^{iq(r-r')} \delta_{y',1} \theta(x'-1)\theta(l-x')$$
$$\times \left\{ [1 - vM(q)]_{11}^{-1} + e^{i\pi/4}[1 - vM_1(q)]_{12}^{-1} \right.$$
$$\left. + e^{-i\pi/4}[1 - vM_1(q)]_{14} \right\} \quad (5.5.17)$$

and

$$P_{33} = 1 - g \int \frac{d^2q}{(2\pi)^2} e^{iq_1} e^{iq(r-r')} \delta_{y',1} \theta(x'-2)\theta(l+1-x')$$
$$\times \left\{ [1 - vM(q)]_{33}^{-1} + e^{i\pi/4}[1 - vM_1(q)]_{32}^{-1} \right.$$
$$\left. + e^{-i\pi/4}[1 - vM_1(q)]_{34} \right\} \quad (5.5.18)$$

To proceed, we need to know the elements of the matrix $[1 - vM_1(q)]_{\alpha\beta}^{-1}$. These are

$$[1 - vu_1(q)]_{11}^{-1} = \frac{[1 + v^2 - 2v\cos(q_2) - v(1-v^2)e^{iq_1}]}{d(q_1, q_2)} \quad (5.5.19a)$$

$$[1 - vu_1(q)]_{12}^{-1} = \frac{e^{-i\pi/4}[-v^2 + ve^{-iq_2} - v^2 e^{iq_1} e^{-iq_2} - v^3 e^{iq_1}]}{d(q_1, q_2)}$$
$$\quad (5.5.19b)$$

$$[1 - vM_1(q)]_{14}^{-1} = \frac{e^{i\pi/4}[-v^2 + ve^{iq_2} - v^2 e^{iq_1} e^{iq_2} - v^3 e^{iq_1}]}{d(q_1, q_2)}$$
$$\quad (5.5.19c)$$

$$[1 - vM_1(q)]_{32}^{-1} = \frac{e^{i\pi/4}[-v^2 + ve^{-iq_2} - v^2 e^{-iq_1} e^{-iq_2} - v^3 e^{-iq_1}]}{d(q_1, q_2)}$$
$$\quad (5.5.19d)$$

$$[1 - vu_1(q)]_{33}^{-1} = \frac{[1 + v^2 - 2v\cos(q_2) - v(1-v^2)e^{-iq_1}]}{d(q_1, q_2)}$$
$$\quad (5.5.19e)$$

$$[1 - vM_1(q)]_{34}^{-1} = \frac{e^{-i\pi/4}[-v^2 + ve^{iq_2} - v^2 e^{-iq_1 + iq_2} - v^3 e^{-iq_1}]}{d(q_1, q_2)}$$
$$\quad (5.5.19f)$$

5.5 SQUARE LATTICE ISING MODEL: CORRELATION FUNCTION

and the determinant of $[1 - vM(q)]$ is

$$d(q_1, q_2) = (1 - v^2)^2 - 2v(1 - v^2)(\cos q_1 + \cos q_2) \quad (5.5.20)$$

Substituting (5.5.18) into (5.5.16), we have

$$P_{11} = 1 - g \int \frac{d^2q}{(2\pi)^2} e^{-iq_1} e^{iq(r-r')} \delta_{y',1} \theta(x'-1)\theta(l-x)$$
$$\times \left\{ \frac{[1 - v^2 - 2v^2 e^{+iq_1} \cos q_2 - v(1+v^2)e^{iq_1}]}{d(q_1, q_2)} \right\} \quad (5.5.21)$$

and by (5.5.18)

$$P_{33} = 1 - g \int \frac{d^2q}{(2\pi)^2} e^{iq_1} e^{iq(r-r')} \delta_{y',1} \theta(x'-2)\theta(l+1-x)$$
$$\times \left\{ \frac{[1 - v^2 - 2v^2 e^{-iq_1} \cos q_2 - v(1+v^2)e^{-iq_1}]}{d(q_1, q_2)} \right\} \quad (5.5.22)$$

Now let us consider the structure of the $N \times N$ matrix P_{11} carefully. If we order the elements of the matrix by $(xy; x'y')$ as shown in (5.5.21), then it is clear that the determinant we need is just the determinant of the small block in the upper left-hand corner of dimension $l \times l$, as shown in Fig. 5.5.2.

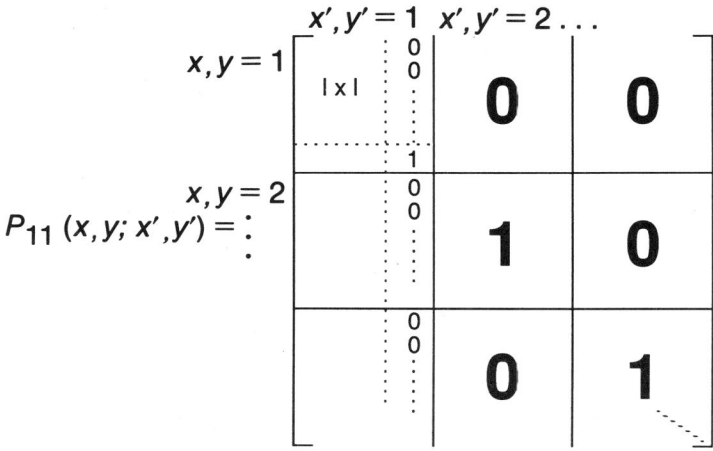

Fig. 5.5.2 Block structure of the matrix $P_{11}(x, y; x', y')$.

174 THE ISING MODEL

Denoting this submatrix by $\tilde{P}_{11}$, we have

$$\tilde{P}_{11} = \int \frac{d^2q}{(2\pi)^2} e^{iq_1(x-x')} \left[1 - g \frac{(1-v^2)e^{-iq_1} - 2v^2\cos q_2 - v(1+v^2)}{(1+v^2)^2 - 2v(1-v^2)(\cos q_1 + \cos q_2)} \right] \tag{5.5.23}$$

P_{33} has a similar structure, and again, the determinant reduces to that of an $l \times l$ subblock $\tilde{P}_{33}$ with elements

$$\tilde{P}_{33} = \int \frac{d^2q}{(2\pi)^2} e^{iq_1(x-x')} \left[1 - g \frac{(1-v^2)e^{iq_1} - 2v^2\cos q_2 - v(1+v^2)}{(1+v^2)^2 - 2v(1-v^2)(\cos q_1 + \cos q_2)} \right] \tag{5.5.24}$$

If we make the change of variable $q_1 \to -q_1$ and let $x - x' \to x' - x$, then $\tilde{P}_{33} = \tilde{P}_{11}$. Therefore, the correlation function has the form

$$G(l+1,1;1,1) = v^l \det|\tilde{P}_{11}| \tag{5.5.25}$$

The integral in (5.5.21) over q_2 can be simplified by first replacing $g = 1/v - v$, to get

$$\tilde{P}_{11} = \int_0^{2\pi} \frac{dq_1}{2\pi} e^{iq_1(x-x')} \int_0^{2\pi} \frac{dq_2}{2\pi} \frac{2(1+v^2) - 2v(1-v^2)\cos q_1 - 1/v(1-v^2)^2 e^{-iq_1}}{(1-v^2)^2 - 2v(1-v^2)\cos q_1 - 2v(1-v^2)\cos q_2} \tag{5.5.26}$$

We now use the fact that

$$\int_0^{2\pi} \frac{1}{A + B\cos\theta} d\theta = \begin{cases} \frac{1}{\sqrt{A^2 - B^2}} & A^2 > B^2 \\ 0 & A^2 < B^2 \end{cases} \tag{5.5.27}$$

to find

$$\tilde{P}_{11} = \int_0^{2\pi} \frac{dq_1}{2\pi} e^{iq_1(x-x')} \frac{2(1+v^2) - 2v(1-v^2)\cos q_1 - 1/v(1-v^2)^2 e^{-iq_1}}{\left\{ \left[(1+v^2)^2 - 2v(1-v^2)\cos q_1\right]^2 - 4v^2(1-v^2)^2 \right\}^{1/2}} \tag{5.5.28}$$

Define

$$\bar{v} = \frac{1-v}{1+v} \tag{5.5.29}$$

5.5 SQUARE LATTICE ISING MODEL: CORRELATION FUNCTION

The numerator in (5.5.29) can then be written in the form

$$2(1 + v^2) - 2v(1 - v^2)\cos q_1 - \frac{1}{v}(1 - v^2)^2 e^{-iq_1}$$

$$= \frac{1}{v(1 + v)^2}\left[(v\bar{v} - e^{-iq_1})(\bar{v} - ve^{iq_1})\right] \qquad (5.5.30)$$

Similarly, the square root in the denominator can be factored as

$$\left[(1 + v^4)^2 - 2v(1 - v^2)\cos q_1\right]^2 - 4v^2(1 - v^2)^2$$

$$= \frac{1}{(1 + v^2)^4}\left[(v\bar{v} - e^{-iq_1})(v\bar{v} - e^{iq_1})(\bar{v} - ve^{iq_1})(\bar{v} - ve^{-iq_1})\right] \qquad (5.5.31)$$

This then gives

$$\tilde{P}_{11} = \frac{1}{v}\int_0^{2\pi} e^{iq_1(x-x')}g(q_1) \qquad (5.5.32)$$

where

$$g(q) = \left[\frac{(v\bar{v} - e^{-iq})(\bar{v} - ve^{iq})}{(v\bar{v} - e^{iq})(\bar{v} - ve^{-iq})}\right]^{1/2} \qquad (5.5.33)$$

We can bring the factor v^l inside the determinant, and the correlation function is given by the determinant

$$\langle S_{i+l,j}S_{i,j}\rangle = \begin{bmatrix} C_0 & C_{-1} & C_{-2} & \cdots & C_{l+1} \\ C_1 & C_0 & C_{-1} & \cdots & \\ C_2 & C_1 & & & \\ \vdots & & & & \\ C_{l-1} & & & & \end{bmatrix} \qquad (5.5.34)$$

where

$$C_{m-n} = \int_0^{2\pi} \frac{dq}{2\pi} e^{iq(m-n)} g(q) \qquad (5.5.35)$$

176 THE ISING MODEL

Note that the (n, m)th element of the determinant in (5.5.34) depends only on $n - m$, that is, the elements along each diagonal are equal. Such a determinant is called a Toeplitz determinant, and in general they are difficult to evaluate. However, in the limit $l \to \infty$ we can evaluate (5.5.32) by use of Szegö's theorem [3]

$$\lim_{l \to \infty} \left(\frac{D_l}{\mu^l} \right) = \exp\left(\sum_{n=1}^{\infty} n g_{-n} g_n \right) \quad (5.5.36)$$

where

$$\ln \mu = \int_0^{2\pi} \frac{dq}{2\pi} \ln g(q) \quad (5.5.37)$$

and

$$g_n = \int_0^{2\pi} \frac{dq}{2\pi} e^{-inq} \ln g(q) \quad (5.5.38)$$

For a detailed discussion and proof of Szegö's theorem, see McCoy and Wu [5]. We can cast $g(q)$ into a slightly more symmetric form by defining

$$\alpha_1 = v\bar{v}$$
$$\alpha_2 = \frac{\bar{v}}{v} \quad (5.5.39)$$

so that

$$g(q) = \frac{\left(1 - \alpha_1 e^{iq}\right)\left(1 - \alpha_2 e^{-iq}\right)^{1/2}}{\left(1 - \alpha_1 e^{iq}\right)\left(1 - \alpha_2 e^{iq}\right)} \quad (5.5.40)$$

In our case, the integral in (5.5.37) vanishes and therefore $\mu = 1$.

To evaluate the g_n, (5.5.36), we must consider the cases $T > T_c$ and $T < T_c$ separately. Note that $\alpha_1 < 1$ for all T, whereas for $T < T_c$, $\alpha_2 < 1$ and for $T > T_c$, $\alpha_2 > 1$.

For $T < T_c$ both α_1 and α_2 are less than 1 and we can expand the logarithmic factors

$$\ln(1 - \alpha e^{\pm iq}) = -\sum_{m=1}^{\infty} \frac{\alpha^m}{m} e^{\pm imq} \quad (5.5.41)$$

5.5 SQUARE LATTICE ISING MODEL: CORRELATION FUNCTION

For $n > 0$ we have

$$g_n = \frac{1}{2n}(\alpha_2^n - \alpha_1^n) \tag{5.5.42}$$

and for $n < 0$ we find $g_{-n} = -g_n$.
Substituting this into (5.5.35), we have

$$\sum_n g_n g_{-n} = -\frac{1}{4}\sum_n \frac{(\alpha_2^n - \alpha_1^n)}{n}$$

$$= \frac{1}{4}\ln\left[\frac{(1-\alpha_2^2)(1-\alpha_1^2)}{(1-\alpha_1\alpha_2)^2}\right] \tag{5.5.43}$$

Therefore, for $T < T_c$ we have

$$\lim_{l\to\infty}\langle S_{l+1,1} S_{1,1}\rangle = \left[\frac{(1-\alpha_1^2)(1-\alpha_2^2)}{(1-\alpha_1\alpha_2)^2}\right]^{1/4} \qquad T < T_c \tag{5.5.44}$$

For $T > T_c$, $\alpha_2 > 1$ and we must take

$$\ln\left[\frac{1-\alpha_2 e^{-iq}}{1-\alpha_2 e^{iq}}\right] = -iq + \ln\left[\frac{1-\alpha_2^{-1}e^{iq}}{1-\alpha_2^{-1}e^{iq}}\right]^{1/2} \tag{5.5.45}$$

The logarithmic terms can now be expanded and we find for $n > 0$,

$$g_n = \frac{1}{n}(-1)^n - \frac{\alpha_2^{-n}}{2n} - \frac{\alpha_1^n}{2n} \tag{5.5.46}$$

and again $g_{-n} = -g_n$. In the exponent of (5.5.35) we have the divergent sum

$$-\sum_n \frac{1}{n} = -\infty \tag{5.5.47}$$

from which we see that

$$\lim_{l\to\infty}\langle S_{l+1,1} S_{1,1}\rangle \to 0 \qquad T > T_c \tag{5.5.48}$$

178   THE ISING MODEL

The magnetization is related to the correlation function by

$$m^2 = \lim_{l \to \infty} \langle S_{l+1,1} S_{1,1} \rangle \qquad (5.5.49)$$

and therefore, for $T < T_c$,

$$m = \left[\frac{(1-\alpha_1^2)(1-\alpha_2^2)}{(1-\alpha_1\alpha_2)^2} \right]^{1/8} \qquad (5.5.50)$$

The magnetization as a function of temperature is shown in Fig. 5.5.3. At $T > T_c$, $\alpha_2 = 1$ and for $T \leq T_c$ we have

$$m \sim \left[\frac{T_c - T}{T} \right]^{1/8} \qquad (5.5.51)$$

From this we see that the exponent β is

$$\beta = \frac{1}{8} \qquad (5.5.52)$$

It is also possible to determine the asymptotic form of the Toeplitz determinant, and the reader who is interested should refer to McCoy and Wu [5]. Here we will only indicate the results:

$$\langle S_{l+1,1} S_{1,1} \rangle$$

$$\begin{cases} \dfrac{\alpha_2^l}{l^{1/2}} & T > T_c \quad (5.5.53a) \\[2ex] \left(\dfrac{1+\alpha_1}{1+\alpha_1}\right)^{1/4} \dfrac{1}{l^{1/4}} & T = T_c \quad (5.5.53b) \\[2ex] \left[\dfrac{(1-\alpha_1^2)(1-\alpha_2^2)}{(1-\alpha_1\alpha_2)^2}\right]^{1/4} \left[1 + \dfrac{1}{2\pi l^2} \dfrac{\alpha_2^{2l}}{\left(\dfrac{1}{\alpha_2}-\alpha_2\right)^2}\right] & T < T_c \quad (5.5.53c) \end{cases}$$

By (5.5.53a) we see that the correlations above T_c decay exponentially

$$\langle S_{l+1,1} S_{1,1} \rangle \sim e^{-l/\xi} \qquad (5.5.54)$$

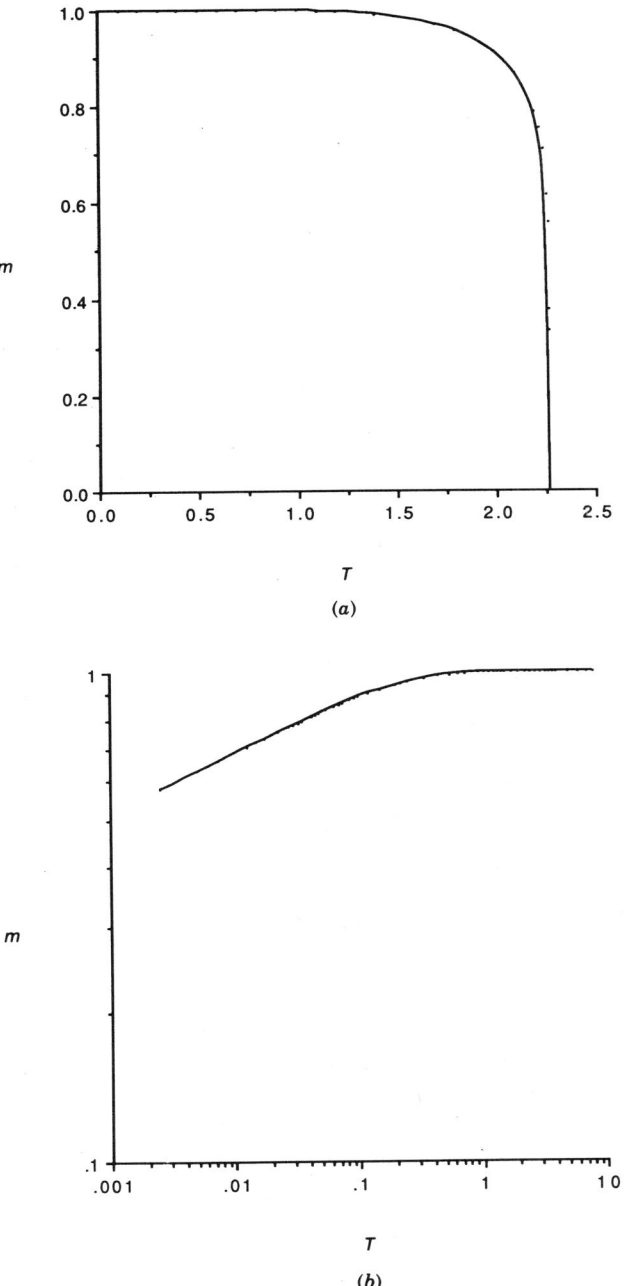

Fig. 5.5.3 (a) The spontaneous magnetization of the Ising model on a square lattice. (b) The magnetization is plotted on a log-log scale. The linear behavior for small t reflects the relation $m \sim (-t)^\beta$ with $\beta = 1/8$.

TABLE 5.5.1 Critical Exponents for the Ising Model in Two and Three Dimensions (Data for $d = 2$ are exact)

d	α	β	γ	ν	η
2	0	1/8	7/4	1	1/4
3	0.133^a	0.324^a	1.238^a	$0.629(4)^b$	$0.031(5)^b$

[a] Calculated using scaling laws.
[b] Reference 7.

where

$$\xi = \frac{1}{\log \alpha_2} \quad (5.5.54)$$

Again, as $T \to T_c$, $\alpha_2 \to 1$, and we have

$$\xi \sim \left[\frac{T - T_c}{T}\right]^{-1} \quad (5.5.56)$$

which gives $\nu = 1$.

At $T = T_c$ the correlation function decays algebraically as $l^{-1/4}$, from which we see that

$$\eta = \frac{1}{4} \quad (5.5.57)$$

In Table 5.5.1 we summarize these results for the critical exponents of the two-dimensional Ising model and in addition we give the exponents calculated for the Ising model on the simple cubic lattice [7].

PROBLEMS

5.1 Calculate the partition function of the one-dimensional ferromagnetic Ising model (use periodic boundary conditions). Show that the specific heat per spin is well behaved in the limit $N \to \infty$ and that therefore there is no phase transition.

5.2 Show that (5.3.10) is true by explicitly evaluating the trace and derive (5.3.13).

5.3 Show that the dual transformation can be expressed in the symmetric form $(\sinh 2\beta J)(\sinh 2\tilde{\beta} J) = 1$. Apply this to the square lattice and calculate J_c.

5.4 Using the dual and star–triangle transformations together, show that the critical coupling for the triangular lattice is given by (5.3.10) and the critical coupling for the hexagonal lattice is given by (5.3.13). Evaluate these expressions for βJ_c in each case.

5.5 (a) Construct the dual lattice to the Kagomé lattice, Fig. 5.P.1.

(b) Show that the partition function on the dual of the Kagomé lattice can be expressed in terms of the partition function of the triangular lattice via two star–triangle transformations.

(c) Using (b) and the new critical coupling for the triangular lattice, find the critical couplings on the Kagomé lattice and its dual.

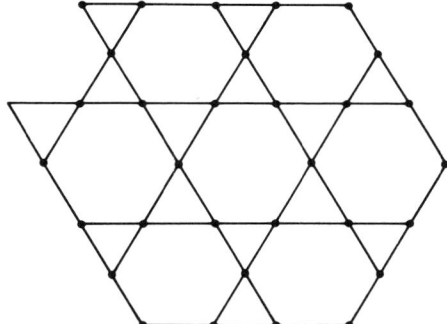

Fig. 5.P.1 The Kagomé lattice (see Prob. 5.5).

REFERENCES

1. Baxter, R. J. (1982) *Exactly Solved Models in Statistical Physics*. Academic, New York (1982).
2. Burgoyne, P. N. (1963) *J. Math. Phys.* **4** 1320.
3. Grenander, U. and Szegö, G. (1958) *Toeplitz Forms and Their Applications*. Univ. California Press, Berkeley.
4. Ising, E. (1925) *Z. Phys.* **31** 253.
5. McCoy, B. M. and Wu, T. T. (1973) *The Two-Dimensional Ising Model*. Harvard Univ. Press, Cambridge, MA.

6. Onsager, L. (1944) *Phys. Rev.* **65** 117.
7. Pawley, G. S., Swendsen, R. H., Wallace, D. J., and Wilson, K. G. (1984) *Phys. Rev. B* **29** 4030.
8. Peierls, R. (1936) *Proc. Cambridge Philos. Soc.* **32** 477.
9. Stanley, H. E. (1971) *Phase Transitions and Critical Phenomena.* Clarendon, Oxford.
10. Vdovichenko, N. V. (1965) *Sov. Phys. JETP* **20** 477.
11. Vdovichenko, N. V. (1965) *Sov. Phys. JETP* **21** (350).

6

RENORMALIZATION GROUP FOR THE ISING MODEL

6.0 INTRODUCTION

In Chap. 4 we outlined in a general way the structure of the RG as applied to critical phenomena, and in this chapter we will make these ideas more concrete by applying them to the Ising model in two dimensions.

There is no recipe for constructing a renormalization group transformation, even within the rather restricted problem we have set for ourselves here, and we will examine several approaches that have proved useful in investigating the Ising model.

All of the transformations considered in this chapter are "real-space" RG transformations in that they are formulated directly in terms of the spins on the lattice.

First, we consider the majority-rule RG developed by Niemeijer and van Leeuwen [9, 10] for the ferromagnetic (FM) Ising model on the triangular lattice. We then extend this method to the problem of the antiferromagnetic Ising model on the triangular lattice and discover that an Ising model is not always as simple as it seems.

6.1 ISING MODEL ON THE TRIANGULAR LATTICE

To begin, we consider a transformation whose physical interpretation is clear, the "block spin" transformation with the majority rule,

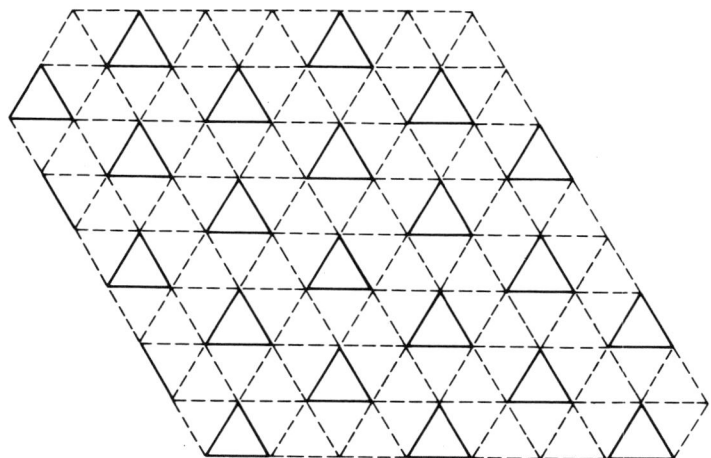

Fig. 6.1.1 Block spins on the triangular lattice. Solid lines indicate intracell couplings, whereas dotted lines show intercell couplings for the nearest-neighbor Hamiltonian.

(4.1.7). This method was introduced by Niemeijer and van Leeuwen [9], and in addition to the original paper, reviews can be found in Niemeijer and van Leeuwen [10] and Hu [8].

As shown in Fig. 6.1.1, we take the sites of the triangular lattice and group them into blocks of 3. Note that the blocks themselves form a triangular lattice with lattice constant $a' = \sqrt{3}\, a$, so the scale factor for the RG transformation is $b = \sqrt{3}$.

The Hamiltonian in the absence of an external field is

$$H = -J \sum_{\langle ij \rangle} s_i s_j \qquad (6.1.1)$$

We now group the terms in H involving the interactions between spins in a given block, which we will label with Greek indices, into $H_\alpha^{(0)}$, and the couplings between blocks α and β into $V_{\alpha\beta}$. Because the original Hamiltonian only incorporates nearest-neighbor couplings, we have

$$H = \sum_\alpha H_\alpha^{(0)} + \sum_{\langle \alpha\beta \rangle} V_{\alpha\beta}^{(1)} \qquad (6.1.2)$$

where $V_{\alpha\beta}^{(1)}$ is the first-order (in J) piece of V. As we will see presently, consistency requires that we include higher-order (in J)

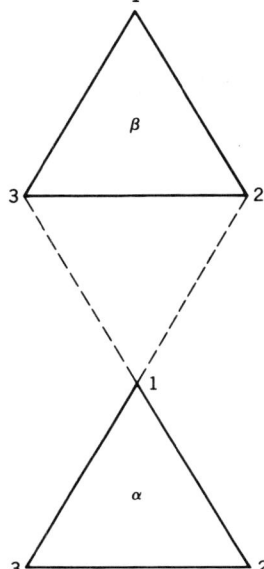

Fig. 6.1.2 Couplings between nearest-neighbor block spins in $V_{\alpha\beta}^{(1)}$.

couplings in the Hamiltonian. Explicitly,

$$H_\alpha^{(0)} = -J\left(s_\alpha^1 s_\alpha^2 + s_\alpha^1 s_\alpha^3 + s_\alpha^2 s_\alpha^3\right) \tag{6.1.3a}$$

and $V_{\alpha\beta}^{(1)}$ is of the form

$$V_{\alpha\beta}^{(1)} = -Js_\alpha^1\left(s_\beta^2 + s_\beta^3\right) \tag{6.1.3b}$$

Note that there are two distinct couplings between a given block and its neighbors as shown in Fig. 6.1.2. The procedure we have introduced does not preserve the whole symmetry of the triangular lattice, and this will lead to problems when we consider higher-order corrections.

In (6.1.3) we have labeled the spins s_α^i, where α is the block index and i runs over the spins within the block. The renormalized Hamiltonian is given by

$$e^{-H'} = \operatorname*{Tr}_{\{s\}} T(\{S\},\{s\})e^{-(H^{(0)}+V)} \tag{6.1.4}$$

where, for the majority rule

$$T(\{S\},\{s\}) = \prod_\alpha \delta(S_\alpha - \mathrm{sgn}(s_\alpha^1 + s_\alpha^2 + s_\alpha^3)) \quad (6.1.5)$$

and $H^{(0)} = \sum_\alpha H_\alpha^{(0)}$ and $V = \sum_{\langle\alpha\beta\rangle} V_{\alpha\beta}^{(1)}$.

We now introduce an average over the original spin variables $\{s_i^\alpha\}$ taken with respect to $H^{(0)}$,

$$\langle A \rangle_0 = \frac{\mathrm{Tr}\, T(\{S\},\{s\})e^{-H^{(0)}}A}{\mathrm{Tr}\, T(\{S\},\{s\})e^{-H^{(0)}}} \quad (6.1.6)$$

With this definition (6.14) can be written as

$$e^{-H'} = \left[\mathrm{Tr}\, T(\{S\},\{s\})e^{-H^{(0)}}\right]\langle e^{-V}\rangle_0 \quad (6.1.7)$$

Exponential averages of this kind can be evaluated by the cumulant expansion

$$\langle e^{-V}\rangle_0 = \exp \sum_{n=1}^\infty \frac{(-1)^n}{n!} C_n \quad (6.1.8)$$

where

$$\begin{aligned}
C_1 &= \langle V \rangle \\
C_2 &= \langle (V - \langle V \rangle)^2 \rangle \\
C_3 &= \langle (V - \langle V \rangle)^3 \rangle \\
C_4 &= \langle (V - \langle V \rangle)^4 \rangle - 3\langle (V - \langle V \rangle)^2 \rangle^2 \quad \text{and so on}
\end{aligned} \quad (6.1.9)$$

For details of the cumulant expansion, see App. A. Note that the cumulant expansion is essentially a power series in the coupling constant J and will therefore give good results near J_c, assuming the series converges, if J_c is small. For the triangular lattice $J_c = 0.275$ and we can hope to achieve reasonable success with just the first few cumulants.

Before we begin evaluating the cumulants, we need to calculate the prefactor in (6.1.7). Because the average over spins in each block can be performed independently, the prefactor factorizes and we have by (6.1.5)

$$\text{Tr} \prod_\alpha \delta(S_\alpha - \text{sgn}(s_\alpha^1 + s_\alpha^2 + s_\alpha^3))e^{-H_\alpha} = z^{N'} \tag{6.1.10}$$

where N' is the number of blocks and

$$z = e^{3J} + 3e^{-J} \tag{6.1.11}$$

There are only three distinct averages that need to be considered for $H^{(0)}$, and they are (Prob. 6.2)

$$\langle s_\alpha^1 \rangle = \langle s_\alpha^2 \rangle = \langle s_\alpha^3 \rangle = \frac{1}{z}(e^{3J} + e^{-J})S_\alpha = a_1 S_\alpha \tag{6.1.12a}$$

$$\langle s_\alpha^1 s_\alpha^2 \rangle = \langle s_\alpha^1 s_\alpha^3 \rangle = \langle s_\alpha^2 s_\alpha^3 \rangle = \frac{1}{z}(e^{3J} - e^{-J}) = a_2 \tag{6.1.12b}$$

$$\langle s_\alpha^1 s_\alpha^2 s_\alpha^3 \rangle = \frac{1}{z}(e^{3J} - 3e^{-J})S_\alpha = a_3 S_\alpha \tag{6.1.12c}$$

6.2 THE FIRST CUMULANT

The first-order contribution to H' is given by the first cumulant

$$H^{(1)} = -\sum_{\langle \alpha\beta \rangle} \langle V_{\alpha\beta}^{(1)} \rangle_0 \tag{6.2.1}$$

Therefore, we have by (6.1.3b)

$$H^{(1)} = -J \sum_{\langle \alpha\beta \rangle} \langle s_\alpha^1(s_\beta^2 + s_\beta^3) \rangle \tag{6.2.2}$$

Because the averages over spins in different blocks are independent, this gives

$$H^{(1)} = -2a_1^2 J \sum_{\langle \alpha\beta \rangle} S_\alpha S_\beta \tag{6.2.3}$$

In lowest order the renormalization equation for the nearest-neighbor coupling is therefore

$$J' = 2J\left(\frac{e^{3J} + e^{-J}}{e^{3J} - e^{-J}}\right)^2 \qquad (6.2.4)$$

This equation has fixed point at $J' = J = J^*$, where

$$J^* = \frac{1}{4}\ln(2\sqrt{2} + 1) = 0.336 \qquad (6.2.5)$$

which is somewhat higher than the exact value $J_c = 0.2746$.

In order to find the thermal eigenvalue y_T, we linearize the RG equation about the fixed point, and in the usual way we have

$$y_T = \frac{\ln\frac{\partial J'}{\partial J}(J = J^*)}{\ln\sqrt{3}} = 0.882 \qquad (6.2.6)$$

which should be compared with the exact value $y_T = 1.00$.

6.3 THE SECOND CUMULANT

The second cumulant is given by (6.1.9), or

$$H^{(2)} = -\sum_{\langle\alpha\beta\rangle}\sum_{\langle\gamma\delta\rangle}\left[\langle V_{\alpha\beta}V_{\delta\gamma}\rangle - \langle V_{\alpha\beta}\rangle\langle V_{\gamma\delta}\rangle\right] \qquad (6.3.1)$$

First, we can see that if the pairs $\langle\alpha\beta\rangle$ and $\langle\gamma\delta\rangle$ have no (block) site in common, that term vanishes, and we only need to consider those terms in which the sites are connected.

Next, we consider the case in which $\langle\alpha\beta\rangle = \langle\gamma\delta\rangle$, which gives, by (6.1.12),

$$\langle V_{\alpha\beta}^2\rangle - \langle V_{\alpha\beta}\rangle^2 = 2J^2(1 + a_2) - 4J^2a_1^4 \qquad (6.3.2)$$

This term is independent of the block spin variables S_α and S_β and does not affect the RG equations. We will see, however, that these "constant" terms, when summed, give the free energy. The contribution to the renormalized nearest-neighbor coupling from the second cumulant arises from terms such as the one shown in Fig. 6.3.1. Note

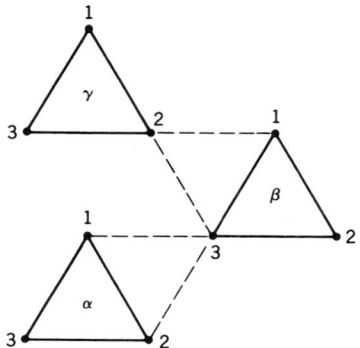

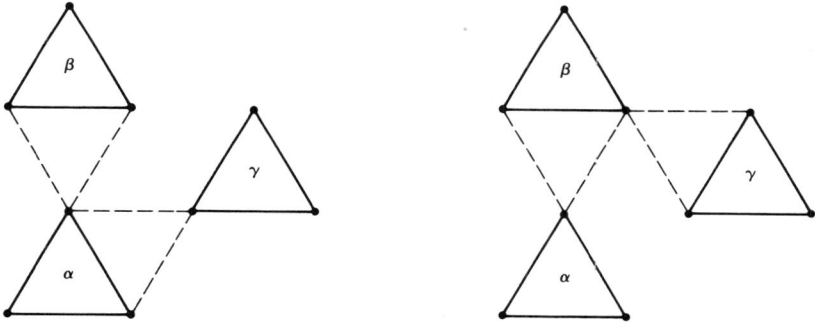

Fig. 6.3.1 Second order contribution to the nearest-neighbor coupling.

that for $\langle \alpha\beta \rangle \neq \langle \gamma\delta \rangle$ each term occurs twice, cancelling the factor of 1/2 in (6.1.8). For each pair of nearest neighbors there are two identical graphs, so we have

$$2J^2\big[\langle(s_\alpha^1 + s_\alpha^2)s_\beta^2(s_\beta^3 + s_\beta^1)s_\gamma^2\rangle - \langle(s_\alpha^1 + s_\alpha^3)s_\beta^2\rangle\langle(s_\beta^3 + s_\beta^1)s_\gamma^2\rangle\big]$$
$$= 2[2a_1^2(1 + a_2) - 4a_1^4]J^2 S_\alpha S_\beta \qquad (6.3.3)$$

In addition to the terms analyzed thus far, Figs. 6.3.2 and 6.3.3 show that entirely new couplings emerge in second order. Figure 6.3.2 shows the two terms that produce a coupling between second-neighbor block spins α and γ, which we denote by $K^{(2)}$,

$$K^{(2)}S_\alpha S_\gamma = J^2\big[\langle(s_\alpha^1 + s_\alpha^2)s_\beta^3 s_\beta^1(s_\gamma^2 + s_\gamma^3)\rangle - \langle(s_\alpha^1 + s_\alpha^2)s_\beta^3\rangle\langle s_\beta^1(s_\gamma^2 + s_\gamma^3)\rangle$$
$$+ \langle s_\alpha^1(s_\beta^2 + s_\beta^3)(s_\beta^1 + s_\beta^2)s_\gamma^3\rangle - \langle s_\alpha^1(s_\beta^2 + s_\beta^3)\rangle\langle(s_\beta^1 + s_\beta^2)s_\gamma^3\rangle\big]$$
$$= [4a_1^2 a_2 - 4a_1^4 + a_1^2(3a_2 + 1) - 4a_1^4]S_\alpha S_\gamma \qquad (6.3.4)$$

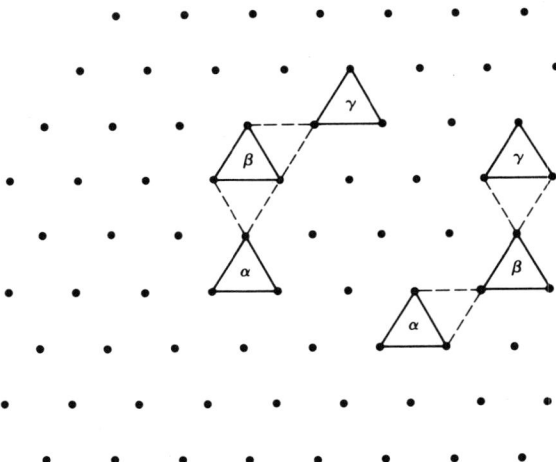

Fig. 6.3.2 Second order contributions to the next-neighbor coupling K.

Similarly, Fig. 6.3.3 corresponds to a third-neighbor coupling L, and we find

$$L^{(2)}S_\alpha S_\gamma = J^2\big[\langle s_\alpha^1(s_\beta^2 + s_\beta^3)s_\beta^1(s_\gamma^2 + s_\gamma^3)\rangle - \langle s_\alpha^1(s_\beta^2 + s_\beta^3)\rangle\langle s_\beta^1(s_\gamma^3 + s_\gamma^3)\rangle\big]$$
$$= 4J^2 a_1^2(a_2 - a_1^2)S_\alpha S_\gamma \qquad (6.3.5)$$

The proliferation of new types of couplings in the Hamiltonian is an inevitable consequence of the RG transformation, and a complete

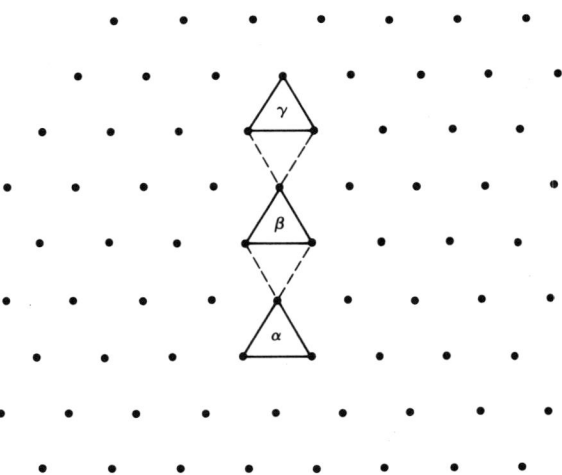

Fig. 6.3.3 Second order contribution to the third-neighbor coupling L.

specification of the fixed-point Hamiltonian requires an infinite number of terms, each with its own coupling constant. The hope is that these new couplings are irrelevant near the fixed point of interest and that sensible results for the critical exponents can be found by considering a finite subset of all couplings. Within the framework of the cumulant expansion, we can see that these new couplings are higher order in J_c, so when J_c is small the truncation of parameter space is a reasonable procedure.

In order to calculate the RG equations consistently to second order, we are forced to incorporate second- and third-neighbor couplings into the initial form of the Hamiltonian. These will not affect $H^{(0)}$, but will simply add to V. By the preceding calculation we may regard K and L as second order in J. We must now consider a Hamiltonian of the form

$$H = -J \sum_{\langle ij \rangle} s_i s_j - K \sum_{\langle\langle ij \rangle\rangle} s_i s_j - L \sum_{\langle\langle\langle ij \rangle\rangle\rangle} s_i s_j \qquad (6.3.6)$$

6.4 CUMULANTS REVISITED

In the light of the previous discussion, we must recalculate the cumulants starting with the Hamiltonian (6.3.6). Our job is made considerably simpler by the observation that the new couplings K and L are already of second order and will therefore contribute to the second cumulant only in higher order in J. Therefore, we do not need to recalculate the second cumulant and only need to determine the contributions of the new couplings to the first cumulant.

The intercell part of the Hamiltonian can be written as

$$V = V^{(1)} + V^{(2)} \qquad (6.4.1)$$

where

$$V^{(2)}_{\alpha\beta} = -K\left(s^1_\alpha s^1_\beta + s^2_\alpha s^2_\beta + s^3_\alpha s^3_\beta\right) - L\left(s^3_\alpha s^2_\beta + s^2_\alpha s^3_\beta\right) \qquad (6.4.2)$$

is the second-order block coupling and $V^{(1)}$ is as given in (6.1.3b)

The contributions to the renormalized first- and second-neighbor couplings from the first cumulant can now be easily calculated, and we find

$$J^{(1)} S_\alpha S_\beta = \left[21 a_1^2 J + 3 K a_1^2 + 2 L a_1^2\right] S_\alpha S_\beta \qquad (6.4.3)$$

and

$$K^{(1)}S_\alpha S_\gamma = a_1^2 L S_\alpha S_\gamma \qquad (6.4.4)$$

Putting these results together with (6.3.4) and (6.3.5) and keeping terms in the first and second cumulants to second order, the RG equations become

$$J' = 2a_1^2 J + 4a_1^2 \left[1 + a_2 - 2a_1^2\right] J^2 + 3a_1^2 K + 2a_1^2 L \qquad (6.4.5)$$

$$K' = a_1^2 \left[7a_2 + 1 - 8a_1^2\right] J^2 + a_1^2 L \qquad (6.4.6)$$

$$L' = 4a_1^2 (a_2 - a_1^2) J^2 \qquad (6.4.7)$$

6.5 FIXED-POINT ANALYSIS

Calculating the RG equations is only the first part of an RG analysis. Once the RG equations have been established, it is necessary to locate the fixed points of the equations and find the linear scaling fields and their associated eigenvalues in the neighborhood of each fixed point. Because these are problems one always faces in RG calculations, we will take this opportunity to introduce some practical methods for finding the fixed points and scaling fields.

In order to locate the fixed points of the RG equations in the three-dimensional parameter space of equations (6.4.5) to (6.4.7), we need to generalize "Newton's method," for finding the zeros of a function.

First, let us introduce a three-component vector J_k, where

$$J_1 = J \qquad J_2 = K \qquad J_3 = L \qquad (6.5.1)$$

The RG equations can be written in the form

$$J'_k = R_k(\{J_j\}) \qquad (6.5.2)$$

where the R_k are the nonlinear functions given in (6.4.5) to (6.4.7). Writing $J_k = J_k^* + J_k - J_k^*$ and linearizing about the fixed point, we have

$$J'_k - J_k^* = \sum_j T_{kj}(J_j - J_j^*) \qquad (6.5.3)$$

where the real matrix T_{kj} is given by

$$T_{kj} = \left. \frac{\partial R_k}{\partial J_j} \right|_{\{J_k = J_k^*\}} \tag{6.5.4}$$

and should not be confused with the projection operator introduced in (6.1.5). Note that this matrix is not generally symmetric. Because nonsymmetric real matrices do not come up as often in physics as symmetric real matrices, we will treat them here in some detail.

We now introduce the right and left eigenvectors of the linearized RG transformation, ϕ_k, and $\tilde{\phi}_k$, respectively, and the associated eigenvalues $\lambda^{(n)}$, which satisfy

$$\sum_j T_{kj} \phi_j^{(n)} = \lambda^{(n)} \phi_k^{(n)} \tag{6.5.5a}$$

and

$$\sum_k \tilde{\phi}_k^{(n)} T_{kj} = \lambda^{(n)} \tilde{\phi}_j^{(n)} \tag{6.5.5b}$$

The completeness relation for the eigenvectors is

$$\sum_n \tilde{\phi}_i^{(n)} \phi_j^{(n)} = \delta_{ij} \tag{6.5.6}$$

and the orthogonality condition is

$$\sum_j \tilde{\phi}_j^{(n)} \phi_j^{(m)} = \delta_{nm} \tag{6.5.7}$$

The right eigenvectors are a convenient basis in which to expand the displacement from the fixed point in parameter space, $J_k - J_k^*$,

$$J_k - J_k^* = \sum_n u_n \phi_k^{(n)} \tag{6.5.8}$$

where by the completeness relation we have

$$u_n = \sum_k \tilde{\phi}_k^{(n)} (J_k - J_k^*) \tag{6.5.9}$$

The linearized RG equations can now be written in the form

$$J_k' = J_k + \sum_n (\lambda^{(n)} - 1) u_n \phi_k^{(n)} \tag{6.5.10}$$

194   RENORMALIZATION GROUP FOR THE ISING MODEL

The procedure for locating the fixed point is now clear. We make a first guess for the fixed point J_i and apply the RG equations (6.4.5) to (6.4.7) to get J'_k. By (6.5.10) we can then estimate the displacement from the fixed point by solving for the u_n,

$$u_n = \frac{1}{(\lambda^{(n)} - 1)} \sum_j \tilde{\phi}_j^{(n)}(J'_j - J_j) \tag{6.5.11}$$

which then yields a new, and hopefully better, estimate for the fixed point

$$J_k^* = J_k - \sum_n \sum_j \frac{\tilde{\phi}_j^{(n)}(J'_j - J_j)\phi_k^{(n)}}{\lambda^{(n)} - 1} \tag{6.5.12}$$

A good initial guess for the fixed point is the result (6.2.5) obtained in first order, $J^* = 0.366$, $K^* = 0.0$, $L^* = 0.0$. So long as there is at least one relevant scaling field, that is, at least one eigenvalue that is greater than 1, the method produces a series of estimates that converge rapidly to the exact solution. When the algorithm outlined previously is put into use, we find the fixed point

$$J^* = 0.2789 \qquad K^* = -0.0142 \qquad L^* = -0.0152 \tag{6.5.13}$$

The left and right eigenvectors at the fixed point are

$$\phi^{(1)} = \begin{pmatrix} 1.0432 \\ -0.0147 \\ -0.0458 \end{pmatrix} \quad \phi^{(2)} = \begin{pmatrix} -0.0726 \\ 0.0690 \\ 0.0291 \end{pmatrix} \quad \phi^{(3)} = \begin{pmatrix} -0.0294 \\ 0.0543 \\ 0.0168 \end{pmatrix} \tag{6.5.14}$$

$$\tilde{\phi}^{(1)} = \begin{pmatrix} 1.000 \\ -0.7584 \\ 0.6974 \end{pmatrix} \quad \tilde{\phi}^{(2)} = \begin{pmatrix} 1.000 \\ -9.861 \\ 25.839 \end{pmatrix} \quad \tilde{\phi}^{(3)} = \begin{pmatrix} 1.000 \\ 6.9034 \\ 20.480 \end{pmatrix} \tag{6.5.15}$$

and the corresponding eigenvalues are

$$\lambda^{(1)} = 1.7729 \qquad \lambda^{(2)} = 0.1948 \qquad \lambda^{(3)} = -0.1364 \tag{6.5.16}$$

It is also of interest to map out the domain of attraction, or "critical surface," of the fixed point, and in particular we are interested in the intersection of this surface with the J axis, because this corresponds to

6.5 FIXED-POINT ANALYSIS   195

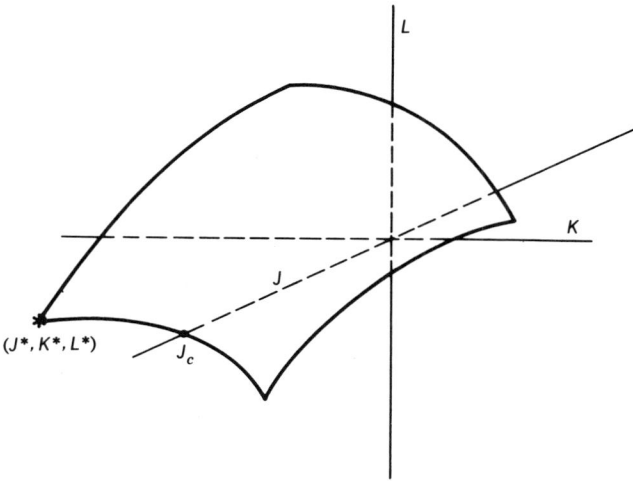

Fig. 6.5.1 The critical surface in the neighborhood of the fixed point, indicated by ∗. The nearest-neighbor critical point is the intersection of this surface with the J axis, labelled J_c.

the critical nearest-neighbor Ising Hamiltonian. The critical surface in the neighborhood of the fixed point is shown schematically in Fig. 6.5.1. The critical point for the nearest-neighbor Ising model is labeled J_c in the figure, and is the intersection of the critical surface with the J axis.

In order to calculate the critical value J_c, we first observe that near the fixed point the critical surface is the set of points for which the amplitude of the relevant scaling field u_1 vanishes. The u_i are given by (6.5.9) within the linear approximation. Setting $K = L = 0$, the critical value J_c is given by

$$J_c = \frac{\sum_k \tilde{\phi}_k^{(1)} J_k^*}{\tilde{\phi}_1^{(1)}}$$

$$= 0.2575 \qquad (6.5.17)$$

Note that this estimate is quite a bit better than the first-order result.

The thermal critical exponent is determined by the largest eigenvalue of T, and we find to second order

$$y_T = \frac{\ln \lambda_1}{\ln \sqrt{3}} = 1.042 \qquad (6.5.18)$$

which is also a great improvement over the first-order result (6.2.6).

6.6 RENORMALIZATION EQUATIONS IN THE PRESENCE OF A MAGNETIC FIELD

In order to obtain the magnetic exponent y_h, we need to include in the Hamiltonian a term corresponding to an external magnetic field h, and so we will now consider

$$H = -J \sum_{\langle ij \rangle} s_i s_j - h \sum_i s_i \tag{6.6.1}$$

The new term modifies the intrablock part of the Hamiltonian $H_\alpha^{(0)}$ of (6.1.3a) which is now

$$H_\alpha^{(0)} = -J\left(s_\alpha^1 s_\alpha^2 + s_\alpha^1 s_\alpha^3 + s_\alpha^2 s_\alpha^3\right) - h\left(s_\alpha^1 + s_\alpha^2 + s_\alpha^3\right) \tag{6.6.2}$$

The averages (6.1.11) and (6.1.12a) to (6.1.12c) must be recalculated, and one finds

$$z_\alpha = a_0 + b_0 S_\alpha \tag{6.6.3}$$

$$\langle s_\alpha^1 \rangle = a_1 S_\alpha + b_1 \tag{6.6.4}$$

$$\langle s_\alpha^2 s_\alpha^3 \rangle = a_2 + b_2 S_\alpha \tag{6.6.5}$$

$$\langle s_\alpha^1 s_\alpha^2 s_\alpha^3 \rangle = a_3 S_\alpha + b_3 \tag{6.6.6}$$

where

$$a_0 = e^{3J} \cosh 3h + 3e^{-J} \cosh h \tag{6.6.7a}$$

$$b_0 = e^{3J} \sinh 3h + 3e^{-J} \sinh h \tag{6.6.7b}$$

$$a_1 = \frac{e^{6J} + 4e^{2J} \cosh 2h + 3e^{-2J}}{e^{6J} + 6e^{2J} \cosh 2h + 9e^{-2J}} \tag{6.6.8a}$$

$$b_1 = \frac{2e^{2J} \sinh 2h}{e^{6J} + 6e^{2J} \cosh 2h + 9e^{-2J}} \tag{6.6.8b}$$

$$a_2 = \frac{e^{6J} + 2e^{2J} \cosh 2h - 3e^{-2J}}{e^{6J} + 6e^{2J} \cosh 2h + 9e^{-2J}} \tag{6.6.9a}$$

$$b_2 = \frac{4e^{2J} \sinh 2h}{e^{6J} + 6e^{2J} \cosh 2h + 9e^{-2J}} \tag{6.6.9b}$$

$$a_3 = \frac{e^{6J} - 9e^{-2J}}{e^{6J} + 6e^{2J} \cosh 2h + 9e^{-2J}} \tag{6.6.10a}$$

$$b_3 = \frac{e^{6J} \sinh 2h}{e^{6J} + 6e^{2J} \cosh 2h + 9e^{-2J}} \tag{6.6.10b}$$

Before we consider the cumulants, first note that the "zeroth-order" term z_α is no longer independent of the block spin S_α and in fact contributes to the renormalized magnetic field. To see this, denote the zeroth-order renormalized field by h'. Then

$$e^{h'S_\alpha} = \cosh h' + S_\alpha \sinh h' = \cosh h'(1 + S_\alpha \tanh h') \quad (6.6.11)$$

The factor z_α from Eq. (6.6.3) can also be cast in this form

$$z_\alpha = a_0\left(1 + \frac{b_0}{a_0}S_\alpha\right) \quad (6.6.12)$$

Comparison of (6.6.11) and (6.6.12) then gives (see also Prob. 6.7)

$$h' = \tanh^{-1}\left(\frac{b_0}{a_0}\right) \quad (6.6.13)$$

Just as in the case of $h = 0$, any overall constant factor does not contribute to the RG equations and can be ignored.

Obviously the zero-field fixed points are still fixed points of the new RG equations, and so in the neighborhood of the $h = 0$ critical point we can take h as small. To lowest order in h, (6.6.13) gives

$$h' = 3a_1(0)h \quad (6.6.14)$$

Evaluating this at the first-order fixed point gives for the magnetic exponent

$$y_h = \frac{\ln 3a_1}{\ln\sqrt{3}} = 1.369 \quad (6.6.15)$$

which should be compared with the exact value $y_h = 15/8 = 1.875$.

6.7 THE FIRST CUMULANT WITH $h \neq 0$

The cumulants are calculated exactly as before, and so just as in (6.2.2) the first-order correction to the renormalized Hamiltonian is

$$H^{(1)'}_{\alpha\beta} = -2J\langle s^1_\alpha\rangle\langle s^2_\beta + s^3_\beta\rangle \quad (6.7.1)$$

Expanding the averages as given in (6.6.4), we have

$$H^{(1)}_{\alpha\beta} = -2Ja_2^2 S_\alpha S_\beta - 2Ja_1 b_1 (S_\alpha + S_\beta) - 2Jb_1^2 \quad (6.7.2)$$

The first term is the first-order contribution to the renormalized nearest-neighbor interaction, the second term is part of the first-order contribution to the renormalized magnetic field, and the last term is independent of the block spins and can be neglected.

In the triangular lattice each spin has six nearest neighbors, and the contribution to the renormalized field calculated in (6.7.2) comes from consideration of just one such neighbor. Therefore, the full renormalized field in first order is given, in the limit $h \to 0$, by

$$h' = 3a_1(0) \left(1 + \frac{16 e^{2J^*} J^*}{\left(e^{3J^*} + 3 e^{-J^*} \right)^2} \right) h \quad (6.7.3)$$

The fixed point is again $J^* = 0.336$, $h^* = 0.0$, and so we find for the magnetic exponent to first order

$$y_h = 2.034 \quad (6.7.4)$$

which is a fair improvement over the zeroth-order result (6.6.15).

6.8 RENORMALIZATION GROUP EQUATIONS TO SECOND ORDER

The second cumulant can be calculated as before, and again we find that new intercell couplings arise. In addition to second- and third-neighbor two-spin interactions, the inclusion of the magnetic field gives rise to new three-spin couplings, which are illustrated in Fig. 6.8.1. Note that the three-spin coupling M, shown in Fig. 6.8.1, fits into the three-spin cell we have been using, so that not only are the intercell couplings modified but the intracell Hamiltonian $H^{(0)}$ must also be modified, and we must take, for the cell α,

$$H^{(0)}_\alpha = -J \left(s_\alpha^1 s_\alpha^2 + s_\alpha^1 s_\alpha^3 + s_\alpha^2 s_\alpha^3 \right) - h \left(s_\alpha^1 + s_\alpha^2 + s_\alpha^3 \right) + M s_\alpha^1 s_\alpha^2 s_\alpha^3 \quad (6.8.1)$$

It is useful to divide the types of terms that can appear in the Hamiltonian into even or odd depending on whether there are an even or odd number of factors of the spin variables. At the FM fixed point all the odd couplings are 0. It then follows by symmetry that the RG equations for even-type couplings are even functions of the odd couplings and the derivatives of the even couplings with respect to the

6.8 RENORMALIZATION GROUP EQUATIONS TO SECOND ORDER

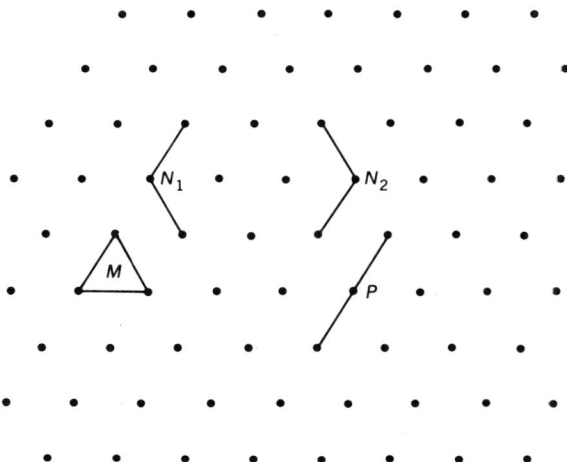

Fig. 6.8.1 The types of three-spin couplings that arise in the second cumulant.

odd couplings evaluated at the fixed point vanish. Similarly, the odd couplings are odd functions of themselves so the derivatives of the odd couplings with respect to the even couplings also vanish at the fixed point. Therefore, the matrix $T_{\alpha\beta}$ has the block-diagonal structure

$$\begin{bmatrix} T_{\alpha\beta}^{e-e} & 0 \\ 0 & T_{\alpha\beta}^{o-o} \end{bmatrix} \quad (6.8.2)$$

Because we only need to keep expressions involving the odd-sector coupling constants in first order, this simplifies our job quite a bit. The calculations are straightforward and the results for the odd-order couplings in the second cumulant approximation are given in Table 6.8.1. Note that there are two inequivalent next-nearest-neighbor three-spin interactions, which we have labeled N_1 and N_2. The fact that these interactions are inequivalent indicates that the renormalized Hamiltonian no longer has the full symmetry of the triangular lattice, and this in turn can be traced back to the way in which the block spins were assigned. We can restore the symmetry of the Hamiltonian in a somewhat ad hoc manner by replacing N_1 and N_2 with their average.

The inclusion of the three-spin term iin $H_\alpha^{(0)}$ modifies the calculation of z_α to second order in J. The new values for a_0 and b_0 are

$$a_0 = e^{3J}\cosh(3h + M) + 3e^{-J}\cosh(h - M) \quad (6.8.3a)$$
$$b_0 = e^{3J}\sinh(3h + M) + 3e^{-J}\sinh(h - M) \quad (6.8.3b)$$

200 RENORMALIZATION GROUP FOR THE ISING MODEL

TABLE 6.8.1 Contributions to Odd-Sector Hamiltonian from the Second Cumulant

Type of term	Multiplicity	Contributions to H'	
	3	$-24J^2 a_1^3 b_1 S_\alpha$	(h)
	3	$6J^2 [b_2 - 4a_1^3 b_1] S_\alpha$	(h)
	6	$24J^2 [a_1 a_2 b_1 - a_1^3 b_1] S_\alpha$	(h)
	1	$4J^2 [a_1^2 b_2 - 2a_1^3 b_1] S_\alpha S_\beta S_\gamma$	(P)
	6	$6J^2 [a_1(3a_2 + 1)b_1 - a_1^3 b_1] S_\alpha$	(h)
	1	$J^2 [3a_1^3 b_2 - 8a_1^3 b_1] S_\alpha S_\beta S_\gamma$	(N_1)
	6	$24J^2 [a_1 a_2 b_1 - a_1^3 b_1] S_\alpha$	(h)
	1	$4J^2 [a_1^2 b_2 - 2a_1^3 b_1] S_\alpha S_\beta S_\gamma$	(N_2)
	6	$24J^2 [a_1(1 + a_2)b_1 - a_1^3 b_1] S_\alpha$	(h)
	1	$4J^2 [a_1^2 b_2 - 2a_1^3 b_1] S_\alpha S_\beta S_\gamma$	(M)
	6	$12J^2 [a_1(1 + a_2)b_1 - 2a_1^3 b_1] S_\alpha$	(h)
	1	$2J^2 [a_1^2 b_2 - 4a_1^3 b_1] S_\alpha S_\beta S_\gamma$	(M)

6.8 RENORMALIZATION GROUP EQUATIONS TO SECOND ORDER

By (6.8.3a) and (6.8.3b), the zeroth-order contribution to h' is

$$h^{(0)} = \frac{e^{3J}(3h + M) + 3e^{-J}(h - M)}{e^{3J} + 3e^{-J}} \tag{6.8.4}$$

The first-order contribution to h is of the same form as we have in (6.7.2) and (6.7.3)

$$h^{(1)} = 12 J a_1 b_1 \tag{6.8.5}$$

The three-spin couplings are of $O(J^2)$ and so in recalculating the first cumulant with the new three-spin interactions, we can drop the three-spin term from H_0.

In Table 6.8.2 we give the various types of terms that arise to second order in the first cumulant.

Before we write out the renormalization equations for the odd-sector coupling constants h, M, N, and P, we must linearize the a's and b's. From (6.6.8) to (6.6.10), it is clear that the a's are all even functions of h and so to first order they can be replaced by their values for $h = 0$.

The b's are all odd functions of h, and except for the zeroth-order term b_0, we can neglect M to $O(J^2)$. We then have

$$a_1 = \frac{e^{4J} + 1}{e^{4J} + 3} \tag{6.8.6a}$$

$$b_1 = \frac{4e^{4J}}{(e^{4J} + 3)^2} h = b'_1 h \tag{6.8.6b}$$

$$a_2 = \frac{e^{4J} - 1}{e^{4J} + 3} \tag{6.8.7a}$$

$$b_2 = \frac{8e^{4J}}{(e^{4J} + 1)^2} h = 2b'_1 h \tag{6.8.7b}$$

$$a_3 = \frac{e^{4J} - 3}{e^{4J} + 3} \tag{6.8.8a}$$

$$b_3 = \frac{12e^{4J}}{(e^{4J} + 3)^2} h = 4b'_1 h \tag{6.8.8b}$$

TABLE 8.6.2 Contributions to the Odd-Sector Renormalized Hamiltonian of $O(J^2)$ from the First Cumulant

Term	Multiplicity	Contribution to H'	
(diagram M)	3	$3Ma_1a_2S_\alpha$	(h)
(diagram M)	1	$Ma_1^2 S_\alpha S_\beta S_\gamma$	(M)
(diagram N)	12	$12Na_1a_2S_\alpha$	(h)
(diagram N)	3	$3Na_1^3 S_\alpha S_\beta S_\gamma$	(M)
(diagram P)	1	$Pa_1^3 S_\alpha S_\beta S_\gamma$	(N)
(diagram P)	2	$2PA_1a_2S_\alpha$	(h)
(diagram K)	6	$6Ka_1b_1S_\alpha$	(h)
(diagram L)	6	$6La_1b_1S_\alpha$	(h)

6.8 RENORMALIZATION GROUP EQUATIONS TO SECOND ORDER

Collecting all terms to $O(J^2)$, the renormalized equations for the odd-sector couplings are

$$h' = 3a_1 h - a_3 M + 12 J a_1 b'_1 h + 3 M a_1 a_2 + 12 N a_1 a_2 + 2 P a_1 a_2$$
$$+ 12 J^2 (1 - 8 a_1^3) b'_1 h + 48 J^2 (a_1 a_2 - a_1^3) b'_1 h$$
$$+ 6 a_1 J^2 (7 + 9 a_2 - 9 a_1^2) b'_1 h + 6 K a_1 b'_1 h + 6 L a_1 b'_1 h \quad (6.8.9a)$$

$$M' = a_1^2 M + 3 a_1^3 N + 4 J^2 (3 a_1^2 - 4 a_1^3) b'_1 h \quad (6.8.9b)$$

$$N' = a_1^3 P + J^2 (7 a_1^2 - 8 a_1^3) b'_1 h \quad (6.8.9c)$$

$$P' = 8 J^2 (a_1^2 - a_1^3) b'_1 h \quad (6.8.9d)$$

Again, note that in (6.8.9c), N is the average of the two inequivalent terms N_1 and N_2 from Table 6.8.1. Because the RG equations for the even-sector couplings are unchanged, we can take for J, K, and L their fixed-point values given in (6.5.13).

Equations (6.8.9a) to (6.8.9d) can be cast as a matrix equation and the eigenvalues are then given by calculating the determinant of the resulting 4×4 matrix. If we are only interested in the largest eigenvalue, this procedure is really not necessary as all the relevant off-diagonal terms are small. (The exact value for the eigenvalue is 2.800, whereas the corresponding diagonal matrix element is 2.815.) We find for the magnetic exponent

$$y_h = \ln \frac{\lambda_h}{\ln \sqrt{3}} = 1.874 \quad (6.8.10)$$

TABLE 6.8.3 Results for Fixed Points and Critical Indices of the Ising Model on the Triangular Lattice

Cell	Order	J^*	K^*	L^*	J_c	y_T	y_h
△	1	0.3356	—	—	0.3356	0.882	2.034
	2	0.2789	0.0142	0.0152	0.2575	1.042	1.874[a]
	3	0.5206	0.0384	0.0516	0.3009	1.146	—
⬡	1	0.3003	—	—	0.3003	0.919	1.9213
	2	0.2980	0.0153	0.0188	0.2647	1.124	1.8787
	3	0.3578	0.0307	0.0310	0.2752	1.072	—
Exact		—	—	—	0.2747	1.000	1.875

[a] This differs from that in [5]. We believe this is the correct value.

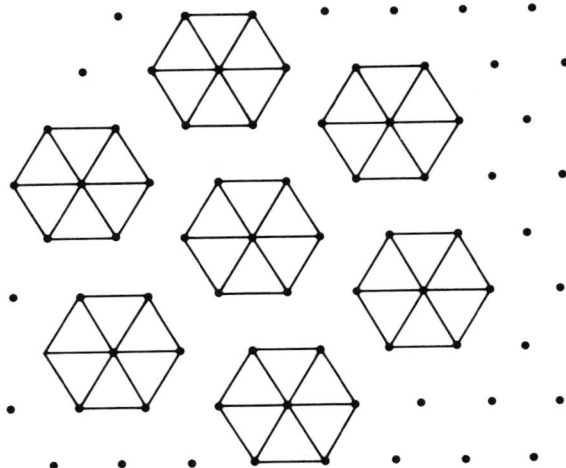

Fig. 6.8.2 The seven-spin cell employed by Sudbo and Hemmer [12].

which differs from the exact result by approximately 0.05%. The results of these calculations, which were first carried out by Braathen and Hemmer [5], are given in Table 6.8.3 along with similar results due to Sudbo and Hemmer [12], who used the seven-spin cell shown in Fig. 6.8.2, and Hemmer and Verlarde [6], who carried out the even-sector calculations to third order.

The remarkable agreement between the value of the magnetic exponent calculated for the three-spin cell to second order and the exact value is probably fortuitous. For the seven-spin cell of Sudbo and Hemmer, the magnetic exponent differs from the exact value by about 0.2%, which is still quite good.

There are a couple of lessons to be learned from the Table 6.8.3. First, the values of the various couplings at the fixed point depend on the particular cell one uses. There is nothing universal about the fixed point itself. The second observation one can make is that increasing the size of the cell does not necessarily improve the accuracy of the calculation. On the other hand, second order is a great improvement over first order. This gives us some hope that the cumulant expansion converges. However, Hsu and Gunton [7] have carried out similar calculations on the square lattice. Their results are listed in Table 6.8.4 and indicate that the third-order results are not uniformly better than those in second order. This sort of behavior leads one to sus-

TABLE 6.8.4 Results for the Cumulant Expansion RG through Third Order for the Ising Model on the Square Lattice [7]

Cell	Order	J_c	y_T	y_h
2×2	1	0.5186	1.006	2.146
	2	0.4300	1.051	1.979
	3	0.4319	1.072	1.975
3×3	1	0.4697	0.927	1.943
	2	0.4302	1.002	1.884
	3	0.4314	1.080	1.899
4×4	1	0.4607	0.932	1.914
	2	0.4330	1.009	1.883
	3	0.4305	1.083	1.914
Exact		—	0.4407	1.000	1.875

pect that the cumulant expansion may be asymptotic rather than convergent.[†]

6.9 ANTIFERROMAGNETIC ISING MODEL ON THE TRIANGULAR LATTICE

It is relatively straightforward to construct an RG transformation for the ferromagnetic Ising model, but when the ordered state of a system is more complex the construction of the RG may not be so obvious. A good example of this is afforded by the Ising model on the triangular lattice with antiferromagnetic nearest-neighbor interactions ($J < 0$).

[†]A series is asymptotic or semiconvergent if [2]

$$\lim_{x \to \infty} x^n R_n(x) = 0 \quad \text{for fixed } n$$

and

$$\lim_{n \to \infty} |x^n R_n(x)| = \infty \quad \text{for fixed } x$$

where $R_n(x)$ is the remainder, $R_n(x) = F(x) - S_n(x)$, and $S_n(x)$ is the sum of the first n terms in the series. The first equation states that for sufficiently large x, the error in approximating the function by a finite number of terms can be made arbitrarily small, whereas the second says that for any x, however large, the series is not convergent. For any given value of x, there is a value of n that minimizes the remainder (see Prob. 6.11).

Fig. 6.9.1 Elementary plaquette on the triangular lattice. For antiferromagnetic nearest-neighbor couplings the order is frustrated.

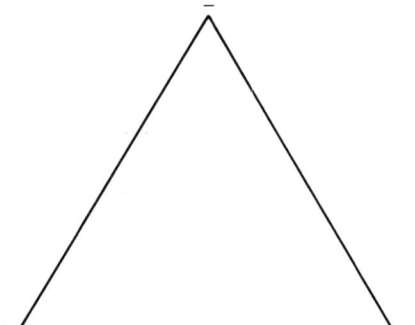

In the absence of an external field the ground state of this model is highly degenerate. In fact, Wannier [13] has shown that the entropy per spin is finite and the system remains paramagnetic even at $T = 0$. To see how this is possible, consider one of the elementary triangles, or plaquettes, of the triangular lattice shown in Fig. 6.9.1.

In the state of lowest energy, all spins should be antiparallel to each of their neighbors, but as one can readily see from the figure, this is impossible. One of the bonds must be broken, a phenomenon called frustration. There is a great deal of freedom in choosing which bonds on the lattice are frustrated, and therefore the ground state is highly degenerate.

If we now apply a weak external field, $h > 0$, the number of ground states is reduced to just three states, one of which is shown in Fig. 6.9.2. The triangular lattice can be decomposed into three triangular sublattices, which we label A, B, and C, as shown in Fig. 6.9.3. In each of the three ground states, the spins on two of the sublattices are parallel to the applied field, whereas those on the third sublattice are antiparallel. The three ground states are conveniently labeled by the sublattice that is antiparallel to the applied field.

It should be clear that the order parameter for this model is not the magnetization as in the ferromagnetic Ising model; the magnetization does not distinguish between the three possible ordered states. The order parameter in this case must be a three-component object and Alexander [1] has argued that this model is equivalent to the three-state Potts model. Therefore, it belongs to an entirely different universality class from the ferromagnetic Ising model. The Hamiltonian for the q-state Potts model is

$$H = -J \sum_{\langle ij \rangle} \delta(\sigma_i, \sigma_j) \qquad (6.9.1)$$

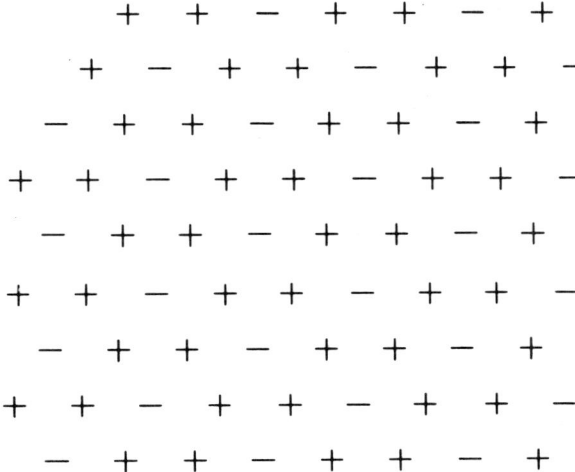

Fig. 6.9.2 One of the three ground states of the AF Ising model on the triangular lattice in a weak positive magnetic field.

where the variables $\sigma_i = 1, 2, 3, \ldots, q$ and $\delta(\sigma, \sigma')$ is the Kronecker delta. The two-state Potts model is equivalent to the Ising model (Prob. 6.8).

To see how tricky constructing an RG transformation can be, let us apply the three-spin-cell majority-rule transformation to one of the ground states shown previously. Because each of the spins in a given

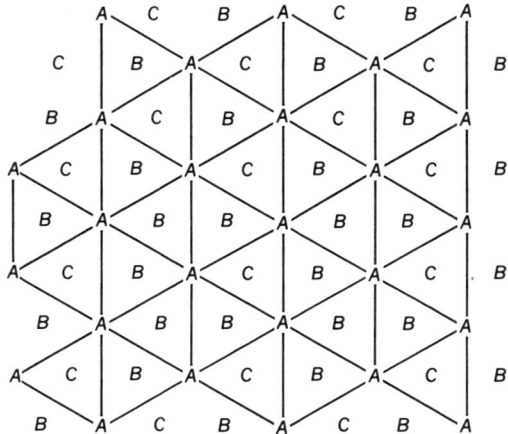

Fig. 6.9.3 Decomposition of the triangular lattice into three triangular sublattices.

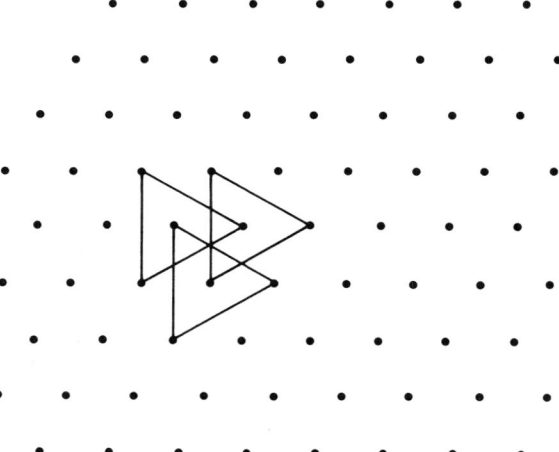

Fig. 6.9.4 Nine-spin cell of Schick, Walker and Wortis [11] composed of 3 interpenetrating 3-spin clusters on each sublattice.

cell belongs to a different sublattice, there will always be two "up" spins and one "down" spin in each cell. By the majority rule, all three ground states will map onto a perfectly ordered ferromagnetic state. This clearly violates the criterion of sec. 4.1 that the RG preserve the nature of the ordered states, and would, if carried blindly through, lead to the erroneous conclusion that the AF Ising model in an external field is in the same universality class as the FM Ising model.

In order to preserve the sublattice structure of the ground states, it is necessary to choose a larger cell, and here we will discuss the simplest choice, which was employed by schick, Walker, and Wortis [11]. The cell consists of nine spins, three from each sublattice, as shown in Fig. 6.9.4. The majority rule can be applied to the spins on each sublattice separately, thereby preserving the structure of the ground states. Of course, there is a price to pay; there are now 512 possible states of the cell to consider. However, if we impose periodic boundary conditions on the cell, then symmetry can be used to greatly reduce the labor involved in constructing the RG equations. The RG transformation will therefore map the nine spins into a single triplet of block spins. Because only one-, two-, and three-spin interactions can exist on such a three-spin cell, we take the initial Hamiltonian to have the form

$$H = -J \sum_{\langle ij \rangle} s_i s_j - h \sum_i s_i - M \sum_{\langle ijk \rangle} s_i s_j s_k + NC \qquad (6.9.2)$$

Note that we have included explicitly a constant term NC, where N is the number of sites, in the Hamiltonian. Such terms do not affect the RG equations for the couplings, but they are generated by the RG. Summing all such terms along the RG trajectory yields the free energy. By using as finite cell with periodic boundary conditions, we avoid the problem of proliferation of new couplings, which we found in the cumulant-expansion method. Instead of truncating in parameter space, we have replaced an infinite lattice by a 3×3 lattice.

The renormalized Hamiltonian is given by

$$e^{-H'} = \operatorname*{Tr}_{\{s_A\}} \operatorname*{Tr}_{\{s_B\}} \operatorname*{Tr}_{\{s_C\}} \prod_i \delta\left(S_A - \operatorname{sgn}\sum_i s_A^i\right) \cdots e^{-H} \quad (6.9.3)$$

If we define the right-hand side of (6.9.3) to be $\exp F(S_A, S_B, S_C)$, then the renormalized couplings can be determined by solving the following set of simultaneous equations:

$$3C' + 9J' + 3h' + 6M' = F(+, +, +) = F_1 \quad (6.9.4a)$$
$$3C' - 3J' + h' - 6M' = F(+, +, -) = F_2 \quad (6.9.4b)$$
$$3C' - 3J' - h' + 6M' = F(+, -, -) = F_3 \quad (6.9.4c)$$
$$3C' + 9J' - 3h' - 6M' = F(-, -, -) = F_4 \quad (6.9.4d)$$

In (6.9.4a) to (6.9.4d), it is important to realize that periodic boundary conditions have been imposed on the three block spins so that each nearest-neighbor coupling is counted three times and the three-spin coupling is counted six times. This "overcounting" ensures that the energy/spin in the ground states is preserved under the RG transformation.[†] The solution to (6.9.4a) to (6.9.4d) is

$$C' = 3C + \frac{1}{24}(F_2 + 3F_2 + 3F_3 + F_4) \quad (6.9.5a)$$

$$h' = \frac{1}{8}(F_1 + F_2 - F_3 - F_4) \quad (6.9.5b)$$

$$J' = \frac{1}{24}(F_1 - F_2 - F_3 + F_4) \quad (6.9.5c)$$

$$M' = \frac{1}{48}(F_1 - 3F_2 + 3F_3 - F_4) \quad (6.9.5d)$$

[†]The authors are indebted to M. Schick for elucidating this point.

Evaluating the trace, (6.9.2) gives

$$e^{F(+++)} = e^{27J+9h+18M} + 9e^{15J+7h+6M} + 27e^{7J+5h+2M}$$
$$+ 18Je^{3J+3h-2M} + 9e^{3J+3h+6M} \tag{6.9.6a}$$

$$e^{F(++-)} = e^{-9J+3h-18M} + 6e^{-J+h-6M} + 3e^{3J+5h-2M}$$
$$+ 9e^{-5J-h-2M} + 18e^{-J+3h-2M}$$
$$+ 18e^{-J+h+2M} + 9e^{-J+h-6M} \tag{6.9.6b}$$

$F(+,-,-)$ and $F(-,-,-)$ are easily found from (6.9.6a) and (6.9.6b) by the observation that under $S \to -S$, $J \to J$, $h \to -h$, and $M \to -M$.

Let us explore for a moment the possible fixed points of the RG equations (6.9.5). First, if we take $h = M = 0$ initially, they remain 0 under the RG as we would hope. In this case $F_1 = F_4$ and $F_2 = F_3$ so that the RG equation for J reduces to

$$J' = \frac{1}{12}\ln\left[\frac{e^{27J} + 9e^{15J} + 27e^{7J} + 27e^{3J}}{7e^{-9J} + 3e^{3J} + 9e^{-5J} + 45e^{-J}}\right] \tag{6.9.7}$$

This equation has a nontrivial fixed point at $J^* = 0.185$ (0.275 exact). In order to find the critical exponents, we must linearize the full RG equations about the fixed point $(0.185, 0, 0)$. We have seen that quite generally the linearized RG in the neighborhood of the ferromagnetic fixed point decomposes into even and odd sectors. In this case there is only one even coupling J, and so the temperature exponent is simply

$$y_T = \frac{\ln\left.\frac{\partial J'}{\partial J}\right|_{J^*}}{\ln\sqrt{3}} = 0.638 \tag{6.9.8}$$

On the other hand, the odd sector of parameter space is two dimensional because we have two odd couplings h and M. If we expand each of the F's defined in (6.9.4a) to (6.9.4d) to first order in h and M, we have

$$h' = \frac{1}{4}\left(\frac{\partial F_1}{\partial h} + \frac{\partial F_2}{\partial h}\right)h + \frac{1}{4}\left(\frac{\partial F_1}{\partial M} + \frac{\partial F_2}{\partial M}\right)M \tag{6.9.9a}$$

$$M' = \frac{1}{24}\left(\frac{\partial F_1}{\partial h} - 3\frac{\partial F_2}{\partial h}\right)h + \frac{1}{24}\left(\frac{\partial F_1}{\partial M} - 3\frac{\partial F_2}{\partial M}\right)M \tag{6.9.9b}$$

where all the derivatives are evaluated at $h = M = 0$. Solving this pair of coupled equations, we find two eigenvalues of the linear RG

$$y_h = 1.454$$
$$y_3 = -1.054 \qquad (6.9.10)$$

One might wonder how we know the relevant eigenvalue is associated with the magnetic field rather than the three-spin interaction. The answer of course is that the linear scaling fields are linear combinations of h and M. If initially we have $M = 0$, $h \neq 0$, the perturbation has a nonzero overlap with both the relevant scaling field and the irrelevant scaling field. Under repeated renormalizations the amplitude of the irrelevant scaling field tends to 0 and the relevant scaling field dominates. The same argument can be made for the case $M \neq 0$, $h = 0$, but because $M = 0$ is in the subspace of physical Hamiltonians, we call the largest eigenvalue in the odd-sector y_h.

Let us now consider the antiferromagnetic case, $J < 0$, in an external field h, which for definiteness we will take to be positive. We have already argued that the ground states in this case have a different symmetry than those of the ferromagnetic Ising model, and on these grounds we expect to find the critical exponents characteristic of the three-state Potts model.

Let us consider for a moment the ground states of the system with $M = 0$, h and $-J$ large. For small h/J the ground state has the symmetry of the Potts phase, and as h/J is increased there will be a transition from the Potts state to a ferromagnetic ground state. The transition will occur (at zero temperature) when the energies of the two states are equal. Comparing the energy per spin in the two ground states, we see that the transition from one to the other will occur for $6J + h = 0$. Therefore, let us look for a fixed point along the line $6J + h = 0$, where again J and h are tending to ∞ in magnitude. Keeping only the highest-order terms in the F's, we have from (6.9.6)

$$F_1 \sim \ln\left[e^{27J + 9h + 18M} + 9e^{15J + 7h + 6M}\right] \qquad (6.9.11\text{a})$$

$$F_2 \sim \ln\left[e^{-9J + 3h - 18M} + 3e^{3J + 5h - 6M}\right] \qquad (6.9.11\text{b})$$

$$F_3 \sim \ln\left[9e^{-5J + h + 2M}\right] \qquad (6.9.11\text{c})$$

$$F_4 \sim \ln\left[18e^{3J - 3h + 2M} + 9e^{3J - 3h + 6M}\right] \qquad (6.9.11\text{d})$$

Ignoring for the moment the three-spin interaction, the RG equations

(6.9.5b) and (6.9.5c) become

$$h' = \frac{4}{3}h \quad (6.9.12a)$$

$$J' = \frac{4}{3}J \quad (6.9.12b)$$

Note that both the renormalized J and h scale in the same way, so their ratio remains constant. Furthermore, because the scale factor, $4/3$, is greater than unity, the fixed point of interest will in fact lie at $J = -\infty$, $h = \infty$. It is somewhat awkward to deal directly with variables that are infinite at the fixed point (see the discussion of Sec. 4.2), so let us introduce new variables that will remain finite as the fixed point is approached. One obvious choice is M itself, whereas for the other we can take

$$z = 6J + h \quad (6.9.13)$$

which measures how far we are from the phase boundary. By (6.9.5a) to (6.9.5d), the renormalized equations for z and M are

$$z' = \frac{1}{8}\ln\left[\frac{(e^{9z+18M} + 9e^{7z+6M})^3 + (18e^{-3z+2m} + 9e^{-3z+6M})}{729(e^{6z-12M} + 3e^{8z})}\right] \quad (6.9.14a)$$

$$M' = \frac{1}{48}\ln\left[\frac{729(e^{12z+24M} + 9e^{10z+12M})}{(e^{3z-18M} + 3e^{5z-6M})^3(18e^{-3z+2M} + 93e^{-3z+6M})}\right] \quad (6.9.14b)$$

These equations have the fixed point

$$z^* = -1.5883 \qquad M^* = 0.1132 \quad (6.9.15)$$

Linearizing equations (6.9.14) about the fixed point, we find that there is one relevant scaling field with scaling exponent $y_h = 0.956$, and one irrelevant scaling field with scaling exponent $y_T = -0.398$. The left eigenvectors of the linearized RG transformation are

$$e_h = (1, 6) \qquad e_T = (1, -10.8) \quad (6.9.16)$$

The original parameter space is three dimensional, and yet we have only solved the RG equations in a two-dimensional subspace. There must be a third parameter, and a useful choice is

$$y = e^{-(3/2)(h-6M)} \qquad (6.9.17)$$

It is easy to see from the RG equations that y decouples from the other RG equations, and satisfies

$$y' = y^{4/3} \qquad (6.9.18)$$

At the fixed point $y^* = 0$, and small perturbations of y about $y = 0$ vanish faster than exponentially. In the language of Sec. 4.2, this is a nonlinear fixed point and strictly speaking equations (6.9.12a) and (6.9.12b) cannot be used to calculate the appropriate scaling exponent. Rather, we should use (6.9.18), which yields

$$y_3 = \frac{\ln \dfrac{\partial y'}{\partial y}}{\ln \sqrt{3}}\bigg|_{y^*=0} = -\infty \qquad (6.9.19)$$

This relatively simple model exhibits yet another fixed point. If we take $J = h = 0$, then the RG equations reduce to a single equation for M. The Ising model with only three-spin interactions is known as the Baxter–Wu model [3, 4] and has been solved exactly. The RG equations in this case are left as an exercise (Prob. 6.10). The exact results are $J_c = 0.4407$, $y_T = 1.5$, and $y_h = 1.875$.

PROBLEMS

6.1 Show that

$$\delta(S_\alpha - \mathrm{sgn}(s_\alpha^1 + s_\alpha^2 + s_\alpha^3)) = \frac{1}{2}\left[1 + \frac{1}{2}S_\alpha(s_\alpha^1 + s_\alpha^2 + s_\alpha^3 - s_\alpha^1 s_\alpha^2 s_\alpha^3)\right]$$

6.2 Derive (6.1.12a) to (6.1.12c).

6.3 Find the fixed point for the RG equation (6.2.4).

6.4 Derive an expression of the form (6.2.4) for a cell consisting of five spins.

6.5 What type of interactions must be included in H if we include the third cumulant in (6.3.6)?

6.6 Any nonsingular function of $S_\alpha = \pm 1$ can be expressed in the form $f(S_\alpha) = a + bS_\alpha$. In particular, show that

$$(a + bS_\alpha)^{-1} = a' + b'S_\alpha$$

and find a' and b' in, terms of a and b.

6.7 The zeroth-order renormalized Hamiltonian is, by (6.6.3), just

$$\sum_\alpha \ln z_\alpha = \sum_\alpha \ln(a_0 + b_0 S_\alpha)$$

Show that $\ln(a + bS_\alpha) = a' + b'S_\alpha$ and that $b' = \tanh^{-1}(b/a)$. This is a neater way of deriving (6.6.13).

6.8 Show that the two-state Potts model

$$H = -J \sum_{\langle ij \rangle} \delta(\sigma_i, \sigma_j) \qquad \sigma_i = 1, 2$$

is equivalent to the Ising model. $\delta(\sigma, \sigma')$ is the Kronecker delta.

6.9 For the triangular lattice derive the scaling law for $x = 6J + h + 6M$ for $-J, h \to \infty$. Find the fixed point and scaling exponent for this scaling field.

6.10 Show that if $J = h = 0$ (Baxter–Wu model) the RG equations (6.9.5) reduce to a single equation $M' = R(M)$. Find the fixed point and scaling exponent.

6.11 Consider the incomplete gamma function (see [2] for further details)

$$I(x, p) = \int_x^\infty e^{-u} u^{-p} \, du$$

(a) Integrating by parts, show that

$$I(x, p) = \frac{e^{-x}}{x^p} - pI(x, p + 1)$$

and therefore that

$$I(x, p) = S_n(x, p) + R_n(x, p)$$

where the nth partial sum is

$$S_n(x, p) = \frac{e^{-x}}{x^p} \sum_{k=0}^{n-1} (-1)^k \frac{(p + k - 1)!}{(p + 1)!} \frac{1}{x^k}$$

and the nth remainder is

$$R_n(x, p) = (-1)^n \frac{(p - n + 1)!}{(p - 1)} I(x, p + n)$$

(b) Show that

$$\lim_{x \to \infty} x^n R_n(x) \to 0 \quad \text{for fixed } n$$

$$\lim_{n \to \infty} |x^n R_n(x)| \to \infty \quad \text{for fixed } x$$

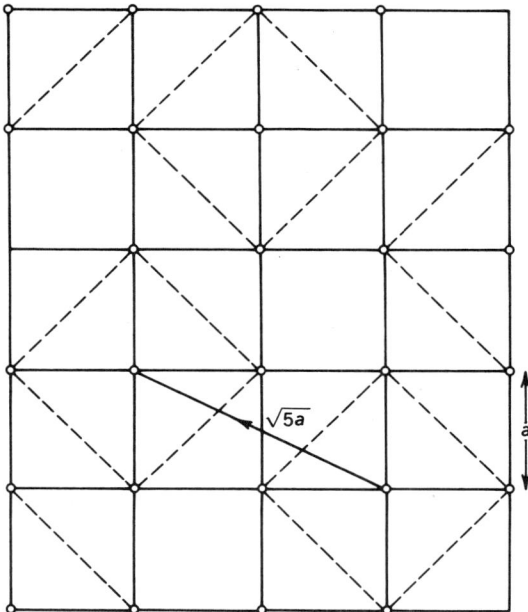

Fig. 6.P.1 5-spin block on the square lattice, $n = \sqrt{5}$. See problem 9.12.

(c) For a given value of x, find the value of n which minimizes the remainder.

6.12 Consider the five-spin blocks on the square lattice ($b = \sqrt{5}$), shown in Fig. 6.P.1.

(a) Find the renormalization equation for the nearest-neighbor coupling in the first cumulant approximation. Find the fixed point and calculate the thermal exponent.

(b) Calculate the zeroth-order approximation to the renormalized magnetic field and the magnetic exponent in the same approximation.

REFERENCES

1. Alexander, S. (1975) *Phys. Lett. A* **54** 353.
2. Arfken, G. (1985) *Mathematical Methods for Physicists* 3rd ed., p. 339. Academic, New York.
3. Baxter, R. J. (1971) *Phys. Rev. Lett.* **26** 832; *Ann. Phys. (N.Y.)* **70** 323 (1972).
4. Baxter, R. J. and Wu, F. Y. (1973) *Phys. Rev. Lett.* **31** 1294; *Austral. J. Phys.* **27** 357 (1974).
5. Braathen, H. J. and Hemmer, P. C. (1975) *Phys. Nor.* **8** 69.
6. Hemmer, P. C. and Verlarde, M. G. (1979) *J. Phys. A* **9** 1713.
7. Hsu, S.-C. and Gunton, J. D. (1977) *Phys. Rev. B* **15** 2688.
8. Hu, B. (1982) *Phys. Rep.* **91** 233.
9. Niemeijer, Th. and van Leeuwen, J. M. J. (1974) *Physica* **71** 17.
10. Niemeijer, Th. and van Leeuwen, J. M. J. (1976) In *Phase Transitions and Critical Phenomena* (C. Domb, and M. S. Green, eds.) Academic Press, N.Y., vol. 6.
11. Schick, M., Walker, J. S., and Wortis, M. (1977) *Phys. Rev. B* **16** 2205.
12. Sudbo, A. S. and Hemmer, P. C. (1976) *Phys. Rev. B* **13** 980.
13. Wannier, G. (1950) *Phys. Rev.* **79** 357.

7

OTHER REAL-SPACE RENORMALIZATION GROUP METHODS FOR THE ISING MODEL

7.0 INTRODUCTION

In this section we investigate a slightly different class of real-space RG transformations, which in a sense are simpler than the majority-rule transformations, but perhaps more removed from the physically appealing idea of coarse graining. These are the decimation transformations.

In a decimation transformation no new degrees of freedom are introduced, but rather a subset of the original degrees of freedom is eliminated. A simple example of the decimation transformation was employed in Chap. 5 to solve the one-dimensional Ising model. In that case the decimation introduced no new couplings into the Hamiltonian, and a nice closed set of RG equations could be found. In higher dimensions decimation transformations suffer from the same proliferation of couplings we found in the majority-rule RGs. There are several strategies for dealing with this problem.

The first is to simply ignore some of the newly generated couplings as was done with the cumulant expansion. Another possibility is to start with a finite lattice, as in the example of Sect. 6.9, so that once the decimation transformation has been accomplished only a few spins are left, and only a small number of couplings need to be retained. It turns out that this kind of transformation is exact on a hierarchical lattice. A third method, called the Migdal–Kadanoff [4, 6], or bond-moving, method, avoids the proliferation of couplings by moving some

218   OTHER REAL-SPACE RENORMALIZATION GROUP METHODS

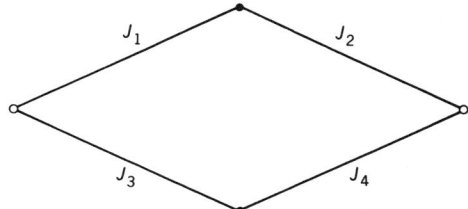

Fig. 7.1.1 Elementary cluster of bonds for the decimation of the diamond hierarchical lattice. The spins on the sites indicated by solid circles are decimated.

of the bonds off of the sites to be decimated so that once the decimation is carried out no new interactions are generated.

7.1 DECIMATION TRANSFORMATION ON A HIERARCHICAL LATTICE

Let us begin this section by constructing the RG on the "diamond" hierarchical lattice. The elementary cell is shown in Fig. 7.1.1, and the decimation procedure involves tracing over the spins at the sites indicated by solid circles. Because the hierarchical lattice is generated by replacing each bond by the elementary cell of Fig. 7.1.1, the decimation works in the opposite sense and yields a single renormalized bond for each cell.

The renormalization equation, which in this case involves only the nearest-neighbor coupling, is

$$e^{J'S_1 S_2} = 4\cosh^2 J(S_1 + S_2) \tag{7.1.1}$$

Taking the possible values ± 1 for S_1 and S_2, we can then find for $v = \tanh J$,

$$v' = \frac{2v^2}{1 + v^4} \tag{7.1.2}$$

This transformation has a fixed point at $v^* = 0.5437$ and the correlation length exponent is $\nu = 1.338$ ($b = \sqrt{2}$).

If we now introduce a magnetic field h, (7.1.1) becomes

$$e^{J'S_1 S_2 + h'(S_1 + S_2)} = Ne^{-h(S_1 + S_2)} \cosh^2[J(S_1 + S_2) + h] \quad (7.1.3)$$

Solving for J' and h' in much the same way as we did in Sect. 5.2, we find

$$e^{2J'} = \frac{\cosh(2J + h)\cosh(2J - h)}{\cosh^2 h} \quad (7.1.4)$$

$$e^{2h'} = e^{2h}\frac{\cosh(2J + h)}{\cosh(2J - h)} \quad (7.1.5)$$

We know that at the fixed point $h^* = 0$ and linearizing about $(J^*, h^* = 0)$, (7.1.4) reproduces (7.1.2), whereas (7.1.5) becomes

$$h' = h(1 + \tanh 2J^*) \quad (7.1.6)$$

Again taking $b = \sqrt{2}$, we find $y_h = 1.758$, just 6% below the exact value of 1.875.

7.2 THE RANDOM-BOND ISING MODEL

An interesting generalization of the Ising model is the case in which the exchange couplings J_{ij} for the Ising model are random variables rather than given constants. Such models have been used to describe random ferromagnets, spin glasses, and alloys of mixed ferromagnetic–antiferromagnetic species. A good review of random Ising models is by Stinchcombe [8].

The bonds J_{ij} are assumed to be statistically independent and governed by the probability distribution $p(J)$. Some special cases of importance are:

Bond-dilute Ising model

$$p(J_{ij}) = x\delta(J_{ij} - J_0) + (1 - x)\delta(J_{ij}) \quad (ij\ nn) \quad (7.2.1)$$

Edwards–Anderson [3] spin glass model

$$p(J_{ij}) = \frac{1}{\sqrt{2\pi J_0^2}} e^{-J_{ij}^2/2J_0^2} \quad (7.2.2)$$

$\pm J$ Model

$$p(J_{ij}) = x\delta(J_{ij} - J_0) + (1 - x)\delta(J_{ij} + J_0) \quad (ij\ nn) \quad (7.2.3)$$

Let us consider again the hierarchical lattice of Fig. 7.1.1 [10] and take for $p(J)$ the bond dilute model, (7.2.1). In place of variables J_{ij} it is, as is often the case, more desirable to work with the variables $v_{ij} = \tanh J_{ij}$.

In terms of the new variable $v = \tanh J$, the probability distribution is (see Prob. 7.2)

$$p(v) = x\delta(v - v_0) + (1 - x)\delta(v) \tag{7.2.4}$$

If we decimate the spins labeled S_1 and S_2 in Fig. 7.1.1 we find a renormalized nearest-neighbor couplings between the undecimated spins S_A and S_B given by

$$v' = \frac{v_1 v_2 + v_3 v_4}{1 + v_1 v_2 v_3 v_4} \tag{7.2.5}$$

Note that this reduces to (7.1.2) when all the v's are set equal.

The probability distribution for the renormalized couplings, which we will refer to as the renormalized probability distribution, is given by

$$p'(v') = \int dv_1 \, dv_2 \, dv_3 \, dv_4 \, \delta[v' - f(v_1, v_2, v_3, v_4)]$$
$$\times p(v_1) p(v_2) p(v_3) p(v_4) \tag{7.2.6}$$

where $f(v_1, v_2, v_3, v_4)$ is the function on the right-hand side of (7.2.5).

Let us consider the bond-dilute Ising model, (7.2.1). In general, $p'(v)$ will not be of the simple "binomial" form (7.2.1), and as the renormalization is repeated by taking $p'(v)$ and calculating $p''(v)$, the renormalized probability distributions become quasicontinuous. That is, the allowed values of v, although still discrete, become very closely spaced.

We may regard (7.2.6) as a renormalization equation for the function $p(v)$, and as usual we are interested in locating fixed distributions $p^*(v)$, which satisfy

$$p^*(v) = \int dv_1 \cdots dv_4 \, \delta[v - f(v_1, \ldots, v_4)] p^*(v_1) \cdots p^*(v_4) \tag{7.2.7}$$

Finding fixed distributions may seem like a formidable task, but we already know of two solutions to (7.2.7). First, for $x = 1$, $v_0 = v^* =$

0.5437... we have again the solution for the perfect lattice. Also, if we take $J_0 \to \infty$ ($T \to 0$), then the problem reduces to bond percolation on the hierarchical lattice, which we have solved in Sect. 3.3. In this case the renormalized probability is

$$p'(v) = x'\delta(v-1) + (1-x')\delta(v) \tag{7.2.8}$$

where x' is given by (3.2.1)

$$x' = 2x^2 - x^4 \tag{7.2.9}$$

This has a nontrivial fixed point for $x^* = (\sqrt{5}-1)/2$. In addition to these critical fixed points, there are the trivial fixed points $x = 1$, $v_0 = 1, 0$, and $x = 0$ (empty lattice).

Let us assume that the renormalized probability distribution is of the form (7.2.4). This reduces the number of parameters from ∞ [the value of $p(v)$ for every v] to just two, x and v_0. We know this approximation is good at both the ferromagnetic and percolation fixed points, and we hope it is not a poor approximation elsewhere along the critical line.

In order to find the renormalized parameters (x', v_0'), we can calculate the first moments of $p(v')$. By (7.2.4)

$$\langle v \rangle_{p'} = x' v_0' \tag{7.2.10}$$

and x' is given by (7.2.9).

To complete the calculation, we need to evaluate (7.2.10) using (7.2.6). We have

$$\langle v \rangle_{p'} = \int dv\, v p'(v)$$

$$= \int dv \int dv_1 \int dv_2 \int dv_3 \int dv_4\, v \delta[v - f(v_1, v_2, v_3, v_4)]$$

$$\times p(v_1) p(v_2) p(v_3) p(v_4) \tag{7.2.11}$$

The integration over v can be done first, giving, by (7.2.5),

$$\langle v \rangle_{p'} = \int dv_1 \int dv_2 \int dv_3 \int dv_4 \left[\frac{v_1 v_2 + v_3 v_4}{1 + v_1 v_2 v_3 v_4}\right] p(v_1) p(v_2) p(v_3) p(v_4) \tag{7.2.12}$$

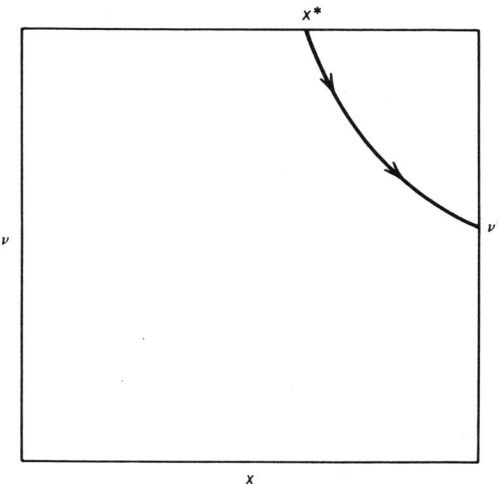

Fig. 7.2.1 Phase diagram of the dilute Ising model. The arrows on the critical line indicate that the RG trajectories flow toward the Ising fixed point at $(1, v^*)$ and away from the percolation fixed point at $(x^*, 1)$.

The remaining integrals are easily performed because they only involve delta functions. Because $p(v)$ is a sum of two terms, expanding in (7.2.12) will lead to 16 terms altogether, but many of them are equal. The final result is

$$\langle v \rangle_{p'} = x^4 \frac{2v_0^2}{1 + v_0^4} + 4x^3(1 - x)v_0^2 + 2x^2(1 - x^2)v_0^2 \quad (7.2.13)$$

Putting (7.2.9) and (7.2.10) together, we have

$$v_0' = \frac{x^4 \dfrac{2v_0^2}{1 + v_0^4} + 4x^3(1 - x)v_0^2 + 2x^2(1 - x^2)^2 v_0^2}{2x^2 - x^4} \quad (7.2.14)$$

This equation, together with (7.2.9), constitute the RG equations in this approximation.

The phase diagram for the dilute Ising model is shown in Fig. 7.2.1. The critical line terminates at the percolation fixed point $(x^*, 1)$, and the ferromagnetic or Ising critical point $(1, v^*)$. For any $x > x^*$ the renormalized values of x tend to unity and so, on the critical line, the percolation fixed point is unstable.

Close to the percolation threshold, $v_c(x)$ approaches unity ($T_c \to 0$) with a finite slope. In terms of the transition temperature this implies that

$$\lim_{x \to x^*} T_c(x) \to -\frac{1}{\ln(x - x^*)} \qquad (7.2.15)$$

7.3 THE MIGDAL–KADANOFF TRANSFORMATION

In physics there are two principal approaches to problems that cannot be solved exactly. The first is a perturbation calculation in which a part of the problem can be solved, and the remaining part can hopefully be treated in successive approximations. In many situations, and in critical phenomena especially, one encounters perturbation series that do not converge. This is probably true of the cumulant expansion RG, and also for the ε expansion, which we will consider in Chap. 9.

The second is a variational approach [2, 4, 6]. The idea here is to first establish that the exact solution of a problem is an extremum of some quantity. For example, the ground state of a quantum system minimizes the expectation value of the energy. We then make a good guess and choose a trial solution, which may depend on one or more adjustable parameters. The values of these parameters are then determined by the condition that the trial solution be extremal with respect to their variation. The Migdal–Kadanoff (MK) approximation is of the second type, although in its most primitive form no adjustable parameters are introduced.

Before we discuss the MK transformation in detail, we first need to establish the appropriate variational principle. Let H be the Hamiltonian whose partition function, or free energy, we wish to know, and suppose that H' is a Hamiltonian whose free energy we do know how to calculate. We denote the difference between H and H' by V, so that

$$H' = H + V \qquad (7.3.1)$$

The variational free energy is then related to the exact free energy by

$$e^{-\beta F'} = e^{-\beta F} \langle e^{-\beta V} \rangle \qquad (7.3.2)$$

from which we have (see Prob. 7.4)

$$e^{-\beta F'} \geq e^{-\beta F} e^{-\beta \langle V \rangle} \qquad (7.3.3)$$

where the average on the right-hand side is taken with respect to H.

We now specialize to the case where $\langle V \rangle$ vanishes identically due to some symmetry of H. In that case (7.3.3) reduces to the desired extremal condition

$$F' \leq F \qquad (7.3.4)$$

Now the question is: What sort of variational potential V should we use? Putting the question another way: What problems do we know how to solve exactly?

Setting aside for the moment the solution of the two-dimensional Ising model outlined in Chap. 5, the only model we have solved completely is the one-dimensional Ising chain. To see how the two-dimensional lattice can be reduced to one-dimensional chains, imagine that half the bonds on the lattice are moved as shown in Fig. 7.3.1. The sites interior to the cells are no longer coupled to anything, and tracing over each of these simply gives a factor of 2. The sites that decorate the edges of the remaining network, indicated by solid circles in Fig. 7.3.1, can now be traced over using the results derived for the one-dimensional chain.

What does it mean to "move a bond" from between sites S_1 and S_2 and put it between sites S_3 and S_4? This process is accomplished by adding to the Hamiltonian a "bond-moving potential"

$$\delta V = J(S_1 S_2 - S_3 S_4) \qquad (7.3.5)$$

The full MK potential is the sum of all such terms

$$V = J \sum_{\langle ij \rangle ; \langle kl \rangle} (S_i S_j - S_k S_l) \qquad (7.3.6)$$

Note that by translational symmetry $\langle V \rangle = 0$, so that the MK approximation to the free energy is a lower bound. Also, because no bonds are lost in the bond-moving procedure, the ground-state energy is preserved in the MK approximation.

Once the bonds have been moved, we only need to calculate the trace over the sites indicated by solid circles in Fig. 7.3.1. This produces (see Sec. 5.1 for details) a new coupling between nearest-neighbor spins on the decimated lattice

$$J' = \tanh^{-1}(\tanh^2 2J) \qquad (7.3.7)$$

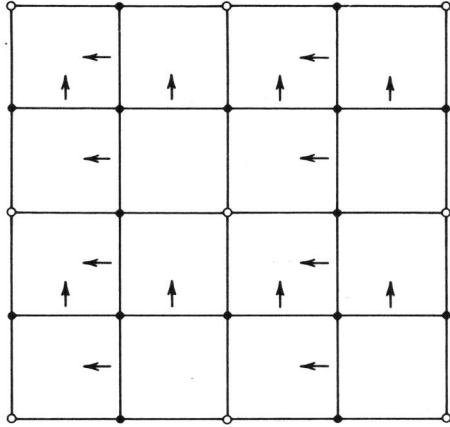

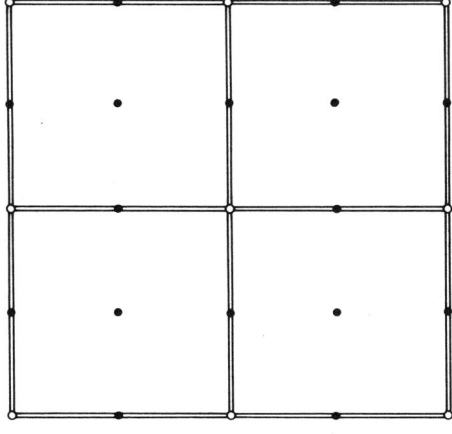

Fig. 7.3.1 Bond-moving procedure for the simple square lattice. Note that the lattice is now reduced to a network of one-dimensional chains. Sites indicated by solid circles can now be decimated without introducing additional couplings.

This equation has a fixed point at $J^* = 0.305$ and the temperature exponent is

$$y_T = \frac{\ln\left(\left.\frac{\partial J'}{\partial J}\right|_{J^*}\right)}{\ln 2} = 0.748 \qquad (7.3.8)$$

At first glance the MK approximation seems to do great violence to the original model. There are two observations that we may make in its defense, apart from the fact that it must give a lower bound to the exact free energy. The first is that the transformation preserves the energy of the ground state (again because $\langle V \rangle = 0$), and the second is

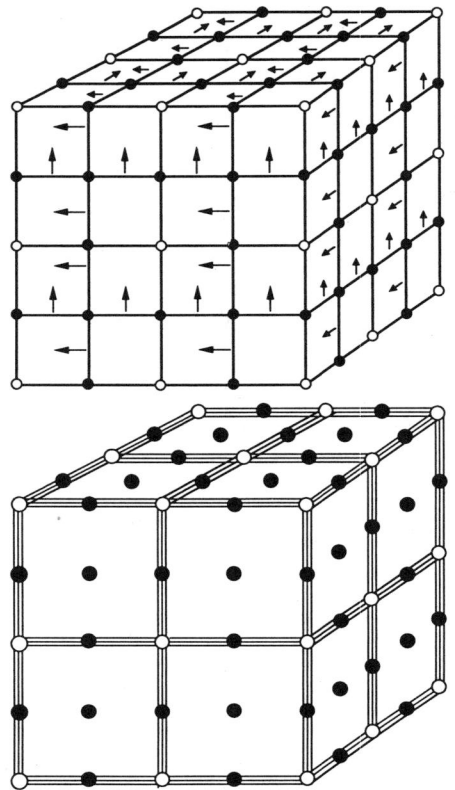

Fig. 7.3.2 Bond-moving procedure for the cubic lattice.

that at high temperatures it again reproduces the free energy of the original model. Therefore, at intermediate temperatures we can hope that it is reasonably good approximation to the exact free energy.

The MK procedure is also nice because it is easily generalized to cells of arbitrary size and to higher dimensions. Taking a cell of b lattice spacings on a (hyper-) cubic lattice in d dimensions, the general MK transformation is

$$J' = \tanh^{-1}\left[\tanh^{b}(b^{d-1}J)\right] \qquad (7.3.9)$$

The bond-moving procedure for the cubic lattice is illustrated in Fig. 7.3.2. Clearly, the fewer bonds we move, that is, the smaller we make b, the closer the MK approximation will be to the exact free energy. In (7.3.9) we can analytically continue from integer values of b to the limit $b \to 1$. If we take $b = 1 + db$, where $db \ll 1$, then the infinitesimal MK transformation is

$$J' = J + \left[(d-1)J + \sinh J \cosh J \ln \tanh J\right] db \qquad (7.3.10)$$

Taking the limit $db \to 0$, we find

$$\frac{dJ'}{db} = (d-1)J + \sinh J \cosh J \ln \tanh J \qquad (7.3.11)$$

The fixed points of the infinitesimal RG transformation are solutions of (7.3.11) for which $dJ'/db = 0$, or

$$(d-1)J^* + \sinh J^* \cosh J^* \ln \tanh J^* = 0 \qquad (7.3.12)$$

Equation (7.3.12) has the usual trivial solutions $J^* = 0$ and $J^* = \infty$, and, in addition, we find a nontrivial fixed point, which for $d = 2$ is $J^* = 0.4407$. This is in fact the exact value for J_c [see Table 5.3.1].

In order to determine the critical exponents we need to linearize (7.3.11) about the fixed point. Writing $J = J^* + j$ to first order in j, we have

$$\frac{dj}{db} = [d + \cosh 2J^* \ln \tanh J^*]j \qquad (7.3.13)$$

This equation has the solution

$$j = j(0)e^{y_T b} \qquad (7.3.14)$$

where the thermal eigenvalue y_T is given by

$$y_T = d + \cosh 2J^* \ln \tanh J^* \qquad (7.3.15)$$

Evaluating (7.3.15) for $d = 2$, we find $y_T = 1.119$, which should be compared to the exact value $y_T = 1.00$.

7.4 THE MIGDAL–KADANOFF TRANSFORMATION IN A WEAK MAGNETIC FIELD

The MK transformation can be generalized to include the effect of a weak external field. First, we will work out the case for $b = 2$ on the square lattice and then tackle the case of arbitrary b and d.

The method is parallel to that of Sec. 5.2 and we can take (5.2.6) over directly if we remember that because of bond moving we need to take $2J$ instead of J for each bond, and that there are now four neighbors that contribute to the renormalized field at each point.

Therefore, we have

$$h' = h + \ln\left[\frac{\cosh(4J + h)}{\cosh(4J - h)}\right] \quad (7.4.1)$$

As in the one-dimensional case, the RG equation for J is not changed so long as $h \ll 1$. To first order in h, (7.4.1) is

$$h' = (1 + 2\tanh 4J)h \quad (7.4.2)$$

which yields for the magnetic exponent y_h,

$$y_h = \frac{\ln(1 + 2\tanh 4J^*)}{\ln 2} \quad (7.4.3)$$

Taking $b = 2$, we find $y_h = 1.529$, which should be compared to the exact result $y_h = 1.875$.

Let us now consider the more general case where b is arbitrary. The dimension d enters only in the number of bonds that are moved to the edges of the cell. The effective coupling is

$$J_e = b^{d-1}J \quad (7.4.4)$$

We now must decimate a chain of $b - 1$ spins terminated by spins S_0 and S_b as shown in Fig. 7.4.1. The partition function for the chain, which is still a function of the terminal spins, is

$$Z(S_0, S_b) = \operatorname*{Tr}_{\{S_1,\ldots,S_{b-1}\}} e^{J_e(S_0 S_1 + \cdots + S_{b-1} S_b)} e^{h(S_0 + S_1 + \cdots + S_b)} \quad (7.4.5)$$

As is often the case, it is useful to express the partition function in terms of the variables $v = \tanh J_e$ and $\tanh(h) \cong h$ for $h \ll 1$. We

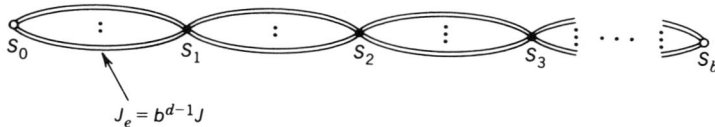

Fig. 7.4.1 One-dimensional chain of spins for the Migdal–Kadanoff transformation with scale factor b. Spins on the interior sites indicated by closed circles are decimated.

have

$$Z(S_0, S_b) = N \operatorname*{Tr}_{\{S_1,\ldots,S_{b-1}\}} (1 + vS_0S_1) \cdots$$
$$\times (1 + vS_{b-1}S_b)(1 + hS_0) \cdots (1 + hS_b) \quad (7.4.6)$$

where N is a constant which need not concern us. Because we are only interested in working to first order in h, we can write

$$Z(S_0, S_b) = Ne^{h(S_0+S_b)}\Bigg\{ \operatorname{Tr}(1 + vS_0S_1) \cdots (1 + vS_{b-1}S_b)$$
$$+ h \sum_{k=1}^{b-1} (1 + vS_0S_1) \cdots (1 + vS_{k-1}S_k)$$
$$\times S_k(1 + vS_kS_{k+1}) \cdots (1 + vS_{b-1}S_b) \Bigg\}$$
(7.4.7a)

Doing the trace starting with S_1 and going up to S_{k-1} and also doing the trace over S_{k+1} up to S_{b-1} gives

$$Z(S_0, S_b) = 2^b N e^{h(S_0+S_b^{-1})} \Bigg[1 + v^b S_0 S_b + \frac{h}{2} \sum_{k=1}^{b-1} \operatorname*{Tr}_{\{S_k\}} (1 + v^k S_0 S_k)$$
$$\times S_k(1 + v^{b-k} S_k S_b) \Bigg] \quad (7.4.7b)$$

Finally, we can perform the trace over S_k with the result

$$Z(S_0, S_b) = 2^b N e^{h(S_0+S_b)} \Bigg[1 + v^b S_0 S_b + h \sum_{k=1}^{b-1} (v^k S_0 + v^{b-k} S_b) \Bigg]$$
(7.4.8)

The geometric series can be summed

$$\sum_{k=1}^{b-1} v^k = \frac{v(1 - v^{b-1})}{1 - v} \quad (7.4.9)$$

and factoring out $(1 + v^b S_0 S_b)$ gives the contribution to the renormalized field due to one such chain

$$h'' = \left[\Lambda(b^{d-1}J, b) - 1\right] \tag{7.4.10}$$

Because there are $2d$ such chains at each terminal site, the renormalized field is

$$h' = \left\{1 + 2d\left[\Lambda(b^{d-1}J, b) - 1\right]\right\}h \tag{7.4.11}$$

where $\Lambda(x, b)$ is given by

$$\Lambda(x, b) = \frac{1 + \tanh x}{1 - \tanh x} \frac{1 - \tanh^b x}{1 + \tanh^b x} \tag{7.4.12}$$

We can again take the limit $b \to 1$, and the infinitesimal RG transformation for the field becomes

$$\frac{dh}{db} = -[2d \sinh J \cosh J \ln \tanh J]h \tag{7.4.13}$$

which of course has the fixed point $h = 0$. The solution to the RG equation for the field near the fixed point $(J^*, 0)$ is simply

$$h = h_0 e^{y_h b} \tag{7.4.14}$$

where the magnetic exponent is

$$y_h = 2d(d - 1)J^* \tag{7.4.15}$$

where we have used (7.3.12) for J^*. For $d = 2$, (7.4.15) gives $y_h = 1.763$, as compared to the exact value 1.875.

7.5 THE MONTE CARLO RENORMALIZATION GROUP

The growth of computing power in recent years has produced a hybrid branch of physics, computer simulation. In a sense a computer simulation is an ideal experiment in which one knows at each moment exactly what the system is doing, and one can calculate whatever observable is of interest. On the other hand, the computer has also become an integral part of the repertoire of the theoretical physicist, allowing the solution of problems that previously were beyond all

7.5 THE MONTE CARLO RENORMALIZATION GROUP

hope. In this section we present a very powerful renormalization group technique based on the Monte Carlo (MC) method for calculating statistical averages, the Monte Carlo renormalization group (MCRG).

For the benefit of those readers not familiar with Monte Carlo methods, we begin with a brief introduction to the standard Metropolis [5] algorithm as it applies to the Ising model. For the student interested in wider applications of the MC method, we recommend the excellent series of books edited by Binder [1].

For definiteness let us consider the problem of calculating the statistical average of a function of the spins A,

$$\langle A \rangle = \frac{\text{Tr } e^{-\beta H} A}{\text{Tr } e^{-\beta H}} \tag{7.5.1}$$

Typically, a Monte Carlo simulation will take place on a lattice of many hundreds or even possibly thousands of sites. Therefore, to calculate the sum in (7.5.1) over every state is out of the question. Fortunately, only a relatively small number of states contribute appreciably to the average, and one should be able to take advantage of this to make the problem tractable. The idea is to generate those states that are the most relevant. We begin with a Master equation for the probability $P(\phi, n)$ that the system will be in state ϕ at the discrete time n,

$$P(\phi, n + 1) = P(\phi, n) + \sum_{\phi'} T(\phi, \phi') P(\phi', n)$$
$$- \sum_{\phi' \neq \phi} T(\phi', \phi) P(\phi, n) \tag{7.5.2}$$

The transition probabilities $T(\phi, \phi')$ can be thought of as arising from the underlying dynamics, or simply as a result of thermal fluctuations of the system in contact with a heat reservoir. Because we are only interested in the equilibrium properties of the system, we take the latter point of view. We further require that in the limit n very large, the distribution $P(\phi, n)$ should approach the canonical distribution

$$\lim_{n \to \infty} P(\phi, n) \to \frac{1}{Z} e^{-\beta E(\phi)} \tag{7.5.3}$$

This can be ensured by requiring that the transition probabilities obey the condition of detailed balance

$$\frac{T(\phi, \phi')}{T(\phi', \phi)} = e^{\beta[E(\phi') - E(\phi)]} \tag{7.5.4}$$

In the Metropolis method the system is allowed to make transitions by flipping one spin at a time according to the following rules:

1. If the energy of the system is lowered by making the flip, then perform the flip.
2. If the energy of the system is raised by the flip, calculate a uniformly distributed random number $0 \leq R \leq 1$. If

$$e^{-\beta[E(\phi') - E(\phi)]} > R$$

then perform the flip, otherwise go on to the next site and repeat.

In this way whatever the initial state of the system, it will eventually evolve into an "equilibrium state." Typically, one allows a few thousand to a few million Monte Carlo steps per spin (MCS) to equilibrate after which the average is taken over a similar number of MCS. In this way a sequence of "important" states is generated which give an accurate estimate of the average (7.5.1).

Now consider [9] a Hamiltonian of the form

$$H = -\sum_{\alpha} J_{\alpha} H_{\alpha} \tag{7.5.5}$$

where the J_{α} are coupling constants ($1/kT$ has been absorbed into the definition of the J's) and the H_{α} are various kinds of interactions. For example, if we take H_1 to be the sum over all spins, then J_1 is the external magnetic field.

Given the Hamiltonian (7.5.5), we can calculate by MC simulation the averages

$$M_{\alpha} = \langle H_{\alpha} \rangle \tag{7.5.6}$$

With each state generated by the MC simulation, we can associate a coarse-grained state according to whatever rule we wish to use. Here, for definiteness, we will consider the majority rule. So long as the coarse-grained lattice is large enough to accommodate all the terms in the Hamiltonian (7.5.5), it is possible to calculate the averages in (7.5.6) over the coarse-grained states as well.

Ultimately, what we want to know are the RG equations that relate the new couplings $\{J_{\alpha}^{(n+1)}\}$ to the old couplings $\{J_{\alpha}^{(n)}\}$. Here the superscript n refers to the level of coarse graining. Clearly, the averages $\langle M_{\alpha}^{(n)} \rangle$ are implicit functions of the $J_{\alpha}^{(n)}$, and in principle we should be able to determine the $J_\alpha^{(n)}$ from the $\langle M_\alpha^{(n)} \rangle$. Unfortunately, we do not have at out disposal the functional relations $M_\alpha^{(n)} = M_\alpha^{(n)}(\{J_\alpha^{(n)}\})$, and this program is not feasible.

Rather than trying to unravel the renormalized couplings from the coarse-grained averages, let us take a slightly different approach and calculate the change in $M_\alpha^{(n+1)}$ produced by an infinitesimal change in the initial couplings $J_\beta^{(n)}$. We have by the chain rule

$$\frac{\partial M_\alpha^{(n+1)}}{\partial J_\beta^{(n)}} = \sum_\gamma \frac{\partial M_\alpha^{(n+1)}}{\partial J_\gamma^{(n+1)}} \frac{\partial J_\gamma^{(n+1)}}{\partial J_\beta^{(n)}} \qquad (7.5.7)$$

Note that on the right-hand side of (7.5.7) we have the matrix of derivatives of the new couplings $J_\alpha^{(n+1)}$) with respect to the old couplings $J_\beta^{(n)}$. If we were at a fixed point, this is just the matrix whose eigenvalues yield the critical exponents (see, e.g., Sec. 6.5). Our object then is to extract this matrix of derivatives from (7.5.7); to do this, we need to know how to calculate $\partial M_\alpha^{(n+1)}/\partial J_\beta^{(n)}$ ad $\partial M_\alpha^{(n+1)}/\partial J_\alpha^{(n)}$.

On the right-hand side of (7.5.7), we have

$$\frac{\partial M_\alpha^{(n+1)}}{\partial J_\gamma^{(n+1)}} = \frac{\partial}{\partial J_\gamma^{(n+1)}} \frac{\operatorname{Tr} H_\alpha \exp \sum_\mu J_\mu^{(n+1)} H_\mu}{\operatorname{Tr} \exp \sum_\mu J_\mu^{(n+1)} H_\mu}$$

$$= \frac{\operatorname{Tr} H_\alpha H_\gamma \exp \sum_\mu J_\mu^{(n+1)} H_\mu}{\operatorname{Tr} \exp \sum_\mu J_\mu^{(n+1)} H_\mu}$$

$$- \left[\frac{\operatorname{Tr} H_\alpha \exp \sum_\mu J_\mu^{(n+1)} H_\mu}{\operatorname{Tr} \exp \sum_\mu J_\mu^{(n+1)} H_\mu} \right] \left[\frac{\operatorname{Tr} H_\gamma \exp \sum_\mu J_\mu^{(n+1)} H_\mu}{\operatorname{Tr} \exp \sum_\mu J_\mu^{(n+1)} H_\mu} \right]$$

$$= \langle H_\alpha^{(n+1)} H_\gamma^{(n+1)} \rangle - \langle H_\alpha^{(n+1)} \rangle \langle H_\gamma^{(n+1)} \rangle \qquad (7.5.8)$$

The left-hand side of (7.5.7) can be simplified in a similar way if we recall (see Sec. 4.1) that

$$e^{-H^{(n+1)}} = \operatorname*{Tr}_{\{s\}} T(S', s) e^{-H^{(n)}}(s) \qquad (7.5.9)$$

We then have

$$\frac{\partial}{\partial J_\beta^{(n)}} e^{-H^{(n+1)}} = \frac{\partial}{\partial J_\beta^{(n)}} \operatorname*{Tr}_{\{s\}} T(\{S\}, \{s\}) \exp \sum_\mu J_\mu^{(n)} H_\mu$$

$$= \operatorname*{Tr}_{\{s\}} T(\{S\}, \{s\}) H_\beta \exp \sum_\mu J_\mu^{(n)} H_\mu \qquad (7.5.10)$$

and therefore

$$\frac{\partial M_\alpha^{(n+1)}}{\partial J_\beta^{(n)}} = \langle H_\alpha^{(n+1)} H_\beta^{(n)} \rangle - \langle H_\alpha^{(n+1)} \rangle \langle H_\beta^{(n)} \rangle \qquad (7.5.11)$$

At first glance the average over operators defined on different levels of coarse graining may look strange. By (7.5.11) what we mean is

$$\langle H_\alpha^{(n+1)} H_\beta^{(n)} \rangle = \frac{\underset{\{S\}}{\text{Tr}} \underset{\{s\}}{\text{Tr}} T(\{S\},\{s\}) H_\alpha(\{S\}) H_\beta(\{s\}) \exp\left[-H^{(n)}(\{s\})\right]}{\underset{\{S\}}{\text{Tr}} \underset{\{s\}}{\text{Tr}} T(\{S\},\{s\}) \exp\left[-H^{(n)}(\{s\})\right]}$$

(7.5.12)

In terms of Monte Carlo averaging, the procedure is to take a given state, and for that state calculate the value of H_β. Next, coarse-grain the state and calculate the value of H_α, and multiply this result by the value of H_β calculated in the first step. Finally, average these products over many states as generated by the Metropolis algorithm.

Let us denote averages of this kind by $C_{\alpha\beta}^{(m,n)}$, where m and n refer to the level of coarse graining applied to the variables associated with H_α and H_β, respectively,

$$C_{\alpha\beta}^{(m,n)} = \langle H_\alpha^{(m)} H_\beta^{(n)} \rangle - \langle H_\alpha^{(m)} \rangle \langle H_\beta^{(n)} \rangle \qquad (7.5.13)$$

and we then have the central result of the MCRG method

$$C_{\alpha\beta}^{(n+1,n)} = \sum C_{\alpha\gamma}^{(n+1,n+1)} T_{\gamma\beta} \qquad (7.5.14)$$

where we have again defined the matrix of derivatives

$$T_{\gamma\beta} = \frac{\partial J_\gamma^{(n+1)}}{\partial J_\beta^{(n)}} \qquad (7.5.15)$$

Finally, multiplying on the left by the inverse of $C_{\alpha\gamma}^{(n+1,n+1)}$, we have

$$T_{\mu\beta} = \sum_\alpha \left[C^{(n+1,n+1)}\right]^{-1}_{\mu\alpha} C_{\alpha\beta}^{(n+1,n)} \qquad (7.5.16)$$

If the correlation functions in (7.5.16) are evaluated at the fixed point, then we can extract the scaling fields and critical exponents from $T_{\mu\beta}$

in the usual way. To complete the program, we must now address the question of how to find the fixed points of the RG transformation.

In order to arrive as close to the fixed point as possible, the initial Hamiltonian must lie close to the critical surface. The problem is that a priori we usually have no idea where the critical surface lies in parameter space. As in the simpler one-parameter problems we have considered, we must find a way of locating this surface.

A further complication arises due to finite-size effects. For definiteness suppose that the scale factor for the RG transformation is $b = 2$ and our initial lattice is of size $L = 2^N$. In each coarse graining the size of the lattice is reduced by a factor of 2. In order to be sure that all the irrelevant scaling fields have been renormalized away, it is necessary to perform as many renormalizations as possible. There is, however, a limit to how far we can go in this direction given by the range of the Hamiltonian. For example, if we include third-neighbor interactions, these cannot be supported on a lattice with $L = 2$. Worse yet, as the size of the lattice is reduced, uncontrollable finite-size effects creep into the calculation.

One way [7] of compensating for these finite-size effects is to simultaneously perform calculations on lattices of initial size $2^N, 2^{N-1}, 2^{N-2}$, and so on. Then we can compare the correlations calculated with the initial Hamiltonian on the lattice with $L = 2^{N-1}$ and correlations calculated on the once-renormalized lattice of size $L = 2^N$. Because these lattices are of the same size, the finite-size effects on the two lattices tend to cancel out. This is illustrated in Fig. 7.5.1.

At the fixed point the averages M_α in (7.5.6) attain their fixed-point values M_α^*. Let us denote the average of the operator H_α taken on the nth coarse-grained lattice of initial size 2^m by $M_\alpha^{(m;n)}$. If the initial Hamiltonian lies on the critical surface, then for large $n < m$ we have

$$M_\alpha^{(m;n)} = M_\alpha^{(m+1;n+1)} = M_\alpha^* \qquad (7.5.17)$$

If we are slightly off of the critical surface so that $J_\alpha = J_\alpha^{(c)} + \delta J_\alpha$, then this error gets magnified on each renormalization and we can write

$$M_\alpha^{(m+1;n+1)} - M_\alpha^{(m;n)} = \sum_\beta \left[\frac{\partial M_\alpha^{(m+1;n+1)}}{\partial J_\beta^{(0)}} - \frac{\partial M_\alpha^{(m,n)}}{\partial J_\beta^{(0)}} \right] \delta J_\beta$$

$$= \sum_\beta \left[C_\alpha^{(m+1;n+1,0)} - C_{\alpha\beta}^{(m;n,0)} \right] \delta J_\beta \quad (7.5.18)$$

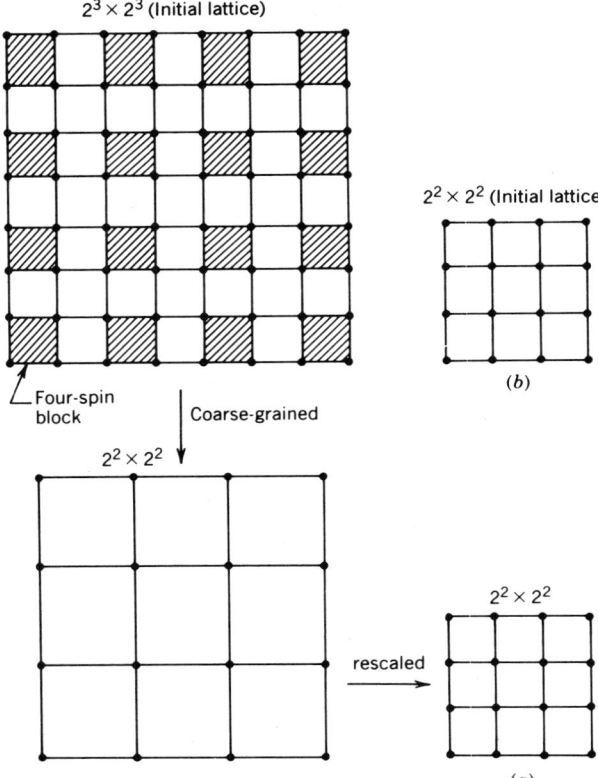

Fig. 7.5.1 At the fixed point, correlations calculated on the once-coarse-grained lattice of original size 2 (*a*) and a lattice of original size 2^2 (*b*) share the same finite-size corrections. Equality of the correlations calculated for (*a*) and (*b*) is therefore a sensitive check for the fixed point.

In (7.5.18) we have extended our notation a bit so that $C_{\alpha\beta}^{m;n,n'}$ refers to correlations calculated on a lattice of initial size 2^m after n and n' coarse grainings. Inverting the matrix on the right-hand side, we have for the deviation from the critical point

$$\delta J_\alpha = \sum_\gamma \left[C_{\alpha\gamma}^{(m+1;n+1,0)} - C_{\alpha\gamma}^{(m;n,0)} \right]^{-1} \left[M_\gamma^{(m+1,n+1)} - M_\gamma^{(m,n)} \right]$$

(7.5.19)

This allows us to correct the parameters in the initial Hamiltonian and to get a better estimate of $J_\alpha^{(c)}$.

Once the critical point has been located with sufficient accuracy, several iterations of the MCRG will bring the system close to its fixed point and the matrix $T_{\alpha\beta}$ of (7.5.16) can be evaluated. The eigenvalues and eigenvectors of this matrix then yield the various critical exponents in the usual way.

PROBLEMS

7.1 Derive (7.1.2) directly by replacing the exponential, $e^{\beta J S_i S_j} \to \cosh \beta J (1 + v S_i S_j)$, for each bond in the basic cell of the hierarchical lattice and calculate the trace over the interior spins.

7.2 Show that (7.2.4) is the correct form for the probability distribution in terms of the variable v.

7.3 Show (7.2.5) is correct by directly calculating the trace over the interior spins.

7.4 Prove the inequality $\langle e^{-\beta V} \rangle \geq e^{-\beta \langle V \rangle}$ which is used in (7.3.3).

7.5 (This is more in the nature of a project.) Implement the MCRG of Sec. 7.5 on the triangular lattice using the majority rule. Keep first-, second-, and third-neighbor pair interactions and compare your results with those of Sec. 6.5.

REFERENCES

1. Binder, K. (1979) *Monte Carlo Methods in Statistical Physics*. Springer, Berlin.
2. Burkhardt, T. W. (1982) In *Real Space Renormalization*, (T. W. Burkhardt and J. M. J. van Leeuwen, eds.). Springer, Berlin.
3. Edwards, S. F. and Anderson, P. W. (1975) *J. Phys. F* **5** 965.
4. Kadanoff, L. P. (1976) *Ann. Phys. (NY)* **100** 359.
5. Metropolis, N., Rosenbluth, A. W., Rosenbluth, M. N., Teller, A. H., and Teller, E. (1953) *J. Chem. Phys.* **21** 1087.
6. Migdal, A. A. (1976) *Sov. Phys. JETP* **42** 413, 743.
7. Pawley, G. S., Swendsen, R. H., Wallace, D. J., and Wilson, K. G. (1984) *Phys. Rev. B* **29** 4030.
8. Stinchcombe, R. B. (1983) In *Phase Transitions and Critical Phenomena* (C. Domb and J. L. Lebowitz, eds.), vol. 7. Academic, New York.
9. Swendsen, R. H. (1979) *Phys. Rev. Lett.* **42** 859.
10. Young, A. P. and Stinchcombe, R. B. (1976) *J. Phys. C* **9** 4419.

8

MEAN FIELD THEORY AND THE GAUSSIAN FIXED POINT

8.0 INTRODUCTION

The earliest theories of phase transitions, the Weiss molecular field theory and the van der Waals theory of the liquid–gas phase transition, are examples of what are now generically called "mean field theories" (MFTs).

Although we will see that MFTs do not usually give quantitatively accurate results close to the critical point, nevertheless they do offer a complete and relatively straightforward description of phase transitions. Very often the first step in the study of a new phase transition is to construct the appropriate mean field theory in order to obtain a global picture of the phase diagram.

There are also cases in which the mean field approximation works quite well quantitatively. For example, the BCS theory of superconductivity is a mean field theory, which, in favorable cases, gives quantitatively accurate results very close to the phase transition. Generally, the accuracy of the MFT depends on the range of the interactions in the system and the dimension of the space in which the system resides. It can, in fact, be shown that the mean field theory is exact for interactions with infinite range or for systems in more than four dimensions.

In this chapter we will discuss MFTs from two points of view. First, we will construct the MFT for the now-familiar Ising model and work out the solution in some detail. We will then show how the partition function for the Ising model can be written in integral form by the Hubbard–Stratonovich transformation. By evaluating the integral in the Gaussian approximation, we recover the previous MFT results and resolve some problems that arise from the neglect of fluctuations in the MF approximation.

We will then look at the Gaussian approximation to the partition function from the point of view of the RG, and introduce the momentum-space RG transformation of Wilson. This section lays the foundation for the systematic theory of critical phenomena generally referred to as the "ε-expansion," which is presented in the next chapter.

8.1 THE MEAN FIELD THEORY FOR THE ISING MODEL

We again begin by writing down the Hamiltonian for the Ising model with ferromagnetic exchange interactions ($J > 0$) in an external field h_0,

$$H = -J \sum_{\langle ij \rangle} S_i S_j - h_0 \sum_i S_i \qquad (8.1.1)$$

Let us now focus our attention on a single spin, say S_k. According to (8.1.1), this spin interacts with its neighbors through a "local field" h_k given by[†]

$$h_k = J \sum_{j(k)} S_j \qquad (8.1.2)$$

so that effectively the Hamiltonian for S_k can be written as

$$H_k = -(h_k + h_0) S_k \qquad (8.1.3)$$

Similar expressions can of course be written down for the neighbors of S_k, and S_k will then contribute to the local field experienced by its neighbors. If we ignore this and simply replace the local field of the neighbors of S_k by its thermal average

$$h_k \to \langle h_k \rangle \equiv h \qquad (8.1.4)$$

[†]The notation $\langle ij \rangle$ in (8.1.1) follows that introduced in Chap. 5 and indicates a sum over nearest-neighbor sites. In (8.1.2) the notation $j(i)$ means sum over all neighbors j of site i.

240 MEAN FIELD THEORY AND THE GAUSSIAN FIXED POINT

we arrive at the mean field approximation to the Hamiltonian for the spin S_k,

$$H_{MF} = -(h + h_0)S_k \qquad (8.1.5)$$

This is the Hamiltonian of an isolated spin interacting with a fixed magnetic field $h + h_0$. The partition function is easily calculated to be

$$Z_1 = 2\cosh \beta(h + h_0) \qquad (8.1.6)$$

and the magnetization per spin m is

$$m = \langle S_k \rangle = \tanh \beta(h + h_0) \qquad (8.1.7)$$

In order to be consistent, the average in (8.1.7) must agree with the average in (8.1.5), which then gives the mean field equation

$$m = \tanh \beta(qJm + h_0) \qquad (8.1.8)$$

In (8.1.8) we have dropped all subscripts because by translational symmetry the same equation holds at all sites. The parameter q is called the coordination number of the lattice, and is simply the number of nearest neighbors of a given site on the lattice.

In general, (8.1.8) must be solved numerically, and the solution for the Ising model on a square lattice is shown in Fig. 8.1.1. The curve of $m(T)$ is characteristic of order parameters. Above the transition temperature the order parameter vanishes identically, whereas at T_c it

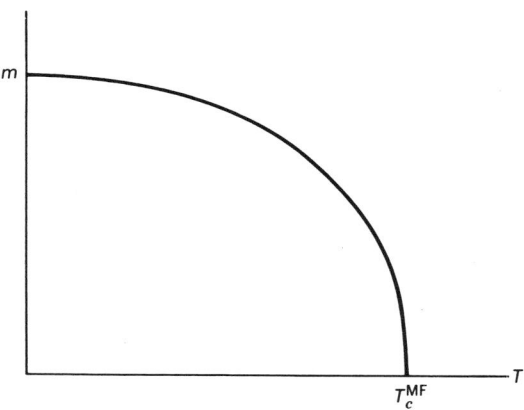

Fig. 8.1.1 Magnetization as a function of temperature for the Ising model on a square lattice in the mean field approximation.

8.1 THE MEAN FIELD THEORY FOR THE ISING MODEL

TABLE 8.1.1 Comparison of Mean Field and Exact Critical Temperatures for the Ising Model

Lattice	d	q	T_m/T_c
Honeycomb	2	3	1.976
Square	2	4	1.763
Triangular	2	6	1.648
Diamond	3	4	1.479
Simple cubic	3	6	1.330
bcc	3	8	1.260
fcc	3	12	1.225

rises continuously from 0, usually with infinite slope, signaling the spontaneous breaking of the symmetry of the disordered phase. In the case of the Ising model, this is the symmetry (with $h_0 = 0$) $S_k \to -S_k$, which is explicitly broken in the MF Hamiltonian (8.1.5) so long as $h \neq 0$.

Very close to T_c the mean field equation (8.1.8) can be expanded for small m and $h_0 = 0$ as

$$m = \beta qJm - \tfrac{1}{3}(\beta qJ)^3 m^3 \qquad (8.1.9)$$

This equation has the trivial solution $m = 0$, and a second real solution

$$m = \sqrt{3}\,(\beta Jq - 1)^{1/2} \qquad (8.1.10)$$

for $T < T_c$, where

$$T_c^{\mathrm{MF}} = qJ \qquad (8.1.11)$$

is the MF transition temperature. In Table 8.1.1 the MF transition temperature is compared to the exact transition temperature for two- and three-dimensional lattices. Note that for each dimension the MF result improves as the number of neighbors increases, and that the results in three dimensions are much better than in two dimensions for lattices with the same coordination number. The fact that the MF result for the critical temperature depends only on the coordination number and not on the dimensionality of the lattice points to the inadequacy of the way in which fluctuations are treated in the MF approximation. The MF theory even predicts a phase transition in one dimension, which we know is incorrect.

By (8.1.10) and setting $Jq = T_c$, it is easy to see that the spontaneous magnetization vanishes as

$$m \sim \left(\frac{T_c - T}{T_c}\right)^{1/2} \qquad (8.1.12)$$

from which we conclude that $\beta = 1/2$.

To find the susceptibility exponent γ, we simply differentiate (8.1.8) with respect to h_0,

$$\chi = \left.\frac{\partial m}{\partial h_0}\right|_{h_0=0} = \frac{\beta(1 - \tanh^2 \beta Jqm)}{1 - \beta Jq(1 - \tanh^2 \beta Jqm)} \qquad (8.1.13)$$

For T close to T_c, we find

$$\chi \sim \left|\frac{T - T_c}{T_c}\right|^{-1} \qquad (8.1.14)$$

and therefore $\gamma = 1$.

To see how the magnetization varies along the critical isotherm, $T = T_c$, expanding (8.1.13) for small h_0 gives

$$h_0 \cong (m + h_0)^3 \qquad (8.1.15)$$

which we may write in the form ($h_0 \ll h_0^{1/3}$)

$$m \sim h_0^{1/3} \qquad T = T_c \qquad (8.1.16)$$

and so the critical isotherm exponent $\delta = 3$.

The MF approximation is fraught with difficulties, as is apparent if we look at the average energy. By (8.1.5), because $h = 0$ for $T > T_c$, the MF average energy vanishes above T_c, which is clearly absurd. The source of this difficulty is that each spin is treated as independent in the MF approximation. There is no correlation at all between even nearest-neighbor spins, so that

$$E = -J \sum_{\langle ij \rangle} \langle S_i S_j \rangle = -J \langle S_i \rangle \langle S_j \rangle \qquad (8.1.17)$$

vanishes. On the other hand, another problem lurks at low temperatures. As T approaches 0, m approaches unity and the energy per spin becomes by (8.1.5)

$$E = -qJ \qquad T \to 0 \qquad (8.1.18)$$

This is twice the correct value, and the reason is that we have summed each term in the Hamiltonian twice. The original Hamiltonian can be written in terms of the local fields as

$$H = -\frac{1}{2} \sum_k h_k S_k - h_0 \sum_k S_k \qquad (8.1.19)$$

The factor of $1/2$ is necessary in order to avoid overcounting each pair of neighboring spins. However, if we now replace h_k by its average value, we do not obtain the correct mean field theory.

As one might expect, we also run into trouble when we try and calculate the specific heat near the critical temperature. By (8.1.17) the specific heat vanishes above T_c, whereas close to, but below, T_c we have (Prob. 8.1)

$$C = \left(\frac{T_c - T}{T_c}\right)^{-1/2} \qquad (8.1.20)$$

Therefore, for $T < T_c$, MF theory gives $\alpha = 1/2$, which is inconsistent with the scaling law (4.6.1d)

$$\alpha + 2\beta + \gamma = 2 \qquad (8.1.21)$$

Taking the MF values for $\beta = 1/2$ and $\gamma = 1$, we should find $\alpha = 0$, that is, a logarithmic divergence in the specific heat. The fact that MF theory predicts critical exponents independent of the dimension of the system is also an indication that something is wrong.

8.2 MEAN FIELD THEORY AS A VARIATIONAL THEORY

Some of these problems can be resolved if we take a slightly different point of view. Let us assume that the fluctuations of the spins about their mean value m are small, and write

$$S_i = m + (S_i - m) \qquad (8.2.1)$$

where $S_i - m$ is the fluctuation of S_i about its mean value. The approximation we are going to make is that $m \gg (S_i - m)$. Inserting this into the Hamiltonian (8.1.1) with $h_0 = 0$ and dropping for a moment the term quadratic in the fluctuations, we have

$$H = \frac{JNq}{2}m^2 - Jqm\sum_i S_i \qquad (8.2.2)$$

where N is the number of spins. Taking the trace over the spins gives for the free energy per spin

$$F = \frac{Jq}{2}m^2 - \frac{1}{\beta}\ln\cosh\beta Jqm \qquad (8.2.3)$$

The value of m is now determined by requiring that the free energy be a minimum with respect to m,

$$\frac{\partial F}{\partial m} = 0 \to m - \tanh\beta Jqm = 0 \qquad (8.2.4)$$

which leads us back to the mean field equation (8.1.8). The average energy is given by

$$E = \frac{\partial}{\partial\beta}(\beta F) \qquad (8.2.5)$$

which still vanishes above T_c, but the problem at $T = 0$ has been fixed, and we recover the proper value, $E_0 = -qJ/2$.

In order to get a more realistic picture of what is going on above T_c, we must retain the term quadratic in the fluctuations in the spins. After all, above T_c there is no mean field and all we have is "fluctuations." Including the terms quadratic in the fluctuations, the partition function is

$$Z = \exp[-\beta JNqm^2/2 + N\ln\cosh\beta Jqm]$$
$$\times \left\langle \exp\beta J\sum_{\langle ij\rangle}(S_i - m)(S_j - m)\right\rangle_{MF} \qquad (8.2.6)$$

where the average in (8.2.6) is taken over the mean field Hamiltonian (8.2.2).

$$\left\langle \exp \beta J \sum_{\langle ij \rangle} (S_i - m)(S_j - m) \right\rangle_{MF}$$
$$= \frac{\operatorname{Tr} \exp \beta Jqm\sum_i S_i \exp \beta J \sum_{\langle ij \rangle}(S_i - m)(S_j - m)}{\operatorname{Tr} \exp Jqm\sum_i S_i} \quad (8.2.7)$$

We have run across averages of this form before, and as before we can apply the cumulant expansion to evaluate (8.2.7). The first cumulant vanishes because $\langle S_i \rangle_{MF} = m$, and the second cumulant is

$$C_2 = \sum_{\langle kl \rangle} \sum_{\langle ij \rangle} \langle (S_i - m)(S_j - m)(S_k - m)(S_l - m) \rangle_{MF} \quad (8.2.8)$$

Because the spins are uncorrelated in the MF Hamiltonian (8.2.7), the terms in the second cumulant vanish unless $\langle ij \rangle = \langle kl \rangle$. In either case we find

$$C_2 = \frac{NqJ^2}{2}(1 - m^2)^2 \quad (8.2.9)$$

and so the partition function within this approximation is

$$Z = \exp\left[-\beta JNqm^2/2 + N\ln\cosh \beta Jm + \beta^2 J^2 Nq/4(1 - m^2)^2\right] \quad (8.2.10)$$

Above T_c the new term is finite and leads to an average energy per spin of

$$\frac{E}{N} = \frac{-\beta J^2 q}{2} \quad T > T_c \quad (8.2.11)$$

The specific heat still exhibits a discontinuity independent of the dimension of the system, but we have a somewhat improved description of the high-temperature phase.

In order to make more progress, we need to go beyond the simple MF theory developed here. There are two directions in which one can seek to improve the theory. The first is to generalize the MF theory and treat exactly not a single spin interacting with the mean field, but rather a finite cluster of spins. The cluster approximation is discussed in the next section. The second approach is to recast the partition function in terms of continuous variables and expand the "thermodynamic action" in powers of the fluctuations about the mean field. This will be discussed in Sect. 8.4.

8.3 CLUSTER APPROXIMATIONS

In the simple mean field approximation, each spin is treated as statistically independent of all the other spins. Clearly, a better approximation would be to take a cluster of spins and treat the interactions among the spins within the cluster exactly, and the interaction of the spins on the boundary of the cluster with spins outside of the cluster in the mean field approximation. This generalization of the mean field approximation is called the cluster approximation. In simple MF theory the cluster consists of a single spin.

To illustrate the method, let us again consider the Ising model on a triangular lattice and take the three-spin cluster shown in Fig. 8.3.1. The effective cluster Hamiltonian is

$$H_c = -J(S_1S_2 + S_1S_3 + S_2S_3) - 4Jm(S_1 + S_2 + S_3) \quad (8.3.1)$$

where, as before, $m = \langle S \rangle$ is independent of the site.

The partition function of the cluster is

$$Z = \text{Tr}\, e^{-\beta H_c} = 2e^{3\beta J}\cosh 12\beta Jm + 6e^{-\beta J}\cosh 4\beta Jm \quad (8.3.2)$$

and, just as in MF theory, we must self-consistently calculate the

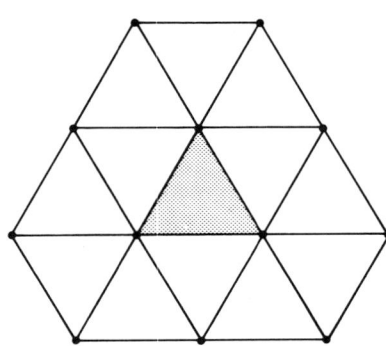

Fig. 8.3.1 Three-spin cluster on the triangular lattice.

magnetization that appears in (8.3.1). We find

$$m = \langle S \rangle = \frac{e^{3\beta J}\sinh 12\beta Jm + e^{-\beta J}\sinh 4\beta Jm}{2e^{3\beta J}\cosh 12\beta Jm + 6e^{-\beta J}\cosh 4\beta Jm} \quad (8.3.3)$$

To locate the critical temperature, we take m small and expand (8.3.3), which yields

$$1 = \frac{2\beta J(3e^{3\beta J} + e^{-\beta J})}{(e^{3\beta J} + 3e^{-\beta J})} \quad (8.3.4)$$

Solving this equation numerically, we find

$$\beta_c J = 0.24235 \quad (8.3.5)$$

The value for the critical temperature is much closer to the exact value of 0.27 than the MF value of 0.17. It is important to note that although this approximation does quite well in giving the critical temperature, it does no better in predicting the critical exponents than ordinary MF theory, and in fact gives exactly the same value for β. To see this, we only need to note that (8.3.3) is an analytic function of m for small m. Expanding to third order will always yield

$$m^2 \sim (\beta J_c - 1) \quad (8.3.6)$$

and so, however large a cluster we choose, the cluster approximation will always give $\beta = 1/2$. By the same token, we will always find $\gamma = 1.0$.

The reason for this somewhat disappointing result lies at the root of critical phenomena. The behavior at the critical point is dominated by very long wavelength fluctuations. No matter how large a cluster we choose, it can only accommodate fluctuations of a finite size and always yields an analytic form for the partition function.

Although the cluster method fails close to the transition temperature, it can give very accurate results for the phase diagram of a system over a large range of temperatures.

8.4 THE HUBBARD–STRATONOVICH TRANSFORMATION

We now develop the second approach to generalizing the MF theory by recasting the partition functions in terms of continuous variables and expanding the "thermodynamic action" in powers of the fluctuations about the mean field. We proceed by expressing the partition function of the Ising model as an integral over a set of continuous fields. The resulting "field theory" bears little resemblance to the original Ising model, and the fact that it must reproduce the critical behavior of the Ising model perhaps lends some weight to the universality hypothesis.

To begin, Let A be a real, $N \times N$ symmetric matrix with positive eigenvalues, λ a vector, and consider the following Gaussian integral:

$$G(\lambda) = \prod_k \int \frac{d\phi_k}{\sqrt{2\pi}} \exp\left(-\frac{1}{2}\sum_{ij} \phi_i A_{ij} \phi_j + \sum_i \lambda_i \phi_i\right) \quad (8.4.1)$$

The integral can be calculated by "completing the square"

$$\frac{1}{2}\sum_{ij} \phi_i A_{ij} \phi_j - \sum_i \lambda_i \phi_i = \sum_{ij} \left(\phi_i - \overline{\phi}_i\right) A_{ij} \left(\phi_j - \overline{\phi}_j\right) - \frac{1}{2}\sum_{ij} \overline{\phi}_j A_{ij} \overline{\phi}_j \quad (8.4.2)$$

where

$$\overline{\phi}_i = \sum A_{ij}^{-1} \lambda_j \quad (8.4.3)$$

Using the formula (see Prob. 8.2)

$$\prod_i \int \frac{d\phi_i}{\sqrt{2\pi}} \exp\left(-\frac{1}{2}\sum_{ij} \phi_i A_{ij} \phi_j\right) = \exp\left(-\frac{1}{2}\mathrm{Tr}\ln A\right) \quad (8.4.4)$$

we finally have

$$\prod_i \int \frac{d\phi_i}{\sqrt{2\pi}} \exp\left(-\frac{1}{2}\sum_{ij} \phi_i A_{ij} \phi_j + \sum_i \lambda_i \phi_i\right)$$

$$= \exp\left(-\frac{1}{2}\mathrm{Tr}\ln A\right) \exp\left(\frac{1}{2}\sum_{ij} \lambda_i A_{ij}^{-1} \lambda_j\right) \quad (8.4.5)$$

Let us now consider the partition function for the Ising model with $h_0 = 0$

$$Z = \mathrm{Tr} \exp\left(\frac{\beta}{2}\sum_{ij} S_i J_{ij} S_j\right) \quad (8.4.6)$$

8.4 THE HUBBARD–STRATONOVICH TRANSFORMATION

where the sum is over all i and j and the factor of $1/2$ is necessary to avoid double-counting. Note that we have generalized the Ising model in (8.4.6) to include interactions between all pairs of spins. If the system has translational symmetry, $J_{ij} = J_{i-j}$ only. By the results obtained previously, the exponential in (8.4.6) can be represented as a Gaussian integral over a set of continuous fields ϕ_i,

$$Z = \frac{\mathrm{Tr}\,\prod_i \int \frac{d\phi}{\sqrt{2\pi}} \exp\left(-\frac{1}{2}\sum_{ij}\phi_i(\beta J)_{ij}^{-1}\phi_j + \sum_i \phi_i S_i\right)}{\prod_i \int \frac{d\phi_i}{\sqrt{2\pi}} \exp\left(-\frac{1}{2}\sum_{ij}\phi_i(\beta J)_{ij}^{-1}\phi_j\right)} \quad (8.4.7)$$

In order to simplify the notation, we define the integration symbol

$$\int D[\phi] = \exp(-\tfrac{1}{2}\mathrm{Tr}\ln\beta J)\prod_i \int d\phi_i \quad (8.4.8)$$

The trace over the spin variables can now be performed with the result

$$Z = \int D[\phi]\exp\left(\frac{1}{2}\sum_{ij}\phi_i(\beta J)_{ij}^{-1}\phi_j + \sum_i \ln\cosh\phi_i\right) \quad (8.4.9)$$

In (8.4.9) no trace remains of the original spin variables. The Ising model has been replaced by an equivalent "field theory." The transformation from the spin variables to continuous real fields is called the Hubbard–Stratonovich or Gaussian transformation.

The auxiliary fields ϕ_i introduced in the Hubbard–Stratonovich transformation are in some sense conjugate to the original spin variables, and in fact they very closely resemble the local field introduced in Sect. 8.1 [see (8.1.2)]. This connection can be made even more precise by calculating the magnetization in the new representation.

Differentiating with respect to ϕ_i inside the integral in (8.4.7), we have

$$\langle S_i \rangle = \frac{1}{Z}\int D[\phi]\exp\left(-\frac{1}{2}\sum_{ij}\phi_i(\beta J)_{ij}^{-1}\phi\right)\frac{\partial}{\partial \phi_j}\exp\left(\sum_k \phi_k S_k\right) \quad (8.4.10)$$

250 MEAN FIELD THEORY AND THE GAUSSIAN FIXED POINT

Now performing the trace over spins, we find that

$$\langle S_i \rangle = \frac{1}{Z} \int D[\phi] \exp\left(-\frac{1}{2}\sum_{ij} \phi_i (\beta J)_{ij}^{-1} \phi_j + \sum_i \ln \cosh \phi_i\right) \tanh \phi_i \quad (8.4.11)$$

We see in (8.4.11) that averages of functions of the original spin variables can be transcribed into averages of new functions of the auxiliary fields ϕ. The effective Hamiltonian for these fields is just the function appearing in the exponent of (8.4.9). Apart from a factor of β, the auxiliary fields are just the local fields at each site. Whereas in the MF approximation the magnetization is simply $\tanh \beta h$, here it is the ensemble average of $\tanh \phi$. As one would expect, the exact result requires that the local fields be allowed to fluctuate.

Now consider the "steepest-descents" approximation to the integral (8.4.9). In the method of steepest descents, one looks for the value of the fields for which the exponent in (8.4.9) is stationary. This most probable field configuration is therefore the solution of the equation

$$\sum_j (\beta J)_{ij}^{-1} \phi_j^0 - \tanh \phi_i^0 = 0 \quad (8.4.12)$$

Assuming that the solution we seek is translationally invariant, we recover the MF equation (8.1.7)

$$\phi^0 = \beta J q \tanh \phi^0 \quad (8.4.13)$$

and again the identification of ϕ with the local field is apparent.

Having found the most probable field, we can expand ϕ_i about ϕ^0 and keep only quadratic (or Gaussian) terms in the exponent. The linear terms in the fluctuations ϕ_i' vanish by virtue of (8.4.12).

Within the Gaussian approximation the partition function for the Ising model now has the form

$$Z = Z_0 \int D[\phi'] \exp\left(-\tfrac{1}{2} \sum_{ij} \phi_i' \left[(\beta J)_{ij}^{-1} - \delta_{ij}(1 - \tanh^2 \phi_i^0)\right] \phi_j'\right) \quad (8.4.14)$$

8.4 THE HUBBARD–STRATONOVICH TRANSFORMATION

where Z_0 is the zeroth-order approximation to the partition function

$$Z_0 = \exp\left(-\tfrac{1}{2}\sum_i \phi_i^0 \tanh \phi_i^0 + \sum_i \ln \cosh \phi_i^0\right) \quad (8.4.15)$$

In order to evaluate the contribution from the Gaussian terms, we take advantage of the translational symmetry of the problem (again assuming ϕ_i^0 is independent of i) and Fourier transform the fields ϕ_i', defining

$$\phi_i' = \frac{1}{\sqrt{N}} \sum_q e^{iqr_i} \phi_q' \quad (8.4.16)$$

Regarding J_{ij} and its inverse as matrices, we have by definition

$$\sum_j J_{ij}(J^{-1})_{jk} = \delta_{ik} \quad (8.4.17)$$

from which it follows by translational symmetry that

$$J^{-1}(q)J(q) = 1 \quad (8.4.18)$$

The partition function is now given by

$$Z = Z_0 \int D[\phi_q'] \exp\left(-\tfrac{1}{2}\sum_q \phi_q'\left[(\beta J(q))^{-1} - (1 - \tanh^2 \phi^0)\right]\phi_{-q}'\right) \quad (8.4.19)$$

Before we proceed, consider the operator appearing in the Gaussian form in (8.4.19)

$$G^{-1}(q) = (\beta J(q))^{-1} - (1 - \tanh^2 \phi_0) \quad (8.4.20)$$

This operator is the inverse of the ϕ–ϕ correlation function

$$G(q) = \langle \phi_q \phi_{-q} \rangle \quad (8.4.21)$$

The average in (8.4.21) is taken with respect to the Gaussian statistical weight in (8.4.19). To see this, we introduce into the integral (8.4.19) a source term and transform the partition function into a generating function for correlations

$$Z(\eta) = \int D[\phi] \exp\left[-\tfrac{1}{2}\sum_q \phi_q G^{-1}(q)\phi_{-q} + \sum \eta_{-q}\phi_q\right] \quad (8.4.22)$$

By differentiating the generating function with respect to the fields η_q and η_{-q}, we have

$$\left.\frac{\partial^2 Z}{\partial \eta_q \partial \eta_{-q}}\right|_{\eta=0} = \int D[\phi]\exp\left[-\tfrac{1}{2}\sum_q \phi_q G^{-1}(q)\phi_{-q}\right]\phi_q\phi_{-q} \quad (8.4.23)$$

On the other hand, the integral in (8.4.23) can be performed by completing the square, with the result

$$Z(\eta) = \exp\left[\tfrac{1}{2}\sum_q \eta_q G(q)\eta_{-q}\right] \quad (8.4.24)$$

Differentiating the preceding expression twice and evaluating it at $\eta = 0$, we arrive at (8.4.21). The Fourier transform of J_{ij} for nearest-neighbor couplings on the cubic lattice is given by

$$J(q) = \sum_{r_{ij}} e^{-iqr_{ij}} J_{ij} = 2J(\cos q_x a + \cos q_y a + \cos q_z a) \quad (8.4.25)$$

so the Fourier transform of the correlation function $G(q)$ is

$$G^{-1}(q) = [2\beta J(\cos q_x a + \cos q_y a + \cos q_z a)]^{-1} - (1 - \tanh^2 \phi^0) \quad (8.4.26)$$

For small q the cosines can be expanded up to order q^2, and we have

$$G(q) = \frac{36\beta J/a^2}{q^2 + r} \quad (8.4.27)$$

where r is given by

$$r = \frac{6}{a^2}[1 - 6\beta J(1 - \tanh^2 \phi^0)] \quad (8.4.28)$$

In order to interpret this result, let us approximate the sum over wave vectors by an integral and perform the inverse Fourier transform for

8.4 THE HUBBARD–STRATONOVICH TRANSFORMATION

$G(\mathbf{x})$, where

$$G(\mathbf{x}) = \frac{36\beta J}{a^2} \int \frac{d^3q}{(2\pi)^3} e^{i\mathbf{q}\cdot\mathbf{x}} \frac{1}{q^2 + r} \tag{8.4.29}$$

The preceding integral is best evaluated in spherical coordinates by taking the z axis along $\mathbf{x}$. After performing the ϕ integration we have

$$G(x) = \frac{36q\beta J}{4\pi^2 a^2} \int_0^\infty q^2 \int_0^\pi \sin\theta \, e^{iqx\cos\theta} \frac{1}{q^2 + r} \, d\theta \, dq \tag{8.4.30}$$

Making the change of variable $\mu = \cos\theta$ and performing the μ integration gives

$$G(x) = \frac{36\beta J}{4\pi^2 a^2} \int_0^\infty dq \frac{q}{ix} \frac{(e^{iqx} - e^{-iqx})}{q^2 + r} \tag{8.4.31}$$

In the second term let $q \to -q$, which allows us to write

$$G(x) = -\frac{9\beta J}{2\pi^2 a_x^2} \int_{-\infty}^\infty dq \, e^{iqx} \left(\frac{1}{q - i\sqrt{r}} + \frac{1}{q + i\sqrt{r}} \right) \tag{8.4.32}$$

The contour for performing this integral in the complex q plane is shown in Fig. 8.4.1. Because $x > 0$ we must close the contour in the upper half-plane, Im $q > 0$. The residue at the simple pole $q = i\sqrt{r}$

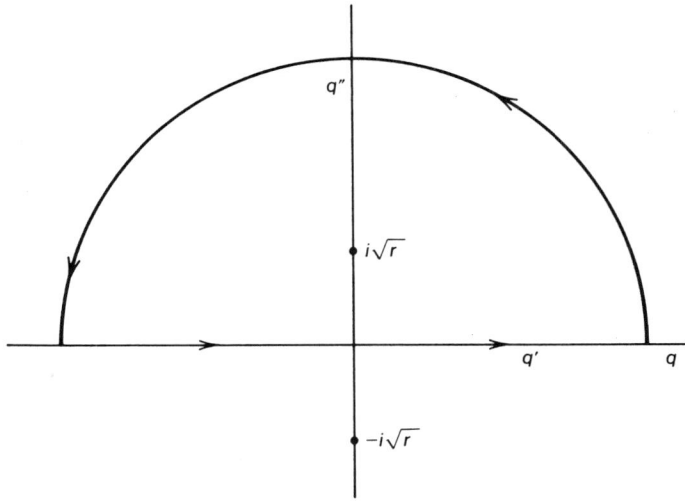

Fig. 8.4.1 Contour in the complex q plane for evaluating the integral (8.4.31).

leads to the result

$$G(x) = \frac{9\beta J}{\pi a^2} e^{-\sqrt{r}x} \tag{8.4.33}$$

If we return to (5.0.10) for the definition of the correlation length, we see that

$$\xi^{-1} = \sqrt{r} \tag{8.4.34}$$

By (8.4.28) it is clear that r tends to 0 linearly as T approaches $T_c = 6J$, and therefore ξ diverges as $T \to T_c$ as

$$\xi \sim \frac{1}{|T - T_c|^{1/2}} \tag{8.4.35}$$

so that the correlation exponent ν is $1/2$. Returning to the expression for the partition function, (8.4.22), if we set all the sources $\eta = 0$, we have

$$Z = \int D[q] \exp\left(\tfrac{1}{2} \sum_q \phi_q G^{-1}(q) \phi_{-q}\right)$$

$$= \exp\left(\tfrac{1}{2} \sum_q \ln G(q)\right) \tag{8.4.36}$$

This gives for the contribution to the free-energy density from the fluctuations (again approximating the sum by an integral)

$$F = -\frac{1}{2} \int \frac{d^d q}{(2\pi)^d} \ln G(q) \tag{8.4.37}$$

[Note in (8.4.37) we have generalized to d dimensions.]

The singular part of the free-energy density comes from the small-q part of the integral (8.4.37). For example, let us calculate the specific heat in the limit $r \to 0$. Recalling that r tends to 0 linearly at T_c, r plays the role of the temperature parameter, so

$$C = -\frac{\partial^2 F}{\partial r^2} \sim \int_0^\Lambda d^d q \frac{1}{(q^2 + r)^2} \tag{8.4.38}$$

where Λ is a cutoff at large momenta needed to make the theory finite. Defining the dimensionless variable $x = q/\sqrt{r}$,

$$C \sim r^{(d-4)/2} \int_0^{\Lambda/\sqrt{r}} dx\, x^{d-1} \frac{1}{(x^2 + 1)^2} \tag{8.4.39}$$

For $d < 4$ the integral in (8.4.39) converges as the upper limit of integration tends to ∞, and we may therefore set the cutoff Λ to ∞. We then have

$$C \sim r^{(d-4)/2} \sim \left|\frac{T - T_c}{T_c}\right|^{d/2 - 2} \tag{8.4.40}$$

From this we see that the specific heat exponent α is

$$\alpha = 2 - \frac{d}{2} \tag{8.4.41}$$

The mean field or "classical exponents" are compared with the exact values for $d = 2$ and the best estimates for $d = 3$ in Table 8.4.1.

8.5 LANDAU–GINZBURG FORM FOR THE FREE-ENERGY FUNCTIONAL AND SPONTANEOUS SYMMETRY BREAKING

The Hubbard–Stratonovich form for the partition function of the Ising model, (8.4.14), does not resemble the initial Ising model very much except in two essential points. First, the symmetry group of the theory ($S \to -S$ or $\phi \to -\phi$) is the same in both, and the dimension of space is the same. We can take one more step toward a generic theory by focusing on the form of the free-energy functional for slowly varying fields. This approach is consistent with our observation that at the critical point it is the large-scale fluctuations of the fields that

TABLE 8.4.1 Comparison of the Critical Exponents for the Ising Model in Two and Three Dimensions with Mean Field Theory

α	β	γ	δ	η	ν	Exponent
$2 - (d/2)$	$\frac{1}{2}$	1	3	0	$\frac{1}{2}$	Mean field
0	$\frac{1}{8}$	$\frac{7}{4}$	15	$\frac{1}{4}$	1	$d = 2$ (exact)
0.113	0.324	1.238	4.82	0.031(5)	0.629(4)	$d = 3$ (MCRG) [2]

determine critical behavior, and the universality hypothesis that asserts that for a given type of order parameter there ought to be a unique critical theory.

We begin with the first term in the free-energy functional on a d-dimensional cubic lattice

$$\frac{1}{2}\sum_{ij}\phi_i(\beta J)_{ij}^{-1}\phi_j = \frac{1}{2}\sum_q \phi_q(\beta J(q))^{-1}\phi_{-q} \tag{8.5.1}$$

where we have introduced the Fourier transform of the fields, (8.4.16), and the generalization of (8.4.25) to d dimensions

$$J(q) = 2J\sum_{i=1}^{d}\cos q_i a \tag{8.5.2}$$

where a is the lattice constant.

For small q we can expand (8.5.2) to $O(q^2)$,

$$J(q) \simeq 2dJ - Ja^2 q^2 + O(q^4) \tag{8.5.3}$$

Inserting this in (8.5.1) gives

$$\frac{1}{2}\sum_q \phi_q[\beta J(q)]^{-1}\phi_{-q} \simeq \frac{1}{2}\sum_q \left(\frac{1}{2\beta\, dJ} + \frac{a^2}{4\beta\, d^2 J}q^2\right)\phi_q\phi_{-q} \tag{8.5.4}$$

We can now go full circle and return to real space by defining a field which is a continuous function of the coordinates

$$\phi(x) = \frac{1}{Na^d}\sum_q e^{i\mathbf{q}\cdot\mathbf{x}}\phi_q \tag{8.5.5}$$

which gives

$$\frac{1}{2}\sum_q \left(\frac{1}{2\, d\beta J} + \frac{a^2}{4\, d\beta^2 J}q^2\right)\phi_q\phi_{-q}$$

$$\to \frac{1}{2}\int d^d x \left[\frac{1}{2\, d\beta J}\phi^2(x) + \frac{a^2}{4\, d\beta^2 J}(\nabla\phi)^2\right] \tag{8.5.6}$$

Next we turn our attention to the second, or potential, term in the energy functional

$$\sum_i \ln\cosh\phi_i \to \int d_x^d \left[\tfrac{1}{2}\phi^2(x) - \tfrac{1}{12}\phi^4(x) + \cdots\right] \tag{8.5.7}$$

where we have expanded the potential in powers of ϕ^2 and replaced the sum over lattice points by an integral of the continuous field $\phi(x)$ over the volume of the system.

In order to cast the free energy in what is called the "standard" Landau–Ginzburg form, let us rescale the fields by

$$\frac{a^2}{4d^2\beta J}\phi^2 \to \phi^2. \qquad (8.5.8)$$

Keeping terms up to ϕ^4, the free energy now has the form

$$F[\phi] = \int d^dx \left[\frac{1}{2}(\nabla\phi)^2 + \frac{r}{2}\phi^2 + \frac{\lambda}{4!}\phi^4\right] \qquad (8.5.9)$$

where we have defined

$$r = \frac{4}{a^2}(1 - 2\, d\beta\, J) \qquad (8.5.10a)$$

and

$$\lambda = 2\left(\frac{8\, d\beta\, J}{a^2}\right)^2 \qquad (8.5.10b)$$

Equation (8.5.9) is called the Landau–Ginzburg form for the free-energy functional, and it has the general form of a term proportional to the square of the gradient of the field plus a polynomial in the fields, which we refer to as the potential $V(\phi)$. The significance of the Landau–Ginzburg model extends beyond its derivation from the Ising model through the Hubbard–Stratonovich transformation. In fact, one can argue from rather general considerations that the Landau–Ginzburg free energy correctly describes the long-wavelength character of all systems. This was the original spirit in which it was proposed, and is in a sense the mathematical embodiment of the idea that near their critical points different systems are characterized by certain universal scaling properties, which depend only on the internal symmetry group of the order parameter.

Within this philosophy any nontrivial model with a given type of order parameter and dimension of space will fall into the same universality class, and therefore the corresponding Landau–Ginzburg form of the free energy is very often taken as the starting point for RG calculations.

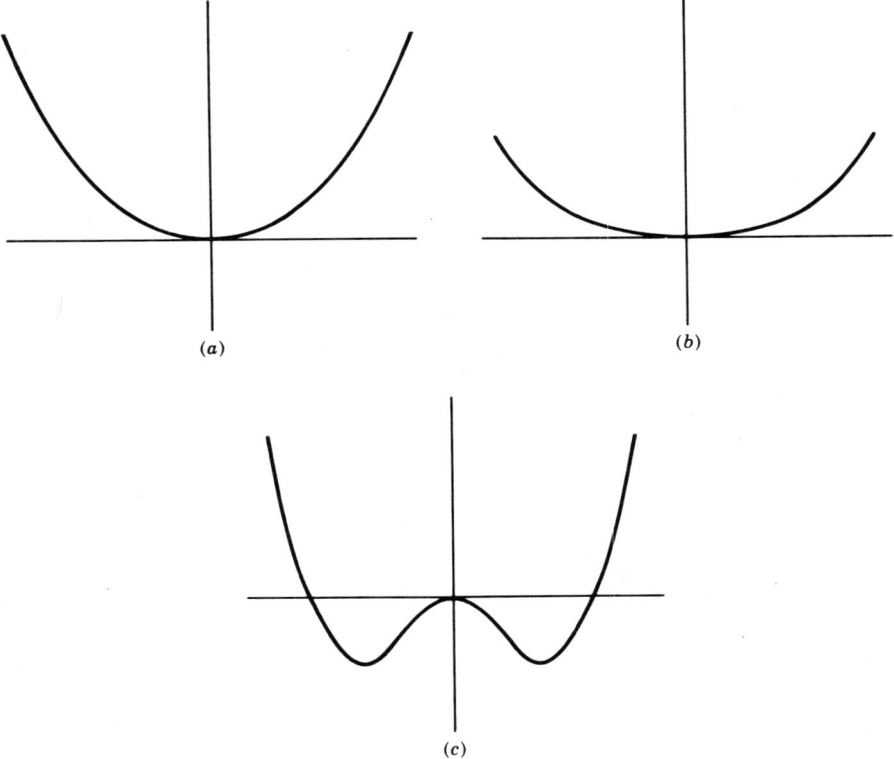

Fig. 8.5.1 The potential $V(\phi)$ for (a) $T > T_0$, (b) $T = T_0$, and (c) $T < T_0$.

The Landau–Ginzburg free-energy functional offers a very simple way of thinking about spontaneous symmetry breaking and critical behavior. Figure 8.5.1 shows the form of the potential as a function of temperature near $T_0 = 2dJ$. Here we wish to distinguish between the mean field critical temperature T_0 and the true critical temperature T_c.

For $T > T_0$, r is positive and $V(\phi)$ has a unique minimum at $\phi = 0$. As $T \to T_0$, r tends to 0 and the potential near $\phi = 0$ flattens out, leading to very strong fluctuations. For $T < T_0$, r becomes negative and minima appear at $\pm \phi_0$. The fact that there are minima at $\pm \phi_0$ reflects the underlying symmetry of the theory.

The states that minimize the potential have lower symmetry than the state $\phi = 0$, but so long as the fluctuations are strong enough the system will visit both minima equally, and the average of ϕ will be 0. As the temperature is lowered below T_0 to T_c, the critical temperature, the system will settle down into one or the other of the broken symmetry states. Thermal fluctuations are no longer strong enough to overcome the free-energy barrier and ϕ assumes a nonzero average.

8.6 THE GAUSSIAN MODEL

We have seen that the long-wavelength properties of the Ising model can be cast in the form of a "field theory" where in this case the field is a real scalar. The representation of the partition function in terms of such a field is generally referred to as the Landau–Ginzburg form for the partition function. If it is true that the critical properties for every system with an order parameter of a particular type and symmetry can be represented by a model of the Landau–Ginzburg form, then we have come a long way toward understanding the universality of critical behavior. The problem of critical phenomena now is reduced to determining the nature of the order parameter in a given transition, writing down the Landau–Ginzburg form for the partition function and then calculating the RG equations. This program will occupy our attention for the next few chapters.

We are going to leave behind the discrete sum over momenta, which is a vestige of the lattice on which the original system lived, and consider the wave vector q to be continuous. The partition function for the Gaussian model is given by the functional integral

$$Z = \int D[\phi] e^{-L[\phi]} \tag{8.6.1}$$

where L is the Landau–Ginzburg free-energy functional

$$L = \int d^d x \left\{ \tfrac{1}{2} |\nabla \phi|^2 + r \phi^2(x) \right\} \tag{8.6.2}$$

In many places, and in particular the text by Amit [1], the free-energy functional is referred to as the "Lagrangian." This notation has its roots in field theory, and we are now at a point where the fields of statistical mechanics and field theory have significant overlap, both formally and conceptually. For the most part we are interested in the RG from the point of view of statistical mechanics, and we will choose our notation accordingly. Because there is no likelihood of confusion, we will refer to L simply as the "free energy."

The simple free-energy functional, (8.6.2), contains only one adjustable parameter r, which we assume vanishes linearly with temperature at the critical point as in (8.5.10a). We will often take advantage of the translational symmetry of the problem and replace the real-space field $\phi(x)$ by its Fourier transform ϕ_q,

$$\phi(x) = \frac{1}{\sqrt{V}} \sum_q e^{iqx} \phi_q \tag{8.6.3}$$

The free energy in terms of the $\phi(q)$ is

$$L = \tfrac{1}{2} \sum_{q=0}^{\Lambda} (q^2 + r) \phi_q \phi_{-q} \tag{8.6.4}$$

A cutoff Λ is imposed on the momentum variables in order to make the theory sensible. Without a cutoff, many of the expressions we will derive would diverge. More physically, in all real problems there is a smallest physically relevant length, and Λ should be thought of as the order of the reciprocal of this length.

We will constantly be called upon to evaluate correlation functions of the Gaussian free energy, and there is a simple way of calculating them. We first introduce sources η_q into the free energy, and so we have

$$L = \tfrac{1}{2} \sum_q^{\Lambda} (q^2 + r) \phi_q \phi_{-q} + \sum_q^{\Lambda} \eta_q \phi_{-q} \tag{8.6.5}$$

The corresponding partition function $Z[\eta]$ is a functional of the sources η_q and is sometimes called the generating functional. We have

$$Z[\eta] = \int D[\phi] e^{-L[\phi;\eta]} \tag{8.6.6}$$

Any average of the fields can now be found by (functionally) differentiating with respect to the appropriate sources and evaluating for all $\eta = 0$. For example,

$$\langle \phi_{-q} \phi_{-q'} \rangle = \frac{1}{Z} \frac{\partial^2 Z}{\partial \eta_q \partial \eta_{q'}} \bigg|_{\eta=0} \tag{8.6.7}$$

We can evaluate (8.5.6) explicitly by "completing the square."

$$Z[\eta] = Z_0 \exp\left[-\frac{1}{2} \sum_q^{\Lambda} \frac{\eta_q \eta_{-q}}{q^2 + r} \right] \tag{8.6.8}$$

From this it is clear that

$$\langle \phi_q \phi_{q'} \rangle = \frac{1}{q^2 + r} \delta(q + q') \quad (8.6.9)$$

where $\delta(q + q')$ is the Kronecker delta function.

The meaning of the parameter r in the free energy can be made clear by considering the susceptibility

$$\chi = \int d^d r\, G(r)$$
$$= G(q = 0)$$
$$= r^{-1} \quad (8.6.10)$$

For the Gaussian model, r is simply the inverse of the susceptibility, and the divergence of the susceptibility at $r = 0$ shows that r is a measure of how far the system is from its critical point. For the Gaussian model to be well defined, we must have $r > 0$.

8.7 RENORMALIZATION GROUP ANALYSIS OF THE GAUSSIAN MODEL

In this simple problem we could just go ahead and do the indicated functional integral as in (8.4.36), but instead let us look at the partition function from the point of view of RG. First, divide the field variables into two groups:

$$\phi_q^{(l)}: 0 \leq q \leq \Lambda/b$$
$$\phi_q^{(s)}: \Lambda/b < q \leq \Lambda \quad (8.7.1)$$

as illustrated in Fig. 8.7.1. The parameter b is, as usual, the scale factor for the RG transformation.

Near the critical point it is the very long wavelength variables that determine the critical properties of the system. The short-wavelength variables $\phi^{(s)}$, which carry information about the microscopic structure of the system, are irrelevant. Therefore, we "integrate out" these

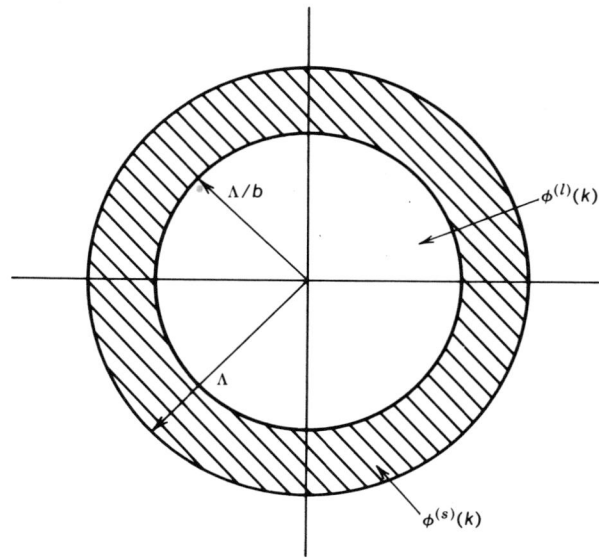

Fig. 8.7.1 Separation of momentum space variables into $\phi^{(l)}$: $0 \leq q \leq \Lambda/b$ and $\phi^{(s)}$: $\Lambda/b \leq q \leq \Lambda$. b is the scale factor of the RG transformation.

degrees of freedom by performing the appropriate integrations in (8.6.1)

$$Z = \int D[\phi^{(l)}] \exp\left[-\tfrac{1}{2} \sum_{q=0}^{\Lambda/b} (q^2 + r)\phi_q \phi_{-q}\right]$$
$$\times \int D[\phi^{(s)}] \exp\left[-\tfrac{1}{2} \sum_{q=\Lambda/b}^{\Lambda} (q^2 + r)\phi_q \phi_{-q}\right] \quad (8.7.2)$$

We then find

$$Z = e^{-F_s} \int D[\phi^{(l)}] \exp\left[-\tfrac{1}{2} \sum_{q=0}^{\Lambda/b} (q^2 + r)\phi_q \phi_{-q}\right] \quad (8.7.3)$$

where F_s, which is given by the second integral in (8.7.2), is the contribution to the free energy from the short-wavelength modes.

The contribution to the free energy from the remaining long-wavelength modes, (8.7.3), is identical to (8.7.2) except that the momentum cutoff Λ has been reduced to Λ/b. Suppose then that we rescale the unit length so that the cutoff is restored to its original value. The

8.7 RENORMALIZATION GROUP ANALYSIS OF THE GAUSSIAN MODEL

cutoff Λ, and anything else with the dimensions of (length)$^{-1}$, must scale as

$$\Lambda' = b\Lambda \qquad (8.7.4)$$

Because the free energy should be invariant, the fields must scale as

$$\phi'_q = b^{(d-2)/2}\phi_q \qquad (8.7.5)$$

In general, as discussed in Sect. 8.4, the fields scale with b as $b^{(d-2+\eta)/2}$, so comparison with (8.7.5) leads to the conclusion that $\eta = 0$.

In terms of the rescaled variables, the partition function has the renormalized form

$$Z = e^{-F_s}\int D[\phi']\exp-\tfrac{1}{2}\sum_q^{\Lambda}(q^2 + r')\phi'_q\phi'_{-q} \qquad (8.7.6)$$

and the only parameter in the problem r obeys the renormalization equation

$$r' = b^2 r \qquad (8.7.7)$$

This equation has two fixed points, $r^* = 0$ and $r^* = \infty$. The first of these choices corresponds to the critical Gaussian fixed point, whereas the second possibility corresponds to the high-temperature disordered phase.

In terms of the correlation length ξ, (8.6.7) becomes

$$\xi(r) = b\xi(b^2 r) \qquad (8.7.8)$$

which produces our earlier result (8.4.34), $\xi = r^{-1/2}$. In terms of the temperature, $r \sim |T - T_c|/T_c$, which gives

$$\xi \sim \left|\frac{T-T_c}{T_c}\right|^{-1/2} \qquad (8.7.9)$$

and again we find $\nu = 1/2$.

8.8 STABILITY OF THE GAUSSIAN FIXED POINT

The Gaussian model arises in a natural way if we expand the free energy about the most probable field and keep just the lowest-order (quadratic) terms in the fluctuations. On the other hand, we have argued that the fluctuations in the critical region are so strong that they completely dominate the behavior of the system. Therefore, it is essential to establish the stability or instability of the Gaussian fixed point. To do this, we need to keep the next term in the expansion of the free energy, and write

$$L = \int d^d x \left\{ \frac{1}{2}(|\nabla \phi|^2 + r\phi^2) + \frac{\lambda}{4!}\phi^4(x) \right\} \quad (8.8.1)$$

We are interested in the effect of the quartic term on the Gaussian fixed point and so we can take λ small. This suggests that we write

$$Z = Z_s \int D[\phi^{(l)}] e^{-L_0[\phi^{(l)}]} \langle e^{-L_4} \rangle_s \quad (8.8.2)$$

where $L_0[\phi^{(l)}]$ is the Gaussian part of the free energy and

$$\langle e^{-L_4} \rangle_s = \frac{\int D[\phi^{(s)}] e^{-L_0[\phi^{(s)}]} e^{-L_4[\phi^{(l)};\,\phi^{(s)}]}}{\int D[\phi^{(s)}] e^{-L_0[\phi^{(s)}]}} \quad (8.8.3)$$

In order to evaluate (8.8.3) we again resort to the cumulant expansion as we did in Sect. 6.1. Because we are exploring the RG in the neighborhood of the Gaussian fixed point, $r = 0$, $\lambda = 0$, we are justified in keeping only the leading-order terms in the cumulant expansion. We begin with the first cumulant.

$$\begin{aligned} C_1 &= \frac{\lambda}{4!} \int d^d x \langle \phi^4(x) \rangle_s \\ &= \frac{\lambda}{4!} \int d^d x (\phi^{(l)}(x))^4 + 4\frac{\lambda}{4!} \int d^d x \langle \phi^{(s)}(x) \phi^{(s)}(x) \rangle_s (\phi^{(l)}(x))^2 \\ &\quad + \frac{\lambda}{4!} \int d^d x \langle (\phi^{(s)}(x))^4 \rangle_s \end{aligned} \quad (8.8.4)$$

where we have used $\phi(x) = \phi^{(l)}(x) + \phi^{(s)}(x)$, expanded the quartic term, and averaged each term. The first term is quartic in the undecimated fields, whereas the second term is quadratic and contributes to the renormalization of r. The third term in (8.8.4) is independent of $\phi^{(l)}$ and has no effect on the RG equations; it is simply part of the free energy contributed by the short-wavelength modes.

Correlations such as $\langle \phi^{(s)}(x)\phi^{(s)}(y)\rangle$ in (8.8.4) occur frequently in RG calculations, so we will take some pains to evaluate this to leading order in r. Reverting to the momentum-space fields,

$$\langle \phi^{(s)}(x)\phi^{(s)}(y)\rangle_s = \int_{\Lambda/b}^{\Lambda} \frac{d^d q}{(2\pi)^d} \frac{e^{iq(x-y)}}{q^2 + r} \tag{8.8.5}$$

Because q is large in the integrand, we can expand and write for $x = y$ as in (8.8.4)

$$\int_{\Lambda/b}^{\Lambda} \frac{d^d q}{(2\pi)^d} \frac{1}{q^2 + r} = \int_{\Lambda/b}^{\Lambda} \frac{d^d q}{(2\pi)^d} \frac{1}{q^2}\left(1 - \frac{r}{q^2} + \cdots\right) \tag{8.8.6}$$

which gives

$$\int_{\Lambda/b}^{\Lambda} \frac{d^d q}{(2\pi)^d} \frac{1}{q^2 + r}$$
$$= \frac{S_d}{(2\pi)^d}\left[\frac{\Lambda^{d-2}(1 - b^{-(d-2)})}{d - 2} - r\frac{\Lambda^{d-4}(1 - b^{-(d-4)})}{d - 4} + \cdots\right] \tag{8.8.7}$$

where S_d is the area of a unit d sphere, which can be expressed in terms of the gamma function as

$$S_d = \frac{2\pi^{d/2}}{\Gamma(d/2)} \tag{8.8.8}$$

To lowest order in the small quantities r and λ, we can neglect the second- (and higher-) order terms in (8.8.7). By the same argument, the higher-order cumulants will contribute terms of $O(\lambda^2)$ and higher, and can also be neglected. Because the field $\phi(x)$ is rescaled by

$b^{(d-2)/2}$ ($\eta = 0$), the renormalization equations for r and λ are

$$r' = b^2 \left[r + \frac{S_d \Lambda^{d/2}}{(2\pi)^d} (1 - b^{2-d}) \frac{\lambda}{3!} \right] \qquad (8.8.9a)$$

$$\lambda' = b^{4-d} \lambda \qquad (8.8.9b)$$

From (8.8.9b) we see that if $d < 4$, $\lambda' > \lambda$ and the quartic term is relevant. Therefore, the Gaussian fixed point is unstable with respect to quartic interactions and does not describe real critical behavior in less than four dimensions. On the other hand, for $d > 4$, the quartic term is irrelevant and the Gaussian fixed point does represent critical behavior. Above four dimensions the critical exponents take on their classical (i.e., Gaussian) values.

Obviously, $d = 4$ is a rather special dimension, and it was the observation that one could regard d as a parameter in equations like (8.7.9a) and (8.7.9b) and analytically continue these equations to include noninteger values for d that led Wilson, Fisher, and others to what is now called the "ε expansion." The idea is simply that if one can work out everything for $d = 4$, then if $d = 4 - \varepsilon$, one can work out everything to a given order in ε. We will carry out this program to second order in ε in the next chapter.

PROBLEMS

8.1 Show that, as in (8.1.20), the specific heat is discontinuous at $T = T_c$ in the mean field approximation.

8.2 Derive (8.4.4).

8.3 Calculate $G(r)$ in $d = 4$ dimensions starting with $G(q)$ as in (8.4.27).

8.4 Show (8.6.9) by explicit differentiation of the generating function (8.6.8).

8.5 Derive the formula (8.8.8) for the area of a unit d sphere.

REFERENCES

1. Amit, D. J. (1978) *Field Theory, the Renormalization Group and Critical Phenomena*. McGraw-Hill, New York.
2. Pawley, G. S., Swendsen, R. H., Wallace, D. J., and Wilson, K. G. (1984) *Phys. Rev. B* **29** 4030.

9

THE ε EXPANSION

9.0 INTRODUCTION

In the previous chapter we saw that the Gaussian fixed point is unstable in less than four dimensions with respect to quartic terms in the free-energy functional. Because such terms are always present in a nontrivial theory, the Gaussian fixed point cannot describe critical behavior in less than four dimensions. There must exist another fixed point, which we shall refer to as the Wilson–Fisher [4, 5] (WF) fixed point, which governs critical behavior for $d < 4$. On the other hand, we have also observed that for $d > 4$, the Gaussian fixed point does give correct critical behavior. If we think of d as a parameter that can vary continuously, then there must be a crossover between the Gaussian and the WF points as d approaches the special value $d = 4$. For $d = 4 - \varepsilon$, where $\varepsilon \ll 1$, the two fixed points should lie close together in parameter space, and this is the central idea behind the ε expansion. If the Gaussian and WF fixed points are close in parameter space and the RG transformation is analytic, then the critical exponents will also differ by ε, or perhaps some higher power of ε, from their mean field, or "classical" values. The hope is that by expanding the RG equations in powers of ε, a convergent series for the critical exponents can be derived. As with other approaches to the RG, this hope is somewhat frustrated and the general feeling is that the ε expansion is asymptotic. From our point of view this is something of a blessing because we are relieved of the onerous task of calculating higher and higher corrections to the critical exponents, and can be reasonably content to work things out to order ε^2.

The construction of the RG equation proceeds exactly as for the Gaussian model (see Sec. 8.7), but instead of exploring the RG in the neighborhood of $r = 0$, $\lambda = 0$, we will assume that

$$r^* \sim O(\varepsilon)$$
$$\lambda^* \sim O(\varepsilon) \tag{9.0.1}$$

and investigate the linear behavior of the RG equations in the neighborhood of the WF fixed point. We also generalize the one-component order parameter introduced in Chap. 8 to an n-component vector of fields

$$\phi_\alpha(x) \qquad \alpha = 1, 2, \ldots, n \tag{9.0.2}$$

In general, the space in which the vector resides is completely separate from the coordinate space of the fields, and we will refer to it as the "internal space" of the fields. We can define symmetry transformations on the internal space, and in fact it is the symmetry group in this internal space which defines the character of the order parameter. Note that the term "group" in this context really refers to a mathematical group, unlike the use of the word in "renormalization group."

The scalar product of two fields is defined in the usual way

$$\psi \cdot \phi = \sum_\alpha \psi_\alpha(x)\phi_\alpha(x)$$
$$= \psi_\alpha(x)\phi_\alpha(x) \tag{9.0.3}$$

In the second line of (9.0.3) we have employed the summation convention in which repeated indices are understood to be summed.

Initially, we will assume that the free energy is a scalar, and in particular a function of $\phi^2 = \phi_\alpha \phi_\alpha$ only. For a quartic interaction term, we have for the Landau–Ginzburg form of the free energy L

$$L = \frac{1}{2}\int d^dx\, \nabla\phi_\alpha \nabla\phi_\alpha + r\phi_\alpha\phi_\alpha + \frac{\lambda}{4!}\int d^dx (\phi_\alpha\phi_\alpha)^2 \tag{9.0.4}$$

Because the fields are now vectors, the two-point correlation function becomes a matrix of correlation functions

$$G_{\alpha\beta}(x - y) = \langle \phi_\alpha(x)\phi_\beta(y)\rangle \qquad (9.0.5)$$

For a scalar free energy like (9.0.4), the matrix of correlation functions takes the particularly simple form in the symmetric phase ($T > T_c$),

$$G_{\alpha\beta}(x - y) = \delta_{\alpha\beta} G(x - y) \qquad (9.0.6)$$

However, below T_c the order parameter

$$m_\alpha = \langle \phi_\alpha(x)\rangle \qquad (9.0.7)$$

is nonzero, which means the internal symmetry of the free-energy functional is spontaneously broken. The order parameter singles out a particular direction in the internal space, and we must now distinguish between correlations along the direction of m_α and perpendicular to this direction.

If we assume that m_α lies along the 1-axis in internal space, then the longitudinal correlation function is

$$G_{11}(x - y) = \langle \phi_1(x)\phi_1(y)\rangle - m^2 \qquad (9.0.8)$$

and the transverse correlations function is

$$\begin{aligned} G_{22}(x - y) &= \langle \phi_2(x)\phi_2(y)\rangle \\ &= G_{33}(x - y) \end{aligned} \qquad (9.0.9)$$

9.1 CONSTRUCTION OF THE RG TRANSFORMATION

In keeping with our philosophy of presenting RG calculations in as straightforward a manner as possible, we will formulate the RG transformation using tools already developed in earlier chapters. However, we do not wish to give the impression that this is the way experts actually carry out RG calculations within the ε expansion. Some of the flavor of such calculations is given in Sec. 9.8. It is beyond the scope of this text to go into the detailed mechanics of RG calculations of this type, and fortunately this task has already been accomplished in the excellent text by Amit [1]. An invaluable source of early references is the review article by Wilson and Kogut [6]. The method employed here, again for pedagogical rather than aesthetic reasons, follows closely the method of Ma [3].

Before we delve into the gory details of the calculation before us, it is a good idea to get an overview of just what we are doing. We begin, as in Sec. 8.7, by dividing the fields into the long- and short-wavelength modes

$$\phi_\alpha(x) = \phi_\alpha^{(l)}(x) + \phi_\alpha^{(s)}(x) \tag{9.1.1}$$

where

$$\phi_\alpha^{(s)}(x) = \int_{\Lambda/b}^{\Lambda} \frac{d^d k}{(2\pi)^d} e^{ikx} \phi_\alpha(k)$$

$$\phi_\alpha^{(l)}(x) = \int_0^{\Lambda/b} \frac{d^d k}{(2\pi)^d} e^{ikx} \phi_\alpha(k) \tag{9.1.2}$$

As usual, b is the scale factor for the RG transformation. By translational symmetry the quadratic, or free, part of the free energy L_0,

$$L_0 = \frac{1}{2} \int d^d x (\nabla \phi_\alpha \nabla \phi_\alpha + r \phi_\alpha \phi_\alpha) \tag{9.1.3}$$

separates ($L_0[\phi^{(l)} + \phi^{(s)}] = L_0[\phi^{(l)}] + L_0[\phi^{(s)}]$), whereas for the interaction term we write

$$V[\phi^{(l)} + \phi^{(s)}] = V[\phi^{(l)}] + V_1[\phi^{(l)}; \phi^{(s)}] \tag{9.1.4}$$

For the quartic interaction (9.0.4) this has the explicit form

$$V_1[\phi^{(l)}; \phi^{(s)}] = \frac{\lambda}{4!} \int d^d x \Big[(\phi_\alpha^{(s)} \phi_\alpha^{(s)})^2 + 4 \phi_\alpha^{(s)} \phi_\alpha^{(l)} \phi_\beta^{(s)} \phi_\beta^{(s)}$$
$$+ 2 \phi_\alpha^{(l)} \phi_\alpha^{(l)} \phi_\beta^{(s)} \phi_\beta^{(s)} + 4 \phi_\alpha^{(s)} \phi_\alpha^{(l)} \phi_\beta^{(s)} \phi_\beta^{(l)}$$
$$+ 4 \phi_\alpha^{(l)} \phi_\alpha^{(l)} \phi_\beta^{(l)} \phi_\beta^{(l)} \Big] \tag{9.1.5}$$

With these definitions the partition function is

$$Z = \int D[\phi^{(l)}] e^{-L[\phi^{(l)}]} \int D[\phi^{(s)}] e^{-L_0[\phi^{(s)}] - V_1[\phi^{(l)}; \phi^{(s)}]} \tag{9.1.6}$$

The integration over the $\phi^{(s)}$ can be put in the form

$$\int D[\phi^{(s)}] e^{-L_0[\phi^{(s)}] - V_1[\phi^{(l)}; \phi^{(s)}]} = Z_s \langle e^{-V_1[\phi^{(l)}; \phi^{(s)}]} \rangle_s \qquad (9.1.7)$$

where Z_s is the contribution to the partition function from the Gaussian part of the free-energy functional

$$Z_s = \int D[\phi^{(s)}] e^{-L_0[\phi^{(s)}]} \qquad (9.1.8)$$

The average in (9.1.7) is of a form we have seen before, and as before we can evaluate it with the cumulant expansion. As we will show, the average of any function of the fields $\phi^{(s)}$ over the Gaussian weight L_0 can be expressed in terms of products of the average of two such fields, which is the correlation function for the short-wavelength modes, or in the language of field theory, the propagator. The free correlation function for the fields $\phi^{(s)}$ is given by

$$\begin{aligned} G^{(s)}_{\alpha\beta}(x - y) &= \langle \phi_\alpha(x) \phi_\beta(y) \rangle \\ &= \delta_{\alpha\beta} G_s(x - y) \end{aligned} \qquad (9.1.9)$$

$G_s(x - y)$ can be evaluated explicitly, and we have

$$G_s(x - y) = \int_{\Lambda/b}^{\Lambda} d^d k \frac{e^{ikr}}{k^2 + r} \qquad (9.1.10)$$

We see that $G_s(x - y)$, like $\phi^{(s)}$, only contains Fourier coefficients in the momentum shell.

Once the average over the short-wavelength modes in (9.1.7) has been carried out, the remaining free energy is a functional of just the long-wavelength modes. The final step in the renormalization procedure is to restore the cutoff, which after decimation lies at Λ/b, to its original value Λ. This is accomplished by rescaling the momentum variables by

$$k = \frac{k'}{b} \qquad (9.1.11)$$

and the fields by

$$\phi = b^{(2-\eta)/2} \phi' \qquad (9.1.12)$$

The exponent η is determined by the condition that the coefficient of the gradient-squared term in L_0, (9.1.3), remains fixed under the RG transformation.

9.2 THE CUMULANT EXPANSION AND FEYNMAN DIAGRAMS

As we will see in a moment, writing down the expressions for the cumulants, expanding out all the factors, and finally evaluating the averages over the $\phi^{(s)}$ leads to a real mess, and it is useful to have a compact notation for representing these expressions. If a picture is worth a thousand words, it is also equal to many lines of algebra, so in this section we present the basics of how to represent terms in the cumulant expansion by Feynman diagrams. More details about Feynman diagrams are given in App. B.

In order that this discussion not proceed completely in a vacuum, let us consider the first cumulant

$$\langle V_1[\phi^{(l)}; \phi^{(s)}]\rangle = \frac{\lambda}{4!}\int d^d x \langle (\phi_\alpha^{(s)}\phi_\alpha^{(s)})^2\rangle + 2\phi_\alpha^{(l)}\phi_\alpha^{(l)}\langle \phi_\beta^{(s)}\phi_\beta^{(s)}\rangle$$
$$+ 4\phi_\alpha^{(l)}\phi_\beta^{(l)}\langle \phi_\alpha^{(s)}\phi_\beta^{(s)}\rangle \qquad (9.2.1)$$

First, we represent the Green function $G_s(x - y)$ by a line extending from y to x (Fig. 9.2.1). The fields $\phi^{(l)}(x)$, which are not averaged over, play the role of fixed functions of x, and we represent these by a dotted line attached to the point x.

The intersection of two or more propagator lines with an interaction (wavy) line is called a vertex. Five types of vertices occur in the present problem, which all contribute to the first cumulant, (9.2.1). They are represented graphically as in Fig. 9.2.2. The potential, which in this case has the simple form $V(x - y) = \lambda \delta(x - y)$, is represented by a wavy line. A factor of $\phi^{(l)}(x)$ is indicated by a dotted line. Although we use a point-like interaction potential, we give the wavy line some extension in space in order to make clear how the indices and coordinate labels are attached. Also, we will see that we must eventually abandon the point-like potential in favor of a more general form when we carry out the cumulant expansion to higher orders.

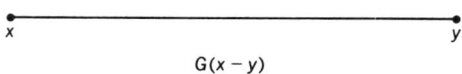

$G(x - y)$

Fig. 9.2.1 The propagator $G(x - y)$ is represented by a solid line.

9.2 THE CUMULANT EXPANSION AND FEYNMAN DIAGRAMS 273

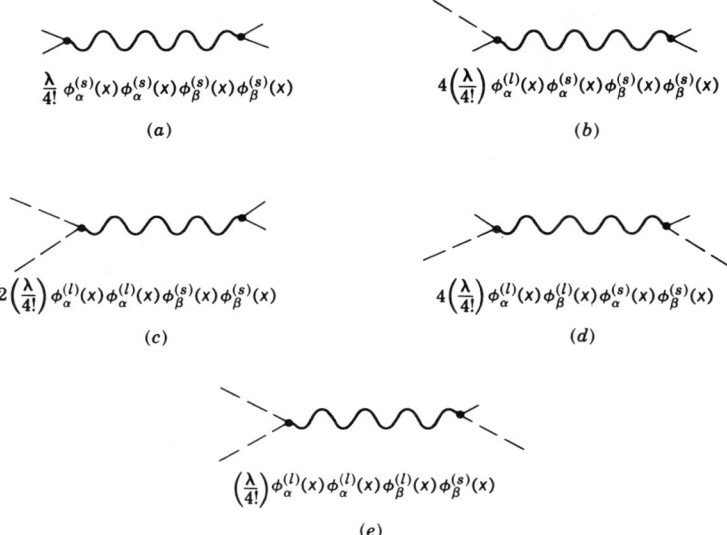

Fig. 9.2.2 Vertex functions for the ϕ^4 interaction. The wiggly line represents a factor of λ. Its extension serves two purposes. In a more general type of quartic interaction, the fields at distant points may actually interact through a two-body potential. Here we use the wiggly line to separate the fields with different indices. This facilitates counting powers of n.

Because the average over the $\phi^{(s)}$ is taken with respect to a Gaussian weight, the average of any odd number of fields vanishes. If we have the average of a product of an even number $2N$ of fields, this is equal to the product of the averages of N pairs of fields summed over all possible ways of pairing the coordinates. For example (again, see App. B for a proof), for four fields we have

$$\langle \phi_\alpha(x_1)\phi_\beta(x_2)\phi_\gamma(x_3)\phi_\delta(x_4) \rangle$$
$$= G_{\alpha\beta}(x_1 - x_2)G_{\gamma\delta}(x_3 - x_4) + G_{\alpha\gamma}(x_1 - x_3)G_{\beta\delta}(x_2 - x_4)$$
$$+ G_{\alpha\delta}(x_1 - x_4)G_{\beta\gamma}(x_2 - x_3) \qquad (9.2.2)$$

Using these rules, the first cumulant is

$$\langle V_1[\phi^{(l)}; \phi^{(s)}] \rangle = \frac{\lambda}{4!} \int d^d x \left[n(n+2)G_s^2(0) \right.$$
$$\left. + 2(n+2)G_s(0)\phi_\alpha^{(l)}(x)\phi_\alpha^{(l)}(x) \right] \quad (9.2.3)$$

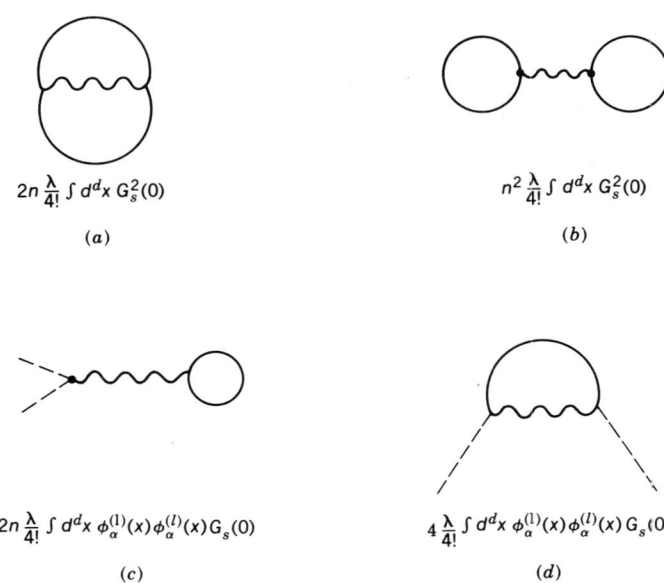

Fig. 9.2.3 Feynman diagrams for the first cumulant, (9.2.3).

The factors of n that appear in (9.2.3) arise from closed propagator loops, because each such loop yields a trace over the internal indices of the identity, $\delta_{\alpha\alpha} = n$. The corresponding graphs are shown in Fig. 9.2.3. Note that the first term in (9.2.3) gives rise to several graphs because, just as in (9.2.2), there are three ways to pair up four fields.

The first term in (9.2.3), which is equal to the sum of graphs (a) and (b) in Fig. 9.2.3, is independent of $\phi^{(l)}$ and is part of the contribution to the free energy from the rapidly varying fields $\phi^{(s)}$. As usual terms of this kind do not play a role in the RG equations and we will not consider them further. The second term in (9.2.3), which is the sum of graphs (c) and (d), yields the first-order (in λ) correction to the quadratic part of the free energy

$$\delta L_2^1 = \frac{\lambda}{4!} 2(n + 2) G_s(0) \int d^d x \, \phi_\alpha^{(l)}(x) \phi_\alpha^{(l)}(x) \qquad (9.2.4)$$

The notation in (9.2.4) is as follows: The subscript refers to the number of factors of $\phi^{(l)}$ appearing in that term of the free energy, whereas the superscript refers to the order in the coupling λ or equivalently, to the order of the cumulant.

An important property of the cumulants, which emerges when we consider the second and higher cumulants, is that they correspond to

9.2 THE CUMULANT EXPANSION AND FEYNMAN DIAGRAMS

connected diagrams. By connected we mean that there is at least one propagator line linking every part of the diagram. For definiteness let us consider the second cumulant

$$C_2[\phi^{(l)}] = \langle V_1^2[\phi^{(l)}; \phi^{(s)}]\rangle - \langle V_1[\phi^{(l)}; \phi^{(s)}]\rangle^2 \qquad (9.2.5)$$

If a diagram has a disconnected part, then the corresponding algebraic expression factorizes. Each of these factors belong to lower-order cumulants, in this case the first cumulant, and are cancelled by the second term in (9.2.5). That this cancellation of disconnected diagrams takes place in all the higher cumulants follows from the observation that the fully renormalized free energy, which is in part the sum of all the cumulants, must be extensive, that is, it must be proportional to the volume of the system. If there were any terms in the renormalized free energy corresponding to disconnected diagrams, that is, diagrams that factorize into $p > 1$ factors, each separate factor would require its own integration over the volume of the system and would lead to a term in the free energy proportional to V_1^p. In the case of the free energy, the connected diagrams are closed loops.

Another way of seeing this [6] is to start with the straightforward expansion of the exponential as a power series and average term by term

$$\langle e^{-V_1[\phi^{(l)}; \phi^{(s)}]}\rangle_s = \sum_{n=0}^{\infty} \frac{(-1)^n}{n!} \langle V_1^n[\phi^{(l)}; \phi^{(s)}]\rangle_s \qquad (9.2.6)$$

The average $\langle V_1^n \rangle$ is equal to the sum of all possible products of connected diagrams with $m_1, m_2, \ldots, m_k$ vertices, where k is the number of connected subdiagrams and the m_j are subject to the restriction that $\sum_j m_j = n$,

$$\langle V_1^n \rangle = \sum_{k=1}^{\infty} \sum_{m_1, \ldots, m_k} \prod_{j=1}^{k} \langle V_1^{m_j}\rangle_c \delta\left(\sum m_i - n\right) \qquad (9.2.7)$$

The subscript c in (9.2.7) indicates that only the connected part is included. Many of the possible partitions of the n factors of V into k connected subdiagrams are equivalent, and the sum can be written as the contribution of one such term weighted by the proper multiplicity. First, we assume that all the m_j are different. The number of ways in which n factors of V_1 can be sorted into k terms of $m_1, m_2, \ldots, m_k$

factors each is

$$f(n; m_1, m_2, \ldots, m_k) = \frac{n!}{m_1! m_2! \cdots m_k!} \qquad (9.2.8)$$

Let us define L_k as

$$L_k = -\frac{(-1)^k}{k!} \langle V_1^k \rangle_c \qquad (9.2.9)$$

We can then write, for the sum of all disconnected diagrams of k loops where all the loops are of different order

$$\langle e^{-V} \rangle_k = (-1)^k L_{m_1} L_{m_2} \cdots L_{m_k} \qquad (9.2.10)$$

Now suppose that several of the m_i are equal so that the number of diagrams of m_1 vertices is p_1 and so on. We can lump all these together if we divide by $p_1!$ to avoid overcounting all the permutations within the group, and so now we have, taking into account diagrams of all possible numbers of loops,

$$\langle e^{-V} \rangle = \sum_{p_1} \frac{(-1)^{p_1} L_1^{p_1}}{p_1!} \sum_{p_2} \frac{(-1)^{p_2} L_2^{p_2}}{p_2!} \cdots$$

$$= e^{-L_1 - L_2 - \cdots} \qquad (9.2.11)$$

By (9.2.11) we see that the average in (9.2.6) is given by the exponential of the sum of all closed connected diagrams. If we now return to (9.2.6), we have

$$\langle e^{-V[\phi^{(l)}; \phi^{(s)}]} \rangle_s = Z_s e^{-L_1 - L_2 - \cdots} \qquad (9.2.12)$$

Therefore, the coarse-grained free energy is

$$L'[\phi^{(l)}] = L_0[\phi^{(l)}] + L_1[\phi^{(l)}] + \cdots \qquad (9.2.13)$$

The fully renormalized free energy is formed by rescaling the length by a factor b^{-1} and the fields by $b^{(d-2+\eta)/2}$, as discussed in Chap. 4.

Because the first cumulant only leads to corrections in the quadratic part of the free energy, the corresponding RG equations will reproduce those at the Gaussian fixed point, which we studied in Chap. 8.

Therefore, in order to go beyond the Gaussian fixed point and discover "nonclassical" critical behavior, we must first tackle the problem of calculating the second cumulant.

9.3 EVALUATION OF THE SECOND CUMULANT

The contributions to the renormalized free energy from the first cumulant are given by (9.2.3). In order to find the WF fixed point we need to include at least the second cumulant, which is given by

$$C_2[\phi^{(l)}] = \left(\frac{\lambda}{4!}\right)^2 \int d^d x \int d^d y$$

$$\times \Big\langle \big[\phi_\alpha^{(s)}(x)\phi_\alpha^{(s)}(x)\phi_\beta^{(s)}(x)\phi_\beta^{(s)}(x)$$

$$+ 4\phi_\alpha^{(l)}(x)\phi_\alpha^{(s)}(x)\phi_\beta^{(s)}(x)\phi_\beta^{(s)}(x) + 4\phi_\alpha^{(l)}(x)\phi_\beta^{(l)}(x)\phi_\alpha^{(s)}(x)\phi_\beta^{(s)}(x)$$

$$+ 2\phi_\alpha^{(l)}(x)\phi_\alpha^{(l)}(x)\phi_\beta^{(s)}(x)\phi_\beta^{(s)}(x) + 4\phi_\alpha^{(l)}(x)\phi_\alpha^{(l)}(x)\phi_\beta^{(l)}(x)\phi_\beta^{(s)}(x)$$

$$- n(n+2)G_s^2(0) - 2(n+2)G_s(0)\phi_\alpha^{(l)}(x)\phi_\alpha^{(l)}(x)\big]$$

$$\times \big[\phi_\gamma^{(s)}(y)\phi_\gamma^{(s)}(y)\phi_\delta^{(s)}(y)\phi_\delta^{(s)}(y) + 4\phi_\gamma^{(l)}(y)\phi_\gamma^{(s)}(y)\phi_\delta^{(s)}(y)\phi_\delta^{(s)}(y)$$

$$+ 2\phi_\gamma^{(l)}(y)\phi_\gamma^{(l)}(y)\phi_\delta^{(s)}(y)\phi_\delta^{(s)}(y) + 4\phi_\gamma^{(l)}(y)\phi_\delta^{(l)}(y)\phi_\gamma^{(s)}(y)\phi_\delta^{(s)}(y)$$

$$+ 4\phi_\gamma^{(l)}(y)\phi_\gamma^{(l)}(y)\phi_\delta^{(l)}(y)\phi_\delta^{(s)}(y) - n(n+2)G_s^2(0)$$

$$- 2(n+2)G_s(0)\phi_\gamma^{(l)}(y)\phi_\gamma^{(l)}(y)\big]\Big\rangle_s \tag{9.3.1}$$

One approach to evaluating this expression is to expand and then collect all terms that are proportional to $\phi^{(l)2}$ and $\phi^{(l)4}$, but this is very tedious. We can make use of the diagram technique and the observation that the free energy contains only connected diagrams to simplify our task somewhat. The extra terms in (9.3.1), which come from $\langle V \rangle$, can be neglected because they only yield disconnected diagrams. [It is a good exercise to go ahead and evaluate (9.3.1) by brute force and see that the disconnected terms all cancel.]

First, we collect all connected diagrams that contribute terms proportional to $\phi^{(l)2}$, that is, terms with two dotted lines. The Feynman diagrams for the contributions to L_2^2 are shown in Fig. 9.3.1. Note that each of these diagrams contains exactly two dotted lines representing the two factors of $\phi^{(l)}$. Figures 9.3.1 g–i vanish because the Green function contains no Fourier components in the range

278 THE ε EXPANSION

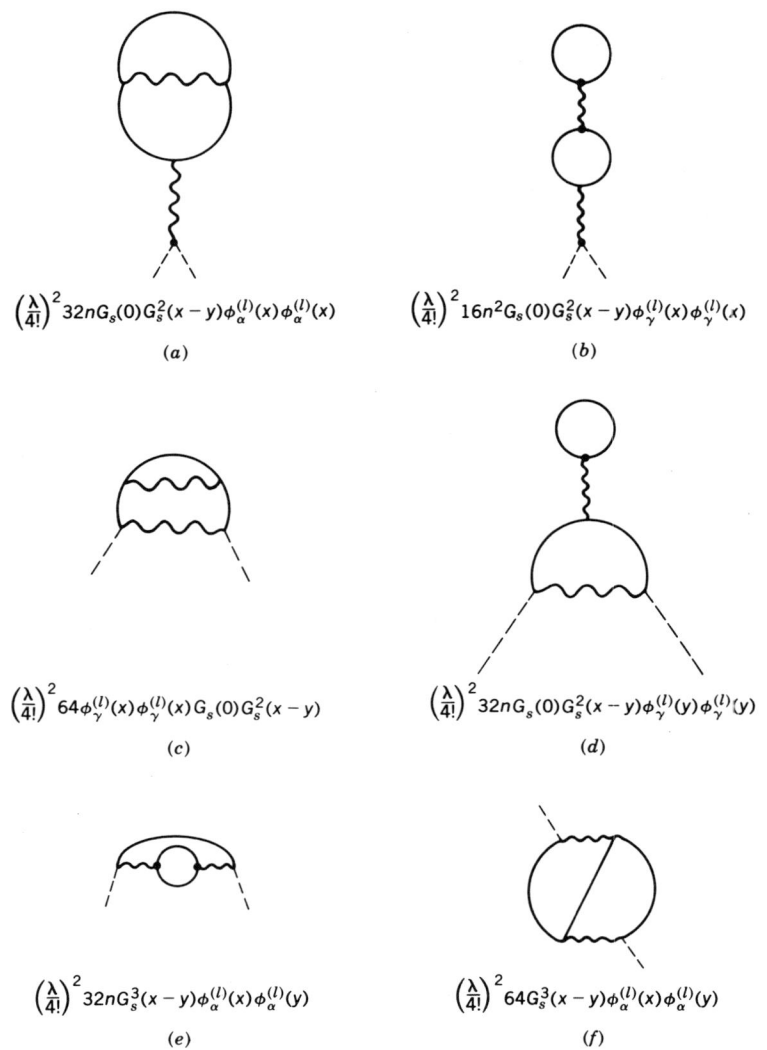

Fig. 9.3.1 Feynman diagrams contributing to the renormalized ϕ^2 term in the second cumulant.

$0 \leq k \leq \Lambda/b$. Summing the remaining terms, we have

$$L_2^2[\phi^{(l)}] = \left(\frac{\lambda}{4!}\right)^2 \int d^dx \int d^dy \left[16(n+2)^2 G_s(0) G_s^2(x-y) \phi_\alpha^{(l)}(x) \phi_\alpha^{(l)}(x) \right.$$
$$\left. + 32(n+2) G_s^3(x-y) \phi_\alpha^{(l)}(x) \phi_\alpha^{(l)}(y) \right] \quad (9.3.2)$$

In the same way, to find the second-order contribution to the fourth-order term in the free energy, we must construct all diagrams with

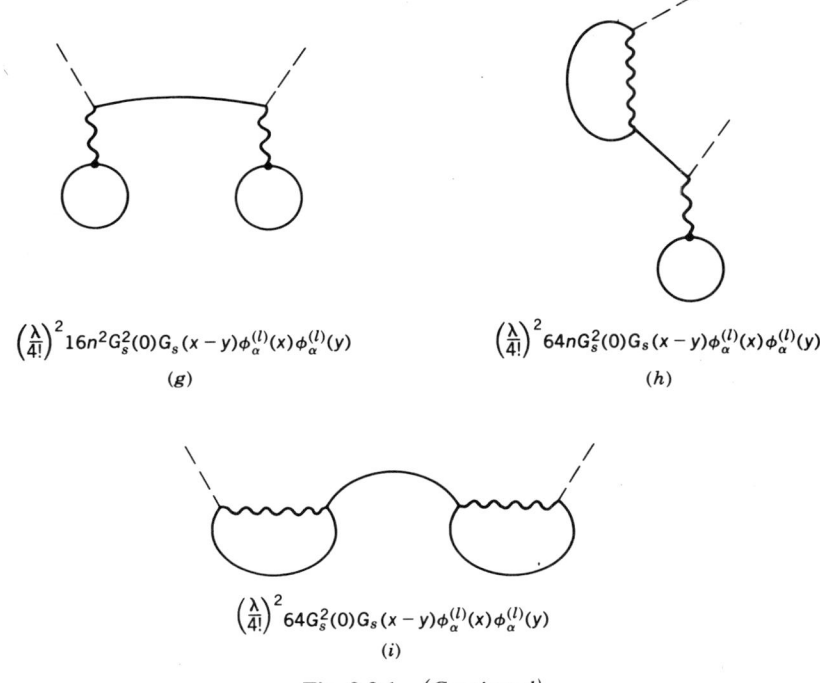

$\left(\frac{\lambda}{4!}\right)^2 16n^2 G_s^2(0) G_s(x-y) \phi_\alpha^{(l)}(x) \phi_\alpha^{(l)}(y)$
(g)

$\left(\frac{\lambda}{4!}\right)^2 64n G_s^2(0) G_s(x-y) \phi_\alpha^{(l)}(x) \phi_\alpha^{(l)}(y)$
(h)

$\left(\frac{\lambda}{4!}\right)^2 64 G_s^2(0) G_s(x-y) \phi_\alpha^{(l)}(x) \phi_\alpha^{(l)}(y)$
(i)

Fig. 9.3.1 *(Continued)*

four dotted lines as shown in Fig. 9.3.2. Summing these, we have

$$L_4^2[\phi^{(l)}] = \left(\frac{\lambda}{4!}\right)^2 \int d^d x \int d^d y$$
$$\times \Big[16(n+2) G_s(0) G_s(x-y) \phi_\alpha^{(l)}(x) \phi_\alpha^{(l)}(x) \phi_\beta^{(l)}(y) \phi_\beta^{(l)}(y)$$
$$+ 8(n+4) G_s^2(x-y) \phi_\alpha^{(l)}(x) \phi_\alpha^{(l)}(x) \phi_\beta^{(l)}(y) \phi_\beta^{(l)}(y)$$
$$+ 32 G_s^2(x-y) \phi_\alpha^{(l)}(x) \phi_\alpha^{(l)}(y) \phi_\beta^{(l)}(x) \phi_\beta^{(l)}(y) \Big] \quad (9.3.3)$$

Although the initial form of free energy, (9.0.4), contains only terms up to fourth order in the fields, the second cumulant generates sixth-order terms, and higher cumulants will of course generate higher-order terms. However, given that λ is the order ε, these terms will be of order ε^2 and higher. For example, the six-field term is of order ε^2 because it emerges in the second cumulant.

If we are interested in calculating the RG equations to $O(\varepsilon^2)$, then we only need to keep terms in the free energy up to sixth order in the fields. In addition, because all contributions to the renormalized free

280 THE ε EXPANSION

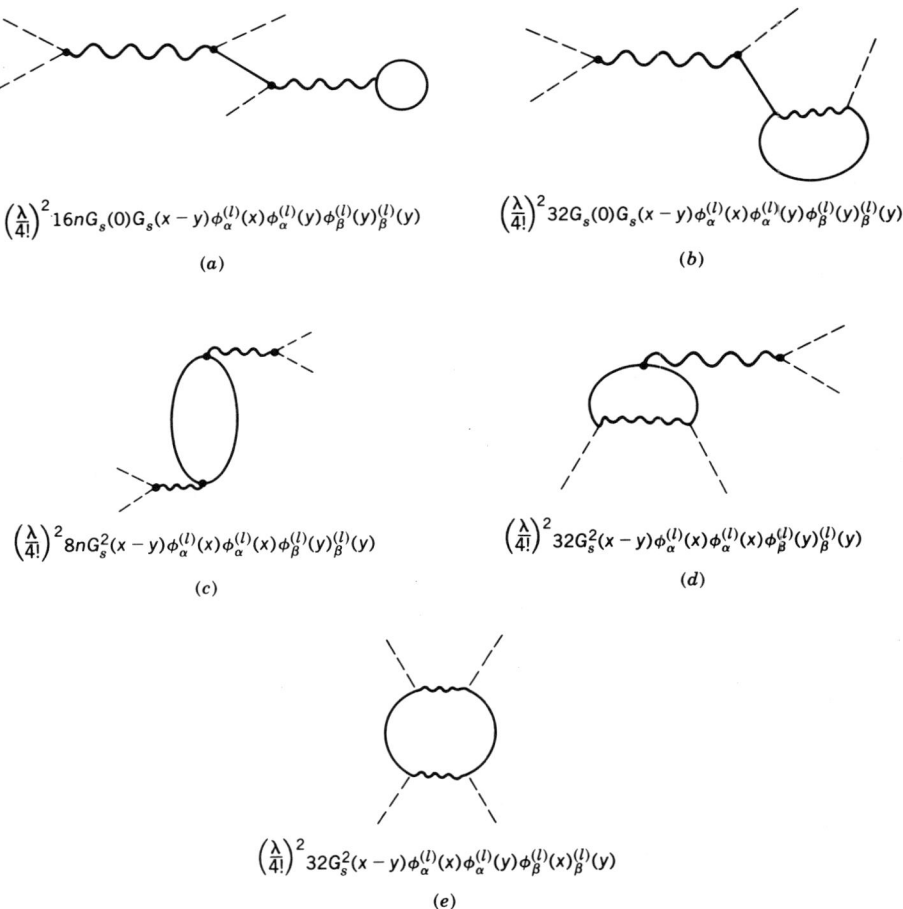

Fig. 9.3.2 Feynman diagrams contributing to the renormalized ϕ^4 term in the second cumulant.

energy from the third and higher cumulants will be at least of $O(\varepsilon^2)$, we only need to calculate the first and second cumulants. The contribution to the six-field term in the free energy from the second cumulant is

$$L_2^6 = -\left(\frac{\lambda}{4!}\right)^2 8 \int d^d x \int d^d y \, \phi_\alpha^{(l)}(x)\phi_\alpha^{(l)}(x)\phi_\beta^{(l)}(x)$$
$$\times G_s(x-y)\phi_\beta^{(l)}(y)\phi_\gamma^{(l)}(y)\phi_\gamma^{(l)}(y) \qquad (9.3.4)$$

and the graph for this term is shown in Fig. 9.3.3.

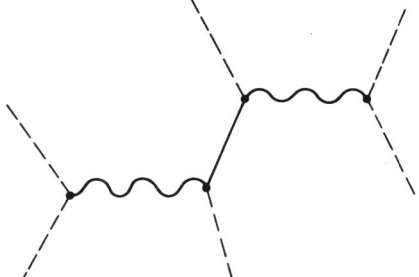

Fig. 9.3.3 Feynman diagrams for the ϕ^6 term in the second cumulant.

In addition to new terms that are higher powers of the fields, the second cumulant also gives rise to terms that are clearly not of the Landau–Ginzburg form, (9.0.4). For example, in the second term of (9.3.3) the fields are evaluated at different periods. We can, however, extract from these expressions terms that are of the Landau–Ginzburg form if we observe that the Green function $G_s(x - y)$ ought to fall off in a distance of the order of Λ^{-1}. Over this interval the relatively slowly varying fields $\phi^{(l)}$ should not change very much, and this suggests what is called the "gradient expansion," in which we write

$$\phi^{(l)}(y) = \phi^{(l)}(x) + (y - x)_i \frac{\partial \phi^{(l)}}{\partial x_i}(x)$$

$$+ \frac{1}{2!}(y - x)_i (y - x)_j \frac{\partial^2 \phi^{(l)}}{\partial x_i \partial x_j}(x) + \cdots \quad (9.3.5)$$

When we do this, for example, in the second term in (9.3.3), we get a series of the form

$$\int d^d x \int d^d y \, G_s^3(x - y) \phi_\alpha^{(l)} \phi_\alpha^{(l)}(y)$$

$$\simeq \left[\int d^d y \, G_s^3(y)\right] \int d^d x \, \phi_\alpha^{(l)}(x) \phi_\alpha^{(l)}(x)$$

$$- \left[\frac{1}{2d} \int d^d y \, y^2 G_s^3(y)\right] \int d^d x \, \nabla \phi_\alpha^{(l)}(x) \nabla \phi_\alpha^{(l)}(x) + \cdots \quad (9.3.6)$$

In (9.3.6) we can use the rotational symmetry of $G(r)$ to show that all integrals of odd powers of r with $G(r)$ vanish, and

$$\int d^d x \, x_i x_j G_s^3(x) = \frac{1}{d} \int d^d x \, x^2 G_s^3(x) \tag{9.3.7}$$

In addition, we have integrated the second term by parts to put it in a more symmetric form.

Unfortunately, the gradient expansion is not only divergent, but the higher-order terms are individually infinite. This divergence can be traced back to the sharp division in momentum space between the fields $\phi^{(s)}$ and $\phi^{(l)}$, which leads to an asymptotic form for $G_s(x)$, with $r/\Lambda^2 \ll 1$, in $d = 4$ dimensions,

$$\begin{aligned} G_s(x) &= \int_{\Lambda/b}^{\Lambda} d^4 k \, \frac{e^{ikx}}{k^2} \\ &= \frac{1}{(2\pi)^2} \frac{J_0(\Lambda x/b) - J_0(\Lambda x)}{x^2} \end{aligned} \tag{9.3.8}$$

where $J_0(x)$ is the usual zero-order Bessel function. For large argument the Bessel function has the behavior

$$\lim_{x \to \infty} J_0(x) = \sqrt{\frac{2}{\pi x}} \cos\left(x - \frac{\pi}{4}\right) \tag{9.3.9}$$

As long as we do not want to go beyond lowest order in ε, we can turn a blind eye to these difficulties and extract results that are independent of exactly how the separation of the fields is achieved. However, in order to calculate the critical exponents to $O(\varepsilon^2)$, we must face up to these difficulties. One approach, due to Wilson [6], is to smoothly interpolate between decimated and undecimated fields. Another approach, which we will examine in detail in Sec. 9.5, is to consider the limit in which the scale factor b is infinitesimally different from unity.

9.4 RENORMALIZATION EQUATIONS TO ORDER ε

If we are interested in locating the WF fixed point just to lowest order in ε, our job is considerably simplified. First, we only need to keep the leading terms in the gradient expansion (9.3.6). In this case, no new terms involving the gradient of the fields are generated by the RG

9.4 RENORMALIZATION EQUATIONS TO ORDER ε

transformation, and in particular the renormalization equation for the gradient-squared term of the free energy is simply

$$\left[\frac{1}{2}\int d^d x (\nabla\phi)^2\right]' = b^\eta \frac{1}{2}\int d^d x (\nabla\phi)^2 \quad (9.4.1)$$

Equation (9.4.1) follows simply from rescaling the coordinates and fields according to (9.1.11) and (9.1.12), respectively. The exponent of η is chosen so that the coefficient of the gradient-squared term is unchanged by the RG transformation, and therefore we must take

$$\eta = 0 \quad (9.4.2)$$

This value of η only holds to $O(\varepsilon)$: In second order in ε, η is nonzero.

If we now collect contributions to the renormalized parameters r and λ and rescale, we arrive at the following RG equations:

$$r' = b^2 \Bigg\{ r + 4(n+2)G_s(0)\frac{\lambda}{4!} - \left(\frac{\lambda}{4!}\right)^2 \int d^d y$$

$$\times \Big[16(n+2)^2 G_s(0) G_s^2(x-y) + 32(n+2) G_s^2(x-y)\Big]\Bigg\} \quad (9.4.3a)$$

$$\lambda' = b^\varepsilon \Bigg\{\lambda - \frac{\lambda^2}{3!}(n+8)\int d^4 y\, G_s^2(x-y)\Bigg\} \quad (9.4.3b)$$

In order to analyze these equations, we must sort out their dependence on ε and on the scale factor b.

First, we consider the factor $G_s(0)$ appearing in (9.4.3a). Setting $d = 4$, we have

$$G_s(0) = \int_{\Lambda/b}^{\Lambda} \frac{d^4 k}{(2\pi)^4} \frac{1}{k^2 + r}$$

$$= \frac{S_4}{(2\pi)^4}\left[\frac{\Lambda^2}{2}(1 - b^{-2}) - r \ln b\right] \quad (9.4.4)$$

Next, we consider the integral of $G_s^2(x-y)$, which appears in (9.4.3b).

Always working to lowest order, we have

$$\int d^4 y\, G_s^2(x-y) = \int_{\Lambda/b}^{\Lambda} \frac{d^4 k}{(2\pi)^4} G_s^2(k)$$

$$\simeq \frac{S_4}{(2\pi)^4} \ln b \qquad (9.4.5)$$

Putting these results together with (9.4.3a) and (9.4.3b), the RG equations to lowest order in ε are

$$r' = b^2 \left\{ r + \frac{n+2}{12} \frac{S_4}{(2\pi)^4} \Lambda^2 (1 - b^{-2}) \lambda - \frac{n+2}{3!} \frac{S_4}{(2\pi)^4} r\lambda \ln b \right\} \qquad (9.4.6a)$$

$$\lambda' = b^\varepsilon \left\{ \lambda - \frac{\lambda^2}{3!} (n+8) \frac{S_4}{(2\pi)^4} \ln b \right\} \qquad (9.4.6b)$$

As usual, our first job is to locate the fixed points of the RG equations. Because (9.4.6b) is a little simpler, we will begin there.

We are assuming that $\varepsilon \ll 1$, so we can express the dependence on the scale factor b in (9.4.6b) as

$$b^\varepsilon = e^{\varepsilon \ln b}$$

$$\simeq 1 + \varepsilon \ln b \qquad (9.4.7)$$

Because λ is, by hypothesis, already of order ε, to lowest order we have

$$\lambda' = \lambda + \left[\varepsilon \lambda - \frac{\lambda^2}{3!} (n+8) \frac{S_4}{(2\pi)^4} \right] \ln b \qquad (9.4.8)$$

Setting $\lambda' = \lambda = \lambda^*$, we have

$$\lambda^* = \frac{3!}{n+8} \frac{(2\pi)^4}{S_4} \varepsilon \qquad (9.4.9)$$

which bears out our original assumption that the fixed-point value of λ is of order ε.

Finding the fixed point of (9.4.6a) is a little less straightforward because the dependence on the scale factor b is not homogeneous.

Inside the brackets we have terms that are not explicit functions of b, terms that go as b^{-2}, and a term that varies as $\ln b$.

Recalling the general discussion of Chap. 4, given initial values of r and λ on the critical line, the fixed point of an RG transformation is reached after many iterations of the RG, or equivalently, in the limit in which $b \to \infty$. Physically, this means we are looking at the system on very large length scales. In this limit we can neglect the terms that depend on b through b^{-2}, and because we are interested in determining the fixed point just to order ε, we can also ignore the term in $\ln b$. Therefore, the fixed-point value for r^* is given by

$$r^* = -\frac{n+2}{12} \frac{S_4}{(2\pi)^4} \lambda^*$$

$$= -\frac{n+2}{n+8} \Lambda^2 \varepsilon \quad (9.4.10)$$

Again, our guess that the fixed-point values of r and λ are of order ε is borne out. The original supposition that a new fixed point (r^*, λ^*) close to the Gaussian fixed point for $4 - d = \varepsilon \ll 1$ has been justified. To summarize, the location of the WF fixed point to $O(\varepsilon)$ is

$$\lambda^* = \frac{3!}{n+8} \frac{(2\pi)^4}{S_4} \varepsilon$$

$$r^* = -\frac{n+2}{n+8} \Lambda^2 \varepsilon \quad (9.4.11)$$

Now that we have located the WF fixed point, the critical exponents can be determined by linearizing the RG equations about the fixed point. Denoting the deviations of r and λ from their fixed-point values by δr and $\delta \lambda$, the linearized RG equations are

$$\delta r' = b^2 \left\{ \delta r + \frac{n+2}{12} \frac{S_4}{(2\pi)^4} \Lambda^2 \, \delta\lambda - \frac{n+2}{n+8} \varepsilon \, \delta r \right.$$

$$\left. + \frac{(n+2)^2}{n+8} \frac{S_4}{(2\pi)^4} \frac{\Lambda^2}{3!} \varepsilon \, \delta\lambda \right\} \quad (9.4.12a)$$

$$\delta\lambda' = (1 + \varepsilon \ln b) \, \delta\lambda - \frac{2}{3!} \frac{(n+8) S_4}{(2\pi)^4} \lambda^* \ln b \, \delta\lambda \quad (9.4.12b)$$

Equation (9.4.12b) is easily solved by using (9.4.7), and we find

$$\delta\lambda' = (1 - \varepsilon \ln b)\,\delta\lambda$$
$$= b^{-\varepsilon}\,\delta\lambda \qquad (9.4.13)$$

The exponent $y_2 = -\varepsilon$ tells us how the quartic interaction transforms under the RG, and we see that, unlike the behavior at the Gaussian fixed point, the WF fixed point is stable with respect to quartic terms in the free energy for $d < 4$. On the other hand, if $d > 4$, so that $\varepsilon < 0$, the WF fixed point is unstable and the Gaussian fixed point is the stable one.

Again making use of (9.4.7), (9.4.12a) can be written in the form

$$\delta r' = b^{2-[(n+2)/(n+8)]\varepsilon}\left[\delta r + \frac{(n+2)S_4\Lambda^2}{12(2\pi)^4}\delta\lambda\right] \qquad (9.4.14)$$

From this we identify the relevant scaling field h_T by

$$h_T = \delta r + \frac{(n+2)S_4\Lambda^2}{12(2\pi)^4}\delta\lambda \qquad (9.4.15)$$

and its associated eigenvalue

$$y_1 = 2 - \frac{n+2}{n+8}\varepsilon \qquad (9.4.16)$$

The "critical surface," which in this case is a line, is the set of points (r_c, λ_c) on which h_T vanishes, so that

$$r_c = -\frac{(n+3)S_4\Lambda^2}{12(2\pi)^4}\lambda_c \qquad (9.4.17)$$

As we have argued previously, the deviation of r from its fixed-point value is a measure of the reduced temperature $t = (T - T_c)/T_c$, and we can identify y_1 with the thermal eigenvalue y_T. In addition, we know that the correlation length is related to r by $\xi = r^{-1/2}$, and so the correlation length exponent ν is

$$\nu = \frac{1}{y_T}$$
$$= \frac{1}{2} + \frac{n+2}{4(n+8)}\varepsilon \qquad (9.4.18)$$

TABLE 9.4.1 Critical Exponents of Order ε

α	β	γ	η	ν
$\dfrac{4-n}{2(n+8)}\varepsilon$	$\dfrac{1}{2} - \dfrac{3\varepsilon}{2(n+8)}$	$1 + \dfrac{(n+2)\varepsilon}{2(n+8)}$	0	$\dfrac{1}{2} + \dfrac{(n+2)\varepsilon}{4(n+8)}$

Now that we have calculated two exponents (η and ν) to order ε, the others can be determined by applying the scaling laws, (4.6.1a) to (4.6.1e). Results to first order in ε are given in Table 9.4.1.

These results for the critical exponents are an eloquent expression of the universality hypothesis. We see that only the dimensionality of space, $d = 4 - \varepsilon$, and the number of components in the order parameter n enter into the expression for the exponents. This happy situation persists beyond the first order in ε, and if the ε expansion were convergent, we would have a proof of the universality hypothesis. Unfortunately, there are strong indications that the ε expansion is asymptotic (see the discussion in Sec. 6.8), and in fact results to second order in ε seem to give better values for the critical exponents in three dimensions than those carried out to third order. In Table 9.4.2 we compare the results for the critical exponents of the ε expansion up to third order with values obtained by other methods for order parameters with $n = 1$ (Ising model) and $n = 3$ (Heisenberg model) components in three dimensions.

In analyzing the behavior of the RG equation, it is convenient to express (9.4.6a) and (9.4.6b) in differential form by taking

$$b = e^{dl} \qquad (9.4.19)$$

where $dl \ll 1$, and we consider r and λ to be functions of the continuous variable l. The renormalization equations can also be

TABLE 9.4.2 Critical Exponents Calculated by ε Expansion

	$n = 1$		$n = 2$	
	γ	η	γ	η
Other methods	1.250(3)	0.56(8)	1.38(1)	0.043(14)
$O(\varepsilon)$	1.167	0.0	1.227	0.0
$O(\varepsilon^2)$	1.244	0.020	1.347	0.021
$O(\varepsilon^3)$	1.197	0.037	1.325	0.039

simplified by rescaling r and λ as

$$r \to \Lambda^2 r$$

$$\lambda \to \frac{3!(2\pi)^4}{(n+8)S_4}\lambda \qquad (9.4.20)$$

When we do this, (9.4.6a) and (9.4.6b) become

$$\frac{dr}{dl} = 2r + \frac{n+2}{n+8}\lambda - \frac{n+2}{n+8}r\lambda \qquad (9.4.21a)$$

$$\frac{d\lambda}{dl} = \varepsilon\lambda - \lambda^2 \qquad (9.4.21b)$$

It is possible to solve this set of coupled nonlinear equations by quadrature. First, (9.4.21b) can be integrated to give

$$\lambda = \left[\frac{1}{\lambda_0}e^{-\varepsilon l} + \frac{1}{\varepsilon}(1 - e^{-\varepsilon l})\right]^{-1} \qquad (9.4.22)$$

Note that in the limit $l \to \infty$, λ always tends to its value at the WF fixed point

$$\lim_{l \to \infty} \lambda(l) \to \lambda^* + e^{y_2 l}\left(\frac{\varepsilon}{\lambda_0} - 1\right) \qquad (9.4.23)$$

with $\lambda^* = \varepsilon$ and $y_2 = -\varepsilon$.

Equation (9.4.21a) can be cast in the form

$$\frac{dr}{dl} = r\left(2 - \frac{n+2}{n+8}\lambda\right) + \frac{n+2}{n+8}\lambda \qquad (9.4.24)$$

Defining

$$\theta(l) = \int_0^l dl' \left[2 - \frac{n+2}{n+8}\lambda(l')\right] \qquad (9.4.25)$$

and

$$r(l) = e^{\theta(l)}\bar{r}(l) \qquad (9.4.26)$$

we have

$$\frac{d\bar{r}}{dl} = e^{-\theta(l)}\left(\frac{n+2}{n+8}\right)\lambda(l) \tag{9.4.27}$$

which can be integrated to give

$$\bar{r}(l) = \bar{r}_0 + \left(\frac{n+2}{n+8}\right)\int_0^l dl'\, e^{-\theta(l')}\lambda(l') \tag{9.4.28}$$

Taking this result and putting it together with (9.4.26), we have

$$r(l) = r_0 e^{\theta(l)} + \left(\frac{n+2}{n+8}\right)e^{\theta(l)}\int_0^l dl'\, e^{-\theta(l')}\lambda(l') \tag{9.4.29}$$

Let us now examine the behavior of this solution for large l. Let $l \gg l_0 \gg 1$, so that

$$\theta(l) \simeq \theta(l_0) + y_2(l - l_0) \tag{9.4.30}$$

and

$$\int_0^l dl'\, e^{-\theta(l')}\lambda(l') \simeq \int_0^{l_0} dl'\, e^{-\theta(l')}\lambda(l') + \frac{\lambda^* e^{-\theta(l_0)}}{y_2}\left[1 - e^{-y_2(l-l_0)}\right] \tag{9.4.31}$$

We now have by (9.4.29) to (9.4.31)

$$r(l) = \left\{r_0 + \left(\frac{n+2}{n+8}\right)\left[\int_0^{l_0} dl'\, e^{-\theta(l')}\lambda(l') + \lambda^* e^{-\theta(l_0)}\right]e^{y_2(l-l_0)}\right\} - \frac{\lambda^*}{y_2} \tag{9.4.32}$$

Evidently, the quantity in the curly brackets in (9.4.32) is the nonlinear generalization of the linear scaling field h_T, (9.4.15), and it scales with the exponent y_1.

If we take the limit $l_0 \to \infty$, the exact critical line (r_c, λ_c) is given by [remember that θ and $\lambda(l)$ depend on λ_c]

$$r_c + \frac{n+2}{n+8}\int_0^\infty dl'\, e^{-\theta(l')}\lambda(l'; \lambda_c) = 0 \tag{9.4.33}$$

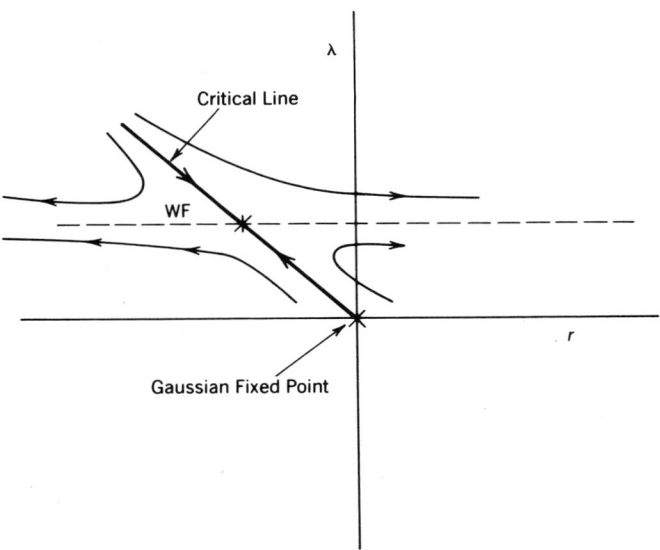

Fig. 9.4.1 RG trajectories to $O(\varepsilon)$. The heavy line is the critical line and connects the Wilson–Fisher (WF) and Gaussian fixed points.

Close to the WF fixed point, (9.4.33) reduces to our previous result, (9.4.17).

Typical RG trajectories are shown in Fig. 9.4.1. The Gaussian and WF fixed points are indicated by an *, and the line connecting them is the critical line.

It is possible for an RG trajectory to pass close to both fixed points, for example, the trajectory on the lower left of the figure. For moderate values of $l < l_0$ the RG equations linearized about the Gaussian fixed point will be valid, and for this limited range of scales one will observe scaling with the classical, or Gaussian, critical exponents. As one approaches the critical line, more closely, or equivalently for $l \gg l_0$, the trajectory enters the linear scaling region of the WF fixed point and the true scaling behavior of the system emerges.

The value of l_0 at which this crossover from the Gaussian to the WF fixed point takes place tells us the width of the critical region. If l_0 is relatively small, then the critical region will be large, whereas if l_0 is large, then there will be a wide range of temperatures over which one can observe mean field behavior, and the true scaling of the system is revealed only very close to the critical line.

Outside the critical region fluctuations are small and mean field theory can be applied.

9.5 THE WEGNER–HOUGHTON INFINITESIMAL RENORMALIZATION GROUP GENERATOR

In order to carry out the ε expansion beyond first order, we need to keep higher-order terms in the gradient expansion, (9.3.6). Unfortunately, when we try to do this we find expressions that diverge: The gradient expansion together with a sharp cutoff in momentum space just does not work.

The motivation of the gradient expansion is to keep the renormalized free-energy functional in the Landau–Ginzburg form, but clearly this is not possible in general. We are forced then to consider a more general class of free-energy functionals. We can also simplify the form of the RG transformation by taking the limit $b = e^{dl}$, as we did in Sec. 9.4. In this way the RG equations become integro-differential equations.

The infinitesimal RG transformation has been formulated in several ways. Here we will consider the derivation of Wegner and Houghton [7], which is a very natural extension of the methods we have employed thus far.

We begin by generalizing the Landau–Ginzburg form of the free energy (9.0.4) to

$$L = \frac{1}{2!} \sum_{k,k'} v_2(k)\phi_\alpha(k)\phi_\alpha(k')\delta(k+k')$$

$$+ \frac{1}{4!V} \sum_{k_1,k'_1;k_2,k'_2} v_4(k_1,k'_1;k_2,k'_2)\phi_\alpha(k_1)\phi_\alpha(k'_1)\phi_\beta(k_2)$$

$$\times \phi_\beta(k'_2)\delta(k_1 + k'_1 + k_2 + k'_2) \qquad (9.5.1)$$

Note that in (9.5.1) we have assumed the free energy is a function of $\phi_\alpha\phi_\alpha$ only, and the momentum vectors for the pairs of fields with the same internal index are labeled (k, k').

The δ's appearing in (9.5.1) are Kronecker deltas and are included explicitly to ensure translational invariance of L. The familiar Landau–Ginzburg free energy (9.0.4) is recovered if we choose

$$v_2(k) = k^2 + r \qquad v_4(k_1,k'_1;k_2,k'_2) = \lambda \qquad (9.5.2)$$

The first step in constructing the RG transformation is to divide the field variables into those with momenta in the shell

$$(1 - \delta)\Lambda < p \leq \Lambda \tag{9.5.3a}$$

and those with momenta in the interior of the sphere

$$0 \leq q \leq (1 - \delta)\Lambda \tag{9.5.3b}$$

In what follows we will adopt the convention that momenta in the shell will be labeled p, and those in the interior will be denoted by q. We will also rescale the momentum variables

$$q \to \frac{q}{\Lambda} \tag{9.5.4}$$

so that the cutoff is at $p = 1$ and $0 \leq q \leq 1 - \delta$.

The partition function is

$$\begin{aligned} Z &= \int D[\phi] e^{-L[\phi]} \\ &= \int D[\phi(q)] e^{-L[\phi(q)]} \int D[\phi(p)] e^{-L_1[\phi(q); \phi(p)]} \end{aligned} \tag{9.5.5}$$

where

$$L_1[\phi(q); \phi(p)] = L[\phi] - L[\phi(q)] \tag{9.5.6}$$

The integration in (9.5.5) can be evaluated by cumulants in the usual way by taking averages with respect to $L_0[\phi]$, where

$$L_0[\phi] = \frac{1}{2} \sum_{p, p'} v_2(p) \phi_\alpha(p) \phi_\alpha(p) \delta(p + p') \tag{9.5.7}$$

If we expand L_1 in a power series in $\phi(p)$, we have

$$L_1[\phi(q); \phi(p)] = \sum_p \frac{\partial L_1[\phi(q)]}{\partial \phi_\alpha(p)} \phi_\alpha(p)$$
$$+ \frac{1}{2!} \sum_{p, p'} \frac{\partial^2 L_1[\phi(q)]}{\partial \phi_\alpha(p) \partial \phi_\beta(p')} \phi_\alpha(p) \phi_\beta(p') + \cdots$$

$$\tag{9.5.8}$$

Each of the derivatives of L_1 is evaluated at $\phi(p) = 0$ and is therefore a function of the slowly varying fields $\phi(q)$ only. A typical term in the cumulant will be of the form

$$\sum_{\{p_{1,j}\}} \cdots \sum_{\{p_{k,j}\}} \left\langle \frac{\partial^{n_1} L_1}{\partial \phi^{n_1}} \phi(p_{1,1}) \cdots \phi(p_{1,n_1}) \frac{\partial^{n_2} L_1}{\partial \phi^{n_2}} \phi(p_{2,1}) \right.$$

$$\left. \times \cdots \phi(p_{2,n_2}) \cdots \frac{\partial^{n_k} L}{\partial \phi^{n_k}} \phi(p_{k,1}) \cdots \phi(p_{k,n_k}) \right\rangle_c$$
(9.5.9)

where the subscript c indicates the cumulant or connected part.

The average of products of fields is the sum of products of all possible averages of pairs, where, as in (9.1.9),

$$\langle \phi_\alpha(p) \phi_\beta(p') \rangle = \delta_{\alpha\beta} \frac{1}{v_2(p)} \delta(p + p') \qquad (9.5.10)$$

The average in (9.5.9) involves $n_1 + n_2 + \cdots + n_k$ sums over wave vectors in the momentum shell, but half of these can be eliminated by the delta functions in (9.5.10). Therefore, there are in general $(n_1 + n_2 + \cdots + n_k)/2$ independent summations over wave vectors in the momentum shell. It is possible that a part of a diagram consisting of k momenta has zero total momentum, that is, there arises a factor $\delta(p_1 + \cdots + p_k)$. This puts an added restriction on the number of independent momentum summations so that if there are M such "zero-momentum" parts we will have

$$N = \tfrac{1}{2}(n_1 + n_2 + \cdots + n_k) - M \qquad (9.5.11)$$

independent momentum summations. Because each sum over the momentum shell is proportional to δ, the width of the shell, and we only want to keep terms of $O(\delta)$, we require that $N = 1$, or

$$2(M + 1) = n_1 + n_2 + \cdots + n_k \qquad (9.5.12)$$

Because $M + 1 \le k$ we have

$$\frac{n_1 + n_2 + \cdots + n_k}{k} \le \frac{n_1 + n_2 + \cdots + n_k}{M + 1} = 2 \qquad (9.5.13)$$

From this it follows that the only types of functional derivatives we need to keep are those of first order (taken in pairs) and second order. Therefore, we can write

$$L = L[\phi(k)] + \sum_p \phi_\alpha(p) \frac{\partial L}{\partial \phi_\alpha(p)}$$
$$+ \frac{1}{2!} \sum_{p,p'} \phi_\alpha(p) \frac{\partial^2 L}{\partial \phi_\alpha(p) \partial \phi_\beta(p')} \phi_\beta(p') \delta(p+p') \quad (9.5.14)$$

and the functional integral we need to evaluate is

$$e^{-\Delta L_1} = \int D[\phi(p)] \exp - \left\{ \sum_p \phi_\alpha(p) \frac{\partial L}{\partial \phi_\alpha(p)} \right.$$
$$\left. + \frac{1}{2} \sum_p \phi_\alpha(p) \frac{\partial^2 L}{\partial \phi_\alpha(p) \partial \phi_\beta(-p)} \phi_\beta(-p) \right\} \quad (9.5.15)$$

This is a Gaussian integration and we can evaluate it by completing the square. The result is

$$\Delta L_1 = -\frac{1}{2} \sum_{|p|=1-\delta} \frac{\partial L}{\partial \phi_\alpha(p)} \left[\frac{\partial^2 L}{\partial \phi(p) \partial \phi(-p)} \right]^{-1}_{\alpha\beta} \frac{\partial L}{\partial \phi_\beta(-p)}$$
$$+ \frac{1}{2} \ln \left[\frac{\partial^2 L}{\partial \phi(p) \partial \phi(-p)} \right]_{\alpha\beta} \quad (9.5.16)$$

The sum over momenta in the shell can be written in the form of an integral as

$$\sum_{|p|=1-\delta} = \delta \frac{V}{(2\pi)^d} \int d\Omega(\hat{p}) \quad (9.5.17)$$

where S_d is the area of a unit d sphere, V_d is the volume, and $d\Omega(\hat{p})$ is the element of area on the unit d sphere.

In addition to changes in the free energy produced by decimating the fields $\phi(p)$, changes are caused by rescaling the fields

$$\phi = b^{(2-\eta)/2} \phi' \quad (9.5.18)$$

the momenta

$$k = b^{-1} k' \quad (9.5.19)$$

and the volume

$$V = b^d V' \quad (9.5.20)$$

9.5 RENORMALIZATION GROUP GENERATOR

First, rescaling the fields produces the change

$$\Delta L_2 = L\left[(1+\delta)^{(2-\eta)/2}\phi\right] - L[\phi]$$
$$= \left(\frac{2-\eta}{2}\right)\delta\sum_k \phi_a(k)\frac{\partial L}{\partial \phi_\alpha(k)} \quad (9.5.21)$$

Rescaling the momenta by (9.5.19) causes the potential functions to change

$$\Delta v_n(k_1, k_2, \ldots, k_n) = -\delta \sum_{i=1}^n k_i \frac{\partial v_n}{\partial k_i}(k_1, k_2, \ldots, k_n) \quad (9.5.22)$$

We can express this change in a compact way as

$$\Delta L_3 = -\delta \sum_k \phi_\alpha(k) k \frac{\partial'}{\partial k} \frac{\partial L}{\partial \phi_\alpha(k)} \quad (9.5.23)$$

where the prime on the derivative indicates that it acts only on the potential function. Finally, rescaling the number of degrees of freedom (9.5.20) leads to a change in the nth-order term in the free energy $L^{(n)}$,

$$\Delta L_4^{(n)} = d\left(\frac{n}{2}-1\right)\delta L^{(n)}$$
$$= \frac{\delta}{2}d\sum_k \phi_\alpha(k)\frac{\partial L^{(n)}}{\partial \phi_\alpha(k)} - \delta\, dL^{(n)} \quad (9.5.24)$$

and so in general

$$\Delta L_4 = \delta\left[\frac{d}{2}\sum_k \phi_\alpha(k)\frac{\delta L}{\partial \phi_\alpha(k)} - dL\right] \quad (9.5.25)$$

Putting all these together, we have

$$\Delta L = \delta\left\{\frac{V}{(2\pi)^d}\int d\Omega(\hat{p})\left[\ln\left[\frac{\partial^2 L}{\partial \phi(p)\partial \phi(-p)}\right]_{\alpha\alpha}\right.\right.$$
$$\left. - \frac{\partial L}{\partial \phi_\alpha(p)}\left[\frac{\partial^2 L}{\partial \phi(p)\partial \phi(-p)}\right]^{-1}_{\alpha\beta}\frac{\partial L}{\partial \phi_\beta(-p)}\right)$$
$$- \sum_k \phi_\alpha(k) k \frac{\partial'}{\partial k}\frac{\partial L}{\partial \phi_\alpha(k)} + \frac{1}{2}(2-\eta-d)$$
$$\left. \times \sum_k \phi_\alpha(k)\frac{\partial L}{\partial \phi_\alpha(k)} + dL\right\} \quad (9.5.26)$$

296 THE ε EXPANSION

Following the method of Sec. 9.4, we can define a continuous variable l related to the scale factor b by

$$b = e^l \tag{9.5.27}$$

Dividing (9.5.26) by dl and taking the limit $dl \to 0$, we have

$$\frac{\partial L}{\partial l} = \frac{V}{(2\pi)^d}\int d\Omega(\hat{p})\left(\left[\ln\frac{\partial^2 L}{\partial\phi(p)\,\partial\phi(-p)}\right]_{\alpha\alpha}\right.$$

$$\left. - \frac{\partial L}{\partial\phi_\alpha(p)}\left[\frac{\partial^2 L}{\partial\phi(p)\,\partial\phi(-p)}\right]_{\alpha\beta}^{-1}\frac{\partial L}{\partial\phi_\beta(-p)}\right)$$

$$- \sum_k{}^{'}\phi_\alpha(k)k\frac{\partial}{\partial k}\frac{\partial L}{\partial\phi_\alpha(k)} + \frac{1}{2}(2-\eta-d)$$

$$\times \sum_k \phi_\alpha(k)\frac{\partial L}{\partial\phi_\alpha(k)} + dL \tag{9.5.28}$$

The complicated expression on the right-hand side of (9.5.28) is called the Wegner–Houghton infinitesimal generator of the RG.

9.6 SOLUTION OF THE RENORMALIZATION EQUATIONS TO $O(\varepsilon)$

As it stands, the Wegner–Houghton generator is an imposing and intractable nonlinear functional differo-integral equation. To actually get anything out of it, we again make use of the ε expansion. First, we observe that $v_2(0)$ and $v_4(0)$ are of $O(\varepsilon)$, whereas v_6 is of $O(\varepsilon^2)$, and all higher potential functions are of $O(\varepsilon^3)$ and higher. Therefore, to $O(\varepsilon^2)$, we only need to keep up to six-field terms in the free energy, and the RG equations can be closed.

The strategy is to expand the nonlinear terms in the generator in powers of $\partial^2 L/\partial\phi^2$ and keep only terms to $O(\varepsilon^2)$. Explicitly, we have

$$\left[\ln\frac{\partial^2 L}{\partial\phi^2}\right]_{\alpha\alpha} = n\ln v_2(p) + \ln\left[1 + \frac{1}{v_2(p)}\frac{\partial^2 L_4}{\partial\phi^2} + \frac{1}{v_2(p)}\frac{\partial^2 L_6}{\partial\phi^2}\right]$$

$$\simeq n\ln v_2(p) + \frac{1}{v_2(p)}\frac{\partial^2 L_4}{\partial\phi^2} + \frac{1}{v_2(p)^2}\frac{\partial^2 L_6}{\partial\phi^2}$$

$$- \frac{1}{2}\left(\frac{\partial^2 L_4}{\partial\phi^2}\right)^2 + O(\varepsilon^3) \tag{9.6.1}$$

9.6 SOLUTION OF THE RENORMALIZATION EQUATIONS TO $O(\varepsilon)$

and

$$\left(\frac{\partial^2 L}{\partial \phi^2}\right)^{-1} = \frac{1}{v_2(p)}\left[1 + \frac{1}{v_2(p)}\frac{\partial^2 L_4}{\partial \phi^2} + \frac{1}{v_2(p)}\frac{\partial L_6}{\partial \phi^2}\right]^{-1}$$

$$\simeq \frac{1}{v_2(p)} - \frac{1}{v_2(p)^2}\frac{\partial^2 L_4}{\partial \phi^2} - \frac{1}{v_2(p)^2}\frac{\partial^2 L_6}{\partial \phi^2}$$

$$+ \frac{1}{2}\frac{1}{v_2(p)^3}\left(\frac{\partial^2 L_4}{\partial \phi^2}\right)^2 + O(\varepsilon^3) \quad (9.6.2)$$

By collecting terms with the same number of powers of ϕ, we can derive RG equations for the potential functions v_2, v_4, and v_6:

$$\frac{\partial L_2}{\partial l} = (2 - \eta)L_2 - \sum_k \phi_\alpha(k) k \frac{\partial'}{\partial k}\frac{\partial L_2}{\partial \phi_\alpha(k)}$$

$$+ \frac{1}{2}\frac{V}{(2\pi)^d}\int d\Omega(\hat{p}) \frac{1}{v_2(p)} \frac{\partial^2 L_4}{\partial \phi_\alpha(p) \partial \phi_\alpha(-p)} \quad (9.6.3)$$

$$\frac{\partial L_4}{\partial l} = (4 - 2\eta - d)L_4 - \sum_k \phi_\alpha(k) k \frac{\partial'}{\partial k}\frac{\partial L_4}{\partial \phi_\alpha(k)}$$

$$+ \frac{1}{2}\frac{V}{(2\pi)^d}\int d\Omega(\hat{p})\left[\frac{1}{v_2(p)} \frac{\partial^2 L_6}{\partial \phi_\alpha(p) \partial \phi_\alpha(-p)}\right.$$

$$- \frac{1}{2}\frac{1}{v_2(p)^2} \frac{\partial^2 L_4}{\partial \phi_\alpha(p) \partial \phi_\beta(-p)}$$

$$\left.\times \frac{\partial^2 L_4}{\partial \phi_\beta(-p) \partial \phi_\alpha(p)}\right] \quad (9.6.4)$$

$$\frac{\partial L_6}{\partial l} = (6 - \eta - 2d)L_6 - \sum_k \phi_\alpha(k) k \frac{\partial'}{\partial k}\frac{\partial L_6}{\partial \phi_\alpha(k)}$$

$$- \frac{1}{2}\frac{V}{(2\pi)^d}\int d\Omega(\hat{p}) \frac{1}{v_2(p)} \frac{\partial L_4}{\partial \phi_\alpha(p)} \frac{\partial L_4}{\partial \phi_\alpha(-p)} \quad (9.6.5)$$

Before we plunge into the solution of the RG equations, it will be helpful to reflect a little on their structure.

First, we know from the results of Sect. 6.4 that at the fixed point, v_2 and v_4 are independent of the momentum variables to $O(\varepsilon)$. This is consistent so long as the momentum-independent part of v_6 vanishes.

Furthermore, by (9.6.5), v_6 is determined to $O(\varepsilon^2)$ solely by v_4 to $O(\varepsilon)$, that is, by the momentum-independent part of v_4. With this in mind, let us first construct the solution to (9.6.3) and (9.6.4) to $O(\varepsilon)$. To this end, let us separate v_4 into a momentum-independent part $v_4(0)$ and a momentum-dependent part $\tilde{v}_4$,

$$v_4(k_1, k_2; k_3, k_4) = v_4(0) + \tilde{v}_4(k_1, k_2; k_3, k_4) \quad (9.6.6)$$

where again $v_4(0)$ is $O(\varepsilon)$ and $\tilde{v}$ is $O(\varepsilon^2)$. To $O(\varepsilon)$ the second derivatives of L_4 are

$$\frac{\partial^2 L_4}{\partial \phi_\alpha(p) \partial \phi_\beta(-p)} = \frac{v_4(0)}{4!V} \left[4\delta_{\alpha\beta} \sum_{k,k'} \phi_\gamma(k)\phi_\gamma(k')\delta(k+k') \right.$$
$$\left. + 8 \sum_{k,k'} \phi_\alpha(k)\phi_\beta(k')\delta(k+k') \right] \quad (9.6.7)$$

If we substitute (9.6.7) into (9.6.4) and set all momenta to 0, we have

$$\frac{\partial v_4(0;l)}{\partial l} = \varepsilon v_4(0;l) - \frac{n+8}{6} \frac{S_d}{(2\pi)^d} v_4^2(0;l) \quad (9.6.8)$$

where we have used

$$v_2(k;l) = k^2 + v_2(0;l) \quad (9.6.9)$$

and we have assumed that $v_2(0)$ is $O(\varepsilon)$ so that to lowest order in ε, $v_2(\hat{p}) = 1$. Also to $O(\varepsilon)$ we take $\eta = 0$. First, we see that there is a fixed point

$$v_4^* = \frac{6\varepsilon}{n+8} \frac{(2\pi)^d}{S_d} \quad (9.6.10)$$

and linearizing we have, with

$$v_4(0;l) = v_4^* + v_4'(l)$$
$$\frac{\partial v_4'(l)}{\partial l} = \varepsilon v_4'(l) - \frac{n+8}{3} v_4^* v_4'(l) \quad (9.6.11)$$
$$= -\varepsilon v_4'(l)$$

9.6 SOLUTION OF THE RENORMALIZATION EQUATIONS TO $O(\varepsilon)$

which has the simple solution

$$v'_4(l) = v'_4(0)e^{-\varepsilon l} \qquad (9.6.12)$$

By (9.6.12) we see that v_4 is irrelevant, and we regain our previous result (9.4.13).

Let us now turn to (9.6.3). Here we must expand $1/v_2$ to $O(\varepsilon)$,

$$\frac{1}{k^2 + v_2(0)} = \frac{1}{k^2} \frac{1}{1 + v_2(0)/k^2}$$

$$= \frac{1}{k^2}\left(1 - \frac{v_2(0)}{k^2} + \cdots\right) \qquad (9.6.13)$$

Because $v_4(0)$ is irrelevant we can set it to its fixed-point value so that, by (9.6.13), we have, setting $\eta = 0$ to $O(\varepsilon)$,

$$\frac{\partial v_2(0)}{\partial l} = 2v_2(0) + \frac{S_d}{(2\pi)^d}\frac{v_4^*}{6}(n+2) - \frac{S_d}{(2\pi)^d}\frac{v_4^*}{6}(n+2)v_2(0) \qquad (9.6.14)$$

Using the fixed-point value (9.6.10) for v_4^*, the fixed-point value for $v_2(0)$ is

$$v_2^*(0) = -\frac{1}{2}\frac{n+2}{n+8}\varepsilon + O(\varepsilon^2) \qquad (9.6.15)$$

Linearizing around the fixed point, $v_2(0) = v_2^*(0) + v'_2$, we find

$$\frac{\partial v'_2(l)}{\partial l} = 2\left[1 - \frac{n+2}{2(n+8)}\varepsilon\right]v'_2(l) \qquad (9.6.16)$$

which has the solution

$$v'_2(l) = v'_2(0)\exp\left[\left(2 - \frac{n+2}{n+8}\varepsilon\right)l\right] \qquad (9.6.17)$$

These are, of course, just the results we obtained to $O(\varepsilon)$ in Sec. 9.4,

$$y_1 = 2 - \frac{n+2}{n+8}\varepsilon \qquad (9.6.18)$$
$$y_2 = -\varepsilon$$

300 THE ε EXPANSION

9.7 SOLUTION OF THE RG EQUATIONS TO $O(\varepsilon^2)$: EVALUATION OF η

Now that we have solved the RG equations to $O(\varepsilon)$, we are in a position to extend the solution to $O(\varepsilon^2)$. We begin with the RG equation for L_6, (9.6.5), and again observe that we only need to know L_4 to $O(\varepsilon)$, and to this order the first derivative is

$$\frac{\partial L_4}{\partial \phi_\alpha(p)} = \frac{v_4(0)}{6V} \sum_{k_2, k_2', k_2'} \phi_\alpha(k_1) \phi_\beta(k_2) \phi_\beta(k_2') \delta(p + k_1 + k_2 + k_2') \quad (9.7.1)$$

The linear part of the RG equation is easily solved, so first we will address the nonlinear term

$$\frac{1}{2} \frac{V}{(2\pi)^d} \int d\Omega(\hat{p}) \frac{1}{v_2(p)} \frac{\partial L_4}{\partial \phi_\alpha(p)} \frac{\partial L_4}{\partial \phi_\alpha(-p)}$$

$$= \frac{1}{2} \frac{V}{(2\pi)^d} \left(\frac{v_4(0)}{6V} \right)^2 \int d\Omega(\hat{p}) \sum_{k_1, k_2, k_2'} \sum_{k_1', k_3, k_3'}$$

$$\times \left[\phi_\alpha(k_1) \phi_\beta(k_2) \phi_\beta(k_2') \phi_\alpha(k_1') \phi_\delta(k_3) \phi_\delta(k_3') \right.$$

$$\left. \times \delta(p + k_1 + k_2 + k_2') \delta(-p + k_1' + k_3 + k_3') \right] \quad (9.7.2)$$

By the properties of the Kronecker delta, we can write

$$\delta(p + k_1 + k_2 + k_2') \delta(-p + k_1' + k_3 + k_3')$$
$$= \delta(k_1 + k_1' + k_2 + k_2' + k_3 + k_3') \delta(p + k_1 + k_2 + k_2') \quad (9.7.3)$$

Furthermore, we want our expression for v_6 to be explicitly symmetric. This can be achieved simply by summing over the appropriate permutations of the arguments and dividing by the number of permutations. There are in all eight permutations of the arguments, but by using the additional symmetry $p \to -p$, together with the constraint that the sum of all the internal momenta must vanish, the number of independent permutations is reduced to 6. Using (9.7.3), the sym-

9.7 SOLUTION OF THE RG EQUATIONS TO $O(\varepsilon^2)$: EVALUATION OF η

metrized version of (9.7.2) is

$$\frac{1}{2} \frac{V}{(2\pi)^d} \int d\Omega(\hat{p}) \frac{1}{v_2(p)} \frac{\partial L_4}{\partial \phi_\alpha(p)} \frac{\partial L_4}{\partial \phi_\alpha(-p)}$$

$$= \frac{1}{2} \frac{V}{(2\pi)^d} \left(\frac{v_4(0)}{6V}\right)^2 \int d\Omega(\hat{p}) \frac{1}{v_2(p)} \frac{1}{6}$$

$$\times [\delta(p + k_1 + k_2 + k_2') + \delta(p + k_1 + k_3 + k_3') + \delta(p + k_2 + k_1 + k_1')$$

$$+ \delta(p + k_2' + k_1 + k_1') + \delta(p + k_3 + k_1 + k_1') + \delta(p + k_1 + k_1' + k_3')]$$

$$\times \delta(k_1 + k_1' + k_2 + k_2' + k_3 + k_3') \tag{9.7.4}$$

We can now write the RG equation for v_6,

$$\frac{\partial v_6}{\partial l} = (6 - \eta - 2d)v_6 - \sum_{i=1}^{6} k_i \frac{\partial v_6}{\partial k_i} - \frac{5}{3} \frac{v}{(2\pi)^d} \int d\Omega(\hat{p}) \frac{1}{v_2(p)}$$

$$\times [v_4(p_1, k_1; k_2, k_2')v_4(-p, k_1; k_3, k_3')\delta(p + k_1 + k_2 + k_2')$$

$$+ v_4(p, k_1; k_2, k_2')v_4(-p, k_1; k_3, k_3')\delta(p + k_1 + k_2 + k_2')$$

$$+ v_4(p, k_2; k_1, k_1')v_4(-p, k_2'; k_3, k_3')\delta(p + k_2 + k_1 + k_1')$$

$$+ v_4(p, k_2'; k_1, k_1')v_4(-p, k_2; k_3, k_3')\delta(p + k_2' + k_1 + k_1')$$

$$+ v_4(p, k_3; k_1, k_1')v_4(-p, k_3'; k_2, k_2')\delta(p + k_3 + k_1 + k_1')$$

$$+ v_4(p, k_3'; k_1, k_1')v_4(-p, k_3; k_2, k_2')$$

$$\times \delta(p + k_3' + k_1 + k_1')] \tag{9.7.5}$$

We have argued that v_6 is $O(\varepsilon^2)$ and we have already found v_4 to $O(\varepsilon)$, so to $O(\varepsilon^2)$ we can set $\eta = 0$, $d = 4$, and $v_4 = v_4(0)$ in (9.7.5). Furthermore, we are ultimately interested in calculating the corrections to v_4^* to $O(\varepsilon^2)$, which means that we only need to know v_6 at the fixed point. Setting the left-hand side of (9.7.5) to 0, we have to $O(\varepsilon^2)$,

$$2v_6^* + \sum_{i=1}^{6} k_i \frac{\partial v_6^*}{\partial k_i} = -\frac{5}{3} v_4(0)^2 \int d\Omega(\hat{p}) \frac{1}{v_2(p)}$$

$$\times [\delta(p + k_1 + k_2 + k_2') + \delta(p + k_1 + k_2 + k_2')$$

$$+ \delta(p + k_2 + k_1 + k_1') + \delta(p + k_2' + k_1 + k_1')$$

$$+ \delta(p + k_3 + k_1 + k_1') + \delta(p + k_3 + k_1 + k_1')]$$

$$\tag{9.7.6}$$

In (9.7.6) we have translated the Kronecker delta to a Dirac delta according to the prescription

$$\frac{V}{(2\pi)^d}\delta_K(k) \to \delta_D(k)$$

First, let us examine the homogeneous part of (9.7.6), that is, the terms on the left-hand side. Because these terms arise purely from rescaling, it should be no surprise that the solution to the homogeneous part of (9.7.6) is a generalized homogeneous function, and in fact

$$f(\{k_i\}) = b^2 f(\{bk_i\}) \tag{9.7.7}$$

To see that this is indeed a solution, let $b = 1 + \delta$, $\delta \ll 1$. Then to $O(\delta)$ we have

$$f(\{k_i\}) = (1 + 2\delta)f(\{(1 + \delta)k_i\})$$

$$= f(\{k_i\}) + \left[2f(\{k_i\}) + \sum_i k_i \frac{\partial f}{\partial k_i}\right]\delta \tag{9.7.8a}$$

or

$$2f(\{k_i\}) + \sum_i k_i \frac{\partial f}{\partial k_i} = 0 \tag{9.7.8b}$$

The inhomogeneous terms in (9.7.6) are integrals of the form

$$\int d\Omega(\hat{p})\frac{1}{v_2(p)}\delta(p + k) = \delta(1 - |k|) \tag{9.7.9}$$

where we have used the fact that $|p| = 1$ and to $O(1)$, $v_2(p) = 1$ on the momentum shell at the cutoff.

Because the homogeneous part of $f(k)$ goes like $1/k^2$ and its first derivative must have a delta-function singularity as in (9.7.9), consider the function

$$f(k) = \frac{1}{|k|^2}\theta_+(|k| - 1) \tag{9.7.10}$$

9.7 SOLUTION OF THE RG EQUATIONS TO $O(\varepsilon^2)$: EVALUATION OF η

where θ_+ is the step function defined by

$$\theta_+(x) = \begin{cases} 1 & x > 0 \\ 0 & x \leq 0 \end{cases} \tag{9.7.11}$$

Evidently, $f(k)$ has the correct homogeneous behavior and the discontinuity at $|k| = 1$ arising from the step function produces a δ function, so that we find

$$2f(k) + k\frac{\partial f}{\partial k} = \delta(|k| - 1) \tag{9.7.12}$$

The fixed-point function v_6^* is therefore

$$\begin{aligned} v_6^* = -\frac{5}{3}v_4^*(0)^2 [\, & f(k_1 + k_2 + k_2') + f(k_1' + k_2 + k_2') \\ & + f(k_2 + k_1 + k_1') + f(k_2' + k_1 + k_1') \\ & + f(k_3 + k_1 + k_1') + f(k_3' + k_1 + k_1')] \end{aligned} \tag{9.7.13}$$

Note that because v_6 is multiplied by $\delta(k_1 + k_1' + k_2 + k_2' + k_3 + k_3')$ and $f(k) = f(-k)$, one can replace, for example, $f(k_1' + k_2 + k_2') = f(k_1 + k_2 + k_2') = f(k_1 + k_3 + k_3')$.

We should also note a rather subtle point which is that $f(|k|) = 0$ for $|k| \leq 1$. This reflects the fact that in the delta functions that appear in (9.7.6), we can take $|p|$ infinitesimally greater than unity (within the momentum shell), whereas each of the k_i must individually be less than unity.

Now that we have found v_6 to $O(\varepsilon^2)$, let us turn to the solution of the RG equation for v_4, (9.6.4). We have already evaluated the nonlinear term in the limit $k_i \to 0$ to obtain (9.6.8), but here it is instructive to keep the full momentum dependence. The RG equation for v_4 is, to $O(\varepsilon^2)$,

$$\begin{aligned} \frac{\partial v_4}{\partial l} = \varepsilon v_4 &- \sum_{i=1}^{4} k_i \frac{\partial v_4}{\partial k_i} + \frac{1}{10}\frac{1}{(2\pi)^d} \int d\Omega(\hat{p}) \frac{1}{v_2(p)} \\ &\times [nv_6(p, -p; k_1, k_1'; k_2, k_2') + 4v_6(p, k_1; -p, k_1'; k_2, k_2')] \\ &- \frac{1}{6}\frac{1}{(2\pi)^d} \int d\Omega(\hat{p}) \frac{1}{v_2^2(p)} \\ &\times [nv_4(p, -p; k_1, k_1')v_4(p, -p; k_2, k_2')\delta(k_1 + k_1') \\ &\quad + 4v_4(p, -p; k_1, k_1')v_4(p, k_2; -p, k_2')\delta(k_1 + k_1') \\ &\quad + 4v_4(p, k_1; -p, k_2)v_4(p, k_2'; -p, k_1')\delta(k_1 + k_2)] \end{aligned} \tag{9.7.14}$$

[The delta functions in (9.7.14) and the following are again Kronecker deltas.]

Using the solution (9.7.13) for v_6 and replacing v_4 by $v_4(0)$ in the last group of terms in (9.7.14), we have

$$\frac{\partial v_4}{\partial l} = \varepsilon v_4 - \sum_{i=1}^{4} k_i \frac{\partial v_4}{\partial k_i} - \frac{1}{(2\pi)^d} \frac{v_4^2(0)}{6} \int d\Omega(\hat{p})$$
$$\times [nf(p + k_1 + k_1') + nf(p + k_2 + k_2') + 4f(p + k_1 + k_1')$$
$$+ 4f(p + k_2 + k_2') + 4f(p + k_1 + k_2) + 4f(p + k_1 + k_2')]$$
$$+ \frac{S_d}{(2\pi)^d} \frac{v_4^2(0)}{12} [n\delta(k_1 + k_1') + n\delta(k_2 + k_2') + 4\delta(k_1 + k_1')$$
$$+ 4\delta(k_2 + k_2') + 4\delta(k_1 + k_2) + 4\delta(k_1 + k_2')] \quad (9.7.15)$$

Note that in (9.7.15) we have symmetrized the arguments in the last term so that we can combine the last two terms into a single term, giving

$$\frac{\partial v_4}{\partial l} = \varepsilon v_4 - \sum_{i=1}^{4} k_i \frac{\partial v_4}{\partial k_i} - \frac{S_d}{(2\pi)^d} \frac{v_4^2(0)}{6}$$
$$\times [ng(k_1 + k_1') + ng(k_2 + k_2') + 4g(k_1 + k_1')$$
$$+ 4g(k_2 + k_2') + 4g(k_1 + k_2) + 4g(k_1 + k_2')] \quad (9.7.16)$$

where we have defined

$$g(k) = \frac{1}{S_d} \int d\Omega(\hat{p}) \left(f(k + p) + \frac{1}{2}\delta(k) \right) \quad (9.7.17)$$

Note that the delta function in (9.7.17) is a Kronecker delta, and as defined, $g(k)$ is a continuous function. The integral of $f(k + p)$ over the unit sphere has a discontinuity at $k = 0$ because $f(1) = 0$. However, for k infinitesimally different from 0 the value of the integral is $1/2$.

If we now set all the momenta to 0, we recover (9.6.8). Ultimately, we are interested in finding the exponent of η to $O(\varepsilon^2)$, and to do this we need to evaluate the momentum-dependent contributions to the RG equation for V_2, (9.6.3), which come from the momentum-depen-

9.7 SOLUTION OF THE RG EQUATIONS TO $O(\varepsilon^2)$: EVALUATION OF η

dent part of $v_4, \tilde{v}_4$. Because $\tilde{v}_4$ is $O(\varepsilon^2)$, we have at the fixed point

$$-\sum_{i=1}^{4} k_i \frac{\partial \tilde{v}_4}{\partial k_i} - \frac{S_d}{(2\pi)^d} \frac{v_4^2(0)}{6} [ng(k_1 + k_1') + ng(k_2 + k_2')$$

$$+ 4g(k_1 + k_1') + 4g(k_1 + k_2) + 4g(k_1 + k_2')] = 0 \quad (9.7.18)$$

Let us define yet another function $h(k)$, which is the solution of

$$-k \frac{\partial h(k)}{\partial k} = g(k) - g(0) \quad (9.7.19)$$

with the initial condition $h(0) = 0$. Then $\tilde{v}_4$ is given by

$$\tilde{v}_4(k_1, k_1'; k_2, k_2') = \frac{6\varepsilon^2}{(n+8)^2} \frac{(2\pi)^d}{S_d} (nh(k_1 + k_1') + nh(k_2 + k_2'))$$

$$+ 4[h(k_1 + k_1') + h(k_2 + k_2') + h(k_1 + k_2)$$

$$+ h(k_1 + k_2')]) \quad (9.7.20)$$

This completes the solution of the RG equation for v_4 to $O(\varepsilon^2)$. Finally, we can apply this result to the evaluation of the exponent η. Written out in full, (9.6.3) is

$$\frac{\partial v_2}{\partial l} = (2 - \eta)v_2 - k\frac{\partial v_2}{\partial k} + \frac{1}{6}\frac{1}{(2\pi)^d}\int d\Omega(\hat{p})\frac{1}{v_2(p)}$$

$$\times [nv_4(p, -p; k, -k) + 2v_4(p, k; -p, -k)] \quad (9.7.21)$$

Again, we make the decomposition $v_2 = k^2 + v_2(0)$. In order for a fixed point to exist, we must choose η to cancel any k^2 dependence that emerges from the third term on the right-hand side of (9.7.21).

Substituting the solution (9.7.20) for $d = 4$, we have

$$\frac{1}{6}\frac{1}{(2\pi)^d}\int d\Omega(\hat{p})\frac{1}{v_2(p)}[nv_4(p, -p; k, -k) + 2v_4(p, k; -p, -k)]$$

$$= \frac{12\varepsilon^2(n+2)}{(n+8)^2}\frac{1}{S_4}\int d\Omega(\hat{p})h(p+k) \quad (9.7.22)$$

Setting $p \to -p$ in the second term, we have for (9.7.22)

$$\frac{\partial v_2}{\partial l} = (2 - \eta)v_2 - k\frac{\partial v_2}{\partial k}$$

$$+ \frac{12\varepsilon^2(n+2)}{(n+8)^2} \frac{1}{S_4} \int d\Omega(\hat{p}) h(p+k) \quad (9.7.23)$$

The third term on the right-hand side of (9.7.23) can be expanded in powers of $|k|$. The constant term shifts the fixed point v_2^* in $O(\varepsilon^2)$, whereas the next term, which is proportional to k^2, we can denote by ak^2. If we denote the coefficient of the k^2 term in $v_2(k)$ by C (C is initially taken to be 1), then the RG equation for C is

$$\frac{\partial C}{\partial l} = (2 - \eta)C - 2C + a$$

$$= -\eta C + a \quad (9.7.24)$$

We require that C be fixed at $C = 1$, and therefore we must have

$$\eta = a \quad (9.7.25)$$

To find η then, all we need to do is evaluate the leading k-dependent term of the integral in (9.7.23). Expanding $h(k + p)$ for small k,

$$\frac{1}{S_4} \int d\Omega(\hat{p}) h(p+k) = h(1) + \frac{k^2}{4}(3h'(1) + h''(1)) \quad (9.7.26)$$

By (9.7.19)

$$k\frac{\partial h}{\partial k} = kh'(k)$$

$$= -g(k) \quad (9.7.27)$$

and therefore

$$3h'(1) + h''(1) = -2g(1) + 2g(0) - g(1) \quad (9.7.28)$$

so that finally we have

$$\eta = \frac{3}{2} \frac{\varepsilon^2(n+2)}{(n+8)^2} [-2g(1) + g(0) - g'(1)] \quad (9.7.29)$$

Evaluating $g(1)$ and $g'(1)$ by (9.7.17), we have, for $d = 4$,

$$g(1) = \frac{1}{3}\frac{\sqrt{3}}{4\pi}$$
$$g'(1) = \frac{\sqrt{3}}{2\pi} \tag{9.7.30}$$

so finally we have

$$\eta = \frac{(n+2)\varepsilon^2}{2(n+8)^2} \tag{9.7.31}$$

9.8 FIELD-THEORETIC APPROACH TO THE ε EXPANSION

We have pushed the simple RG approach to order ε^2 and have already encountered severe difficulties. It is possible to improve the RG to avoid some of these difficulties, but in practice the ε expansion to higher order is generally carried out within the framework of field theory and Feynman diagrams. At this point the reader who is not familiar with Feynman diagrams should read App. B carefully. Some of what we will do here is repeated in more detail in App. B. The approach outlined here follows the review of Barber [2].

Let us assume that $T > T_c$ and consider the correlation function

$$G_{\alpha\beta}(r) = \frac{1}{Z}\int D[\phi] e^{-L[\phi]} \phi_\alpha(r) \phi_\beta(0) \tag{9.8.1}$$

where we again have

$$L[\phi] = \int d^dx \left[\frac{1}{2}\nabla\phi_\alpha \nabla\phi_\alpha + \frac{r}{2}\phi_\alpha\phi_\alpha + \frac{\lambda}{4!}(\phi'_\alpha\phi_\alpha)^2\right] \tag{9.8.2}$$

If we expand the average in (9.8.1) in powers of λ, we can represent each term by a Feynman diagram. If the diagram contains a disconnected part, it can be factorized so that we have

$$G_{\alpha\beta} = \frac{1}{Z}\left(\sum \text{disconnected diagrams}\right) \times \left(\sum \text{connected diagrams}\right) \tag{9.8.3}$$

Fig. 9.8.1 First-order Feynman diagrams for $G_{\alpha\beta}$.

The sum of disconnected diagrams is identical to the diagrammatic expansion for Z itself and so the correlation function is given as the sum of connected diagrams. (See App. B for details.) Figure 9.8.1 shows the first-order diagrams for G, whereas Figs. 9.8.2a–j show all the second-order diagrams for G. The general structure of a Feynman diagram for the correlation function is shown in Fig. 9.8.3. The shaded blob is called the "self-energy" Σ, and in coordinate space we can write

$$G_{\alpha\beta}(x-y) = G^{(0)}_{\alpha\beta}(x-y)$$

$$+ \int d^d z \int d^d z'\, G^{(0)}_{\alpha\gamma}(x-z)\Sigma_{\gamma\delta}(z-z')G^{(0)}_{\delta\beta}(z'-y) \tag{9.8.4}$$

Taking advantage of translation invariance by Fourier transforming (9.8.4), we have

$$G(k) = G_0(k) + G_0(k)\Sigma(k)G_0(k) \tag{9.8.5}$$

Figures 9.8.4a and b show the first-order self-energy insertions, whereas Figs. 9.8.5a–j show the second-order self-energy insertions. Note that the self-energy diagrams (c)–(f) can be decomposed into two first-order diagrams connected by a single line. Such diagrams are called improper self-energy diagrams. Diagrams that cannot be separated into two self-energy diagrams by cutting a single line are called proper self-energy diagrams. Figure 9.8.6 shows the proper self-energy diagrams in second order. We denote the sum of all proper self-energy diagrams by $\Sigma^*(k)$. The total self-energy is given in terms

9.8 FIELD-THEORETIC APPROACH TO THE ε-EXPANSION

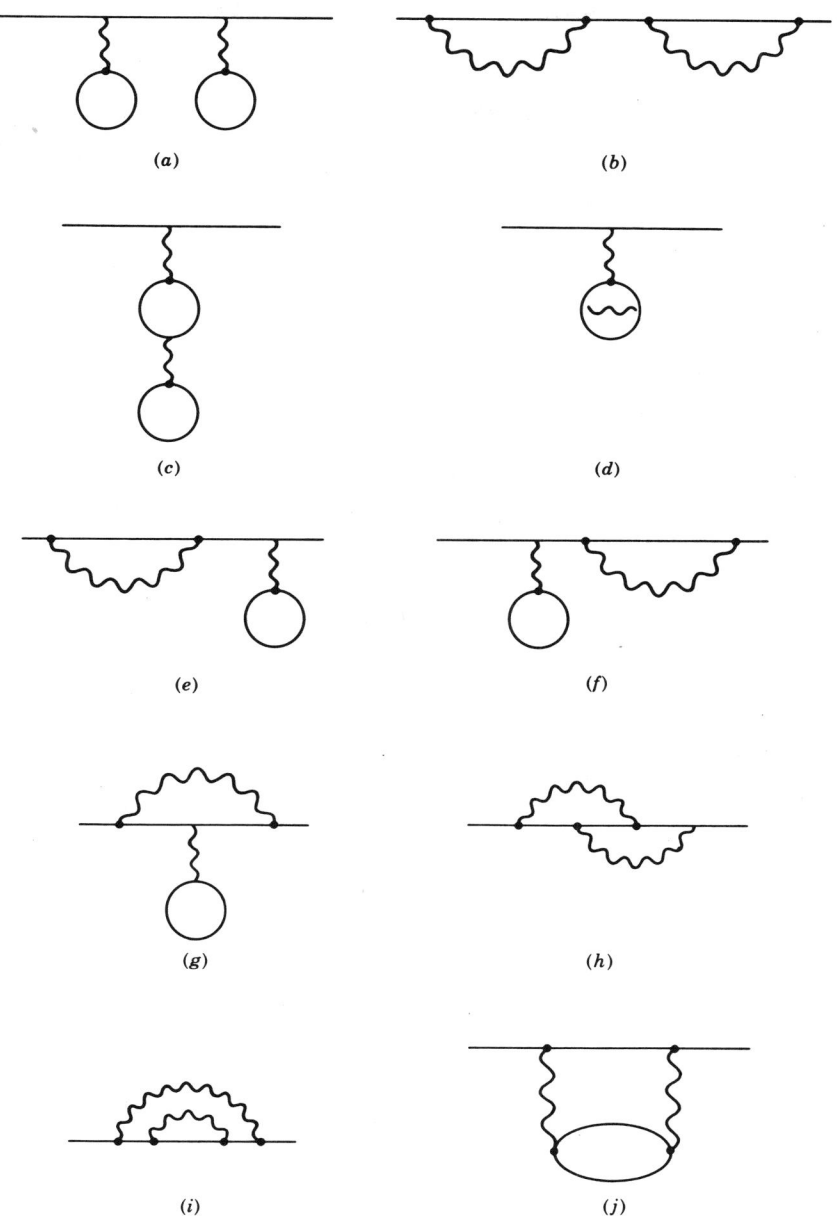

Fig. 9.8.2 Second-order Feynman diagrams for $G_{\alpha\beta}$.

310 THE ε EXPANSION

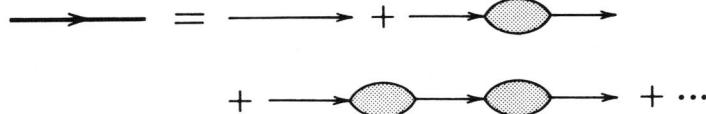

Fig. 9.8.3 Diagrammatic equation for the Green function. The shaded blob is called the self-energy insertion.

Fig. 9.8.4 First-order self-energy insertions.

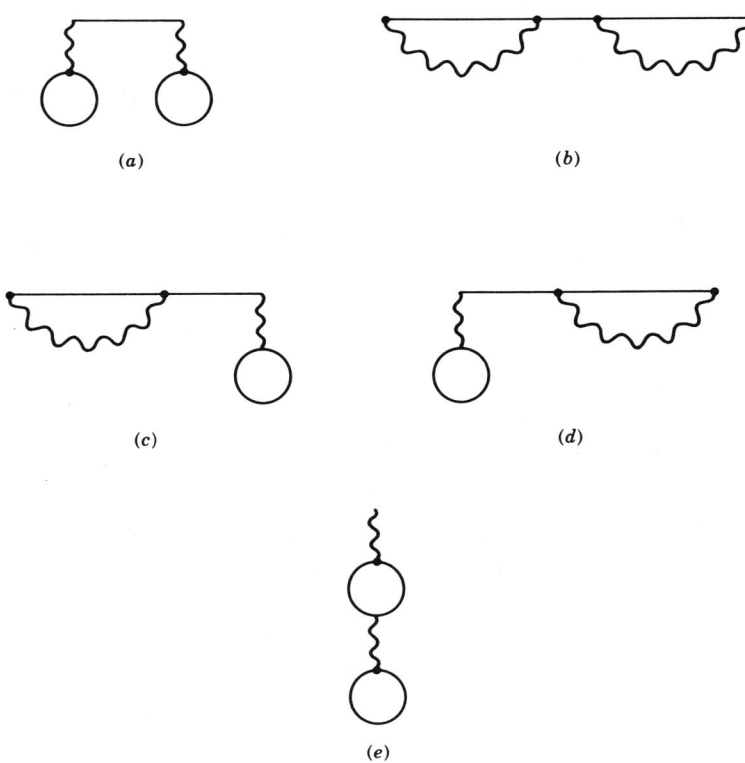

Fig. 9.8.5 Second-order self-energy insertions.

9.8 FIELD-THEORETIC APPROACH TO THE ε-EXPANSION

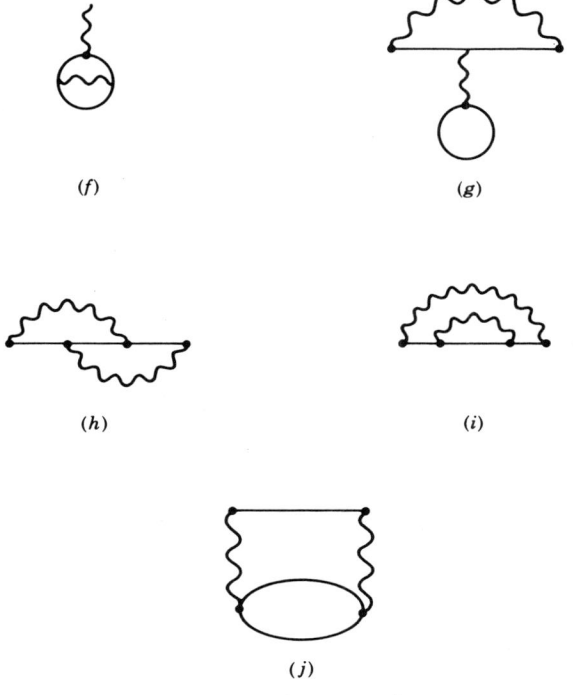

Fig. 9.8.5 *(Continued)*

Fig. 9.8.6 Second-order proper self-energy insertions.

of the proper self-energy by

$$\Sigma(k) = \Sigma^*(k) + \Sigma^*(k)G_0(k)\Sigma^*(k)$$
$$+ \Sigma^*(k)G_0(k)\Sigma^*(k)G_0(k)\Sigma^*(k) + \cdots$$
$$= \frac{\Sigma^*(k)}{1 - G_0(k)\Sigma^*(k)} \tag{9.8.6}$$

where we have formally summed the infinite series.

Taking (9.8.6) together with (9.8.5), we have Dyson's equation for the Green function,

$$G^{-1}(k) = G_0^{-1}(k) - \Sigma^*(k) \tag{9.8.7}$$

Equation (9.8.7) is deceptively simple. It is deceptive because the proper self-energy depends on G and vice versa. To illustrate this, suppose we take a subset of self-energy diagrams Σ_1^* and define

$$G_1^{-1}(k) = G_0^{-1}(k) - \Sigma_1^*(k) \tag{9.8.8}$$

we then have

$$G^{-1}(k) = G_1^{-1}(k) - \Sigma_2^*(k) \tag{9.8.9}$$

where Σ_2^* are all the remaining self-energy insertions. This equation has the same structure as (9.8.7), but we must keep two things in mind. First, we have already taken into account all self-energy insertions of the type Σ_1^*, and so no diagrams of this kind can appear in Σ_2^*. Second, in constructing the diagrams for Σ_2^*, the solid lines now represent G_1 rather than G_0. The power of the Feynman–Dyson approach is that a finite-order evaluation of Σ^* leads to an expression for $G(k)$ to all orders.

At the critical point the correlation length diverges and

$$G^{-1}(k = 0) = 0 \qquad T = T_c \tag{9.8.10}$$

Defining

$$\tilde{r} = r - \Sigma^*(k = 0) \tag{9.8.11}$$

9.8 FIELD-THEORETIC APPROACH TO THE ε-EXPANSION

and

$$W(k) = \Sigma^*(k) - \Sigma^*(k = 0) \tag{9.8.12}$$

Dyson's equation (9.8.7) takes the form

$$G^{-1} = k^2 + \tilde{r} - W(k) \tag{9.8.13}$$

We must now evaluate the proper self-energy up to second order in λ. To first order we have Figs. 9.8.4a and b, which yield (see App. B)

$$\Sigma_1^*(k) = \frac{1}{6}(n + 2)\int d^d q\, G^{(0)}(q) \tag{9.8.14}$$

Note that there is no dependence on the external momentum k, so Σ_1^* does not contribute to $W(k)$, but only to $\tilde{r}$.

Once we absorb the first-order diagrams into G_1, we automatically take into account some of the second-order self-energy insertions. The only diagrams that are not taken into account are Figs. 9.8.5e and f. These diagrams are most easily evaluated in momentum space, and Fig. 9.8.7 shows the definitions of the momentum variables. The sum

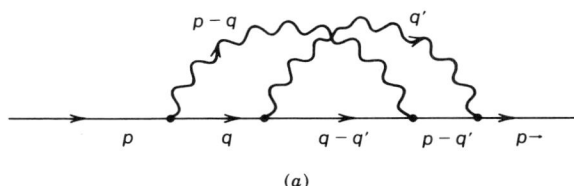

(a)

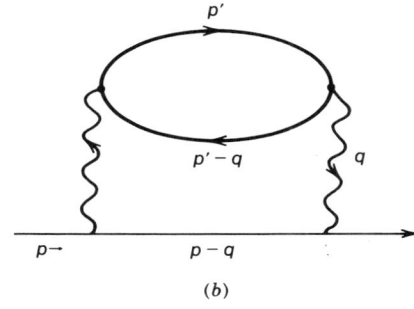

(b)

Fig. 9.8.7 Second-order contributions to the proper self-energy remaining after the first-order terms are taken into account.

of these two terms is

$$W(p) = \frac{(n+2)\lambda^2}{18} \int \frac{d^4q}{(2\pi)^4} \int \frac{d^4q'}{(2\pi)^4} \frac{1}{q^2} \frac{1}{|q-q'|^2} \left(\frac{1}{|p-q'|^2} - \frac{1}{q^2} \right)$$
(9.8.15)

We need to evaluate $W(p)$ to $O(\varepsilon^2)$. Because λ is of $O(\varepsilon)$ at the fixed point, the coefficient of the integral is already of order ε^2 and we can set $d = 4$ in the integral (9.8.15).

We first consider the integral

$$R(q) = \int \frac{d^4q'}{(2\pi)^4} \frac{1}{|q'|^2} \frac{1}{|q-q'|^2}$$

$$= \frac{1}{4\pi^3} \int_0^\Lambda dq' \, q' \int_0^\pi d\phi \, \frac{\sin^2\phi}{q^2 + q'^2 - 2qq'\cos\phi} \quad (9.8.16)$$

The angular integral can be performed first

$$\int_0^\pi d\phi \, \frac{\sin^2\phi}{|q|^2 + |q'|^2 - 2qq'\cos\phi} = \begin{cases} \dfrac{\pi}{2|q'|^2} & q' > q > 0 \\[2mm] \dfrac{\pi}{2|q^2|} & q > q' > 0 \end{cases} \quad (9.8.17)$$

Now performing the remaining integral over q,

$$R(q) = \frac{1}{8\pi^2} \int_0^q \frac{q'}{q^2} + \frac{1}{8\pi^2} \int_q^\Lambda \frac{dq'}{q}$$

$$= \frac{1}{16\pi^2} \left(1 - \ln \frac{q^2}{\Lambda^2} \right) \quad (9.8.18)$$

Substituting (9.8.18) into (9.8.15), we have

$$W(p) = \frac{(n+2)\lambda^2}{(3 \cdot 4!)^2} \int \frac{d^4q}{(2\pi)^4} R(q) \left[\frac{1}{|p-q|^2} - \frac{1}{|q|^2} \right] \quad (9.8.19)$$

The angular integration for the first term on the left-hand side is again given by (9.8.17), which gives

$$W(p) = \frac{(n+2)\lambda^2}{(3 \cdot 4!)^2} \frac{1}{256\pi^4} \int_0^\Lambda dq\, q^3 \left[1 - \ln \frac{q^2}{\Lambda^2}\right]$$
$$\times \left[\frac{1}{q^2}\theta(q-p) + \frac{1}{p^2}\theta(p-q) - \frac{1}{q^2}\right]$$
$$= \frac{(n+2)\lambda^2}{(3 \cdot 4!)^2} \frac{1}{256\pi^4} \int_0^\Lambda dq\, q^3 \left(1 - \ln \frac{q^2}{\Lambda^2}\right)\left(\frac{1}{p^2} - \frac{1}{q^2}\right) \quad (9.8.20)$$

Performing the q integration, we have

$$W(p) = \frac{(n+2)\lambda^2}{(3 \cdot 4!)^2} \frac{1}{512\pi^4} \left(p^2 \ln \frac{p}{\Lambda} - \frac{5}{4}p^2\right) \quad (9.8.21)$$

If we return to (9.8.13), setting $\tilde{r} = 0$, we have

$$p^2 G(p) \simeq 1 + \frac{W(p)}{p^2}$$
$$= 1 + \frac{(n+2)\lambda^2}{(3 \cdot 4!)^2} \frac{1}{512\pi^4} \ln p \quad (9.8.22)$$

In order to find η, we note first that at $T = T_c$ (see Sec. 4.0)

$$G(p) \sim p^{-2+\eta} \quad (9.8.23)$$

Therefore, expanding $p^2 G(p)$ in a series in η, which we know to be small,

$$p^2 G(p) \sim p^\eta$$
$$= 1 + \eta \ln p + \frac{1}{2}(\eta \ln p)^2 + \cdots \quad (9.8.24)$$

Comparison of (9.8.22) with (9.8.24) shows that η is just the coefficient of the $\ln p$ term. Taking for λ its value at the fixed point, (9.8.9),

$$\eta = \frac{1}{2}\frac{n+2}{(n+8)^2}\varepsilon^2 \quad (9.8.25)$$

which agrees with the value calculated in the previous section.

With just a little more effort, we can use the results we have obtained thus far to obtain the susceptibility exponent γ. First, we note that the susceptibility χ is given in terms of the correlation function by

$$\chi = \int d^d r \, G(r) \tag{9.8.26}$$

or, in terms of the Fourier transform of $G(r)$,

$$\chi = G(k = 0) \tag{9.8.27}$$

By (9.8.7) we have

$$\chi^{-1} = \chi_0^{-1} - \Sigma^*(k = 0)$$
$$= \tilde{r} \tag{9.8.28}$$

where χ_0 is the susceptibility of the noninteracting system

$$\chi_0^{-1} = r \tag{9.8.29}$$

At the critical temperature χ^{-1} vanishes, and recalling that r is a function only of the temperature, the critical temperature is given by the implicit equation

$$\tilde{r} = r_c - \Sigma^*(k = 0)$$
$$= 0 \tag{9.8.30}$$

Near T_c we expect $\chi \sim |T - T_c|^{-\gamma}$, or

$$\tilde{r} = r - r_c - (\Sigma^*(0, \tilde{r}) - \Sigma^*(0, 0))$$
$$\sim |r - r_c|^{\gamma} \tag{9.8.31}$$

Dividing by $\tilde{r}$, we have

$$1 + \frac{\Sigma^*(0, \tilde{r}) - \Sigma^*(0, 0)}{\tilde{r}} \sim \tilde{r}^{(1/\gamma - 1)} \tag{9.8.32}$$

To first order in ε,

$$\Sigma^*(0, \tilde{r}) - \Sigma^*(0, 0) = -\frac{(n+2)\lambda}{6} \int_0^\Lambda \frac{d^4k}{(2\pi)^4} \left[\frac{1}{\tilde{r} + k^2} - \frac{1}{k^2}\right] \quad (9.8.33)$$

For small $\tilde{r}$ the dominant term in (9.8.33) is

$$\Sigma^*(0, \tilde{r}) - \Sigma^*(0, 0) \sim -\frac{1}{32\pi^2} \frac{\lambda}{3}(n+2)\tilde{r} \ln \tilde{r} \quad (9.8.34)$$

Substituting this into (9.8.32) gives, after setting $\lambda = \lambda^*$ [see (9.4.9)],

$$1 - \frac{1}{2}\frac{n+2}{n+8}\varepsilon \ln \tilde{r} = \tilde{r}^{(1/\gamma - 1)} \quad (9.8.35)$$

Now in the mean field approximation $\gamma = 1$, so we can expect that $1/\gamma - 1$ is of the order ε. We can then expand the right-hand side of (9.8.35) as

$$\tilde{r}^{(1/\gamma - 1)} \simeq 1 + \left(\frac{1}{\gamma} - 1\right)\ln \tilde{r} + \cdots \quad (9.8.36)$$

Comparing coefficients of $\ln \tilde{r}$ in (9.8.35) and (9.8.36), we see that to $O(\varepsilon)$,

$$\frac{1}{\gamma} = 1 - \frac{1}{2}\frac{n+2}{n+8}\varepsilon \quad (9.8.37)$$

or

$$\gamma = 1 + \frac{1}{2}\frac{n+2}{n+8}\varepsilon \quad (9.8.38)$$

9.9 CRITICAL BEHAVIOR FOR $d = 4$: THE UPPER CRITICAL DIMENSION

It is clear that $d = 4$ plays a special role in the theory of critical phenomena. Above $d = 4$ the critical exponents of the n-vector model take on their classical, or Gaussian, values, whereas for $d < 4$ we find nontrivial critical behavior. An important and interesting question we should ask is: What exactly happens at $d = 4$?

318   THE ε EXPANSION

From the point of view of the RG, at $d = 4$ the Gaussian and WF fixed points merge into a single fixed point. By (9.6.8) and (9.6.14) the RG equations for v_2 and v_4 become, for $d = 4$,

$$\frac{\partial r}{\partial l} = 2r + \frac{n+2}{6}\lambda \qquad (9.9.1)$$

$$\frac{\partial \lambda}{\partial l} = -\frac{n+8}{6}\lambda^2 \qquad (9.9.2)$$

Equation (9.9.2) can be integrated at once to give

$$\lambda(l) = \frac{1}{1 + \lambda_0\left(\dfrac{n+8}{6}\right)l} \qquad (9.9.3)$$

So as $l \to \infty$, λ tends to 0 as we would expect for the Gaussian fixed point.

By (9.9.1) the dominant behavior of r is $r \sim e^{2l}$. For large l, $r \gg 1$ and (9.9.1) can be simplified to

$$\frac{dr}{dl} = \left(2 - \frac{n+2}{6}\lambda\right)r \qquad (9.9.4)$$

Dividing by r and using the large-l limit of λ,

$$r(l) = r_0 e^{2l} l^{-(n+2)/(n+8)} \qquad (9.9.5)$$

In terms of the scale factor $b = e^l$,

$$r(b) = b_0 b^2 (\ln b)^{-(n+2)/(n+8)} \qquad (9.9.6)$$

The leading term is what we would expect for the Gaussian fixed point in the limit $\varepsilon \to 0$ for the Wilson–Fisher fixed point. The logarithmic correction occurs, because, although λ tends to 0 with l, it does so as a power of l rather than exponentially, which means that λ is a marginal variable. It is characteristic of marginal variables that they lead to corrections to the dominant scaling behavior.

To see how the logarithmic corrections affect the familiar scaling laws, consider the case of the correlation length ξ. Because r is

9.9 CRITICAL BEHAVIOR FOR d = 4: THE UPPER CRITICAL DIMENSION

proportional to the reduced temperature t, we have

$$\xi(t) = b\xi\left[b^2(\ln b)^{-(n+2)/(n+8)}t\right] \qquad (9.9.7)$$

Choosing b such that the argument on the right-hand side of (9.9.7) is unity, we have (for large b)

$$\xi(t) = t^{-1/2}|\ln t|^{(n+2)/2(n+8)} \qquad (9.9.8)$$

The leading behavior is again what one would expect from the Gaussian fixed point, and again we see that the leading singular behavior is modified by a logarithmic term.

Except for the specific heat, which in $d = 4$ has an additional logarithmic factor arising from the fluctuations, the logarithmic corrections to the thermodynamic quantities of interest are given by the scaling hypothesis and (9.9.8).

The free energy in the Gaussian model is

$$F = \frac{1}{2}\int_0^\Lambda \frac{d^dk}{(2\pi)^d} \ln(k^2 + r) \qquad (9.9.9)$$

The specific heat can be found by differentiating the free energy twice with respect to r, which is proportional to the reduced temperature. This gives

$$C = \frac{1}{2}\int_0^\Lambda \frac{d^dk}{(2\pi)^d} \frac{1}{(k^2+r)^2} \qquad (9.9.10)$$

Defining a dimensionless variable $x = k^2/r$, we have for $d = 4$,

$$C = \frac{\Lambda^2}{r}\int_0^{\Lambda^2/r} dx \frac{x^3}{(1+x^2)^2} \qquad (9.9.11)$$

As $T \to T_c$, $r \to 0$ and the integral diverges logarithmically as

$$C \sim \frac{S_4}{4(2\pi)^4} \ln \frac{\Lambda^2}{r} \qquad (9.9.12)$$

If we now write down the scaling form of the free energy

$$F = b^{-d} f(\xi(b) t) \tag{9.9.13}$$

where we generalize the usual power-law behavior of $\xi(b)$ to the function of b in (9.9.6), then

$$C \sim b^{-d} \xi^2(b) f''(\xi(b) t) \tag{9.9.14}$$

Substituting in form $\xi(b)$, we have

$$C \sim b^{4-d} (\ln b)^{-2(n+2)/(n+8)} f''[\xi(b) t] \tag{9.9.15}$$

Choosing b so that $\xi(b) t = 1$ and keeping in mind the general prescription that

$$\lim_{d \to 4} b^{4-d} \to \ln b \tag{9.9.16}$$

we find

$$C \sim (\ln t)^{(4-n)/(n+8)} \tag{9.9.17}$$

We see from these calculations that in $d = 4$ dimensions the leading singular behavior is given by the mean field theory, modified by logarithmic factors. Interestingly, the power of the logarithm, which depends only on n, the number of components of the order parameter, is exactly the coefficient of ε in the expression for the critical exponent calculated in the ε expansion, see Table 9.4.1.

PROBLEMS

9.1 For the quartic interaction of (9.0.4), expand $V_l[\phi^{(l)}; \phi^{(s)}]$ in powers of $\phi^{(l)}$ and verify (9.1.5).

9.2 Derive (9.2.3) from (9.2.1).

9.3 Expand (9.3.1) and average term by term to get (9.3.2) and (9.3.3).

9.4 Calculate the six-field term in the free energy which arises from the second cumulant.

9.5 Evaluate the integral in (9.3.8) and derive the form for $G_s(x)$ for $d = 4$.

9.6 Calculate the integrals in (9.4.4) and (9.4.5).

9.7 Given the values for y_T and η to $O(\varepsilon)$, calculate the remaining exponents in Table 9.4.1.

9.8 Verify equations (9.5.21), (9.5.23), and (9.5.25).

9.9 Derive (9.6.8).

9.10 Work out (9.7.5) in detail.

9.11 Verify that (9.7.10) is a solution to (9.7.12) by direct substitution.

9.12 Evaluate $g(1)$ and $g'(1)$ from (9.7.17), see (9.7.30).

REFERENCES

1. Amit, D. J. (1978) *Field Theory, the Renormalization Group, and Critical Phenomena*.
2. Barber, M. N. (1977) *Phys. Rep.* **29C** 1.
3. Ma, S. K. (1976) *Modern Theory of Critical Phenomena*. W. A. Benjamin, Reading, MA.
4. Wilson, K. G. (1972) *Phys. Rev. Lett.* **28** 248.
5. Wilson, K. G. and Fisher, M. E. (1972) *Phys. Rev.* **28** 248.
6. Wilson, K. G. and Kogut, J. (1974) *Phys. Rep.* **12C** 75.
7. Wegner, F. J. and Houghton, A. (1973) *Phys. Rev. A* **8** 401.

10

THE SPHERICAL MODEL AND THE $1/n$ EXPANSION

10.0 INTRODUCTION

We have seen that the Landau–Ginzburg model can be solved exactly in $d = 4$ dimensions, and that a systematic expansion in four dimensions can be derived for the critical exponents. There is another exactly solvable model, the spherical model [3], which has been shown to be equivalent to the $n \to \infty$ limit of the n-vector model [7–9]. The observation that the critical behavior of the $1/n$ model can be determined exactly in the limit $1/n \to 0$ led to the idea of a perturbation theory in powers of $1/n$, which is called the $1/n$ expansion.

In this section we introduce the spherical model and derive the equation of state following the method of Baxter [2]. A more complete discussion including alternate approaches to the problem can be found in Joyce [4]. Apart from serving as the basis of the $1/n$ expansion, these results are of interest in their own right because the spherical model is the only model that has, thus far, been solved in the presence of a magnetic field. The original motivation for the spherical model was to cook up a model that was like the Ising model, but instead of having spins at each site with the restriction

$$s_i^2 = 1 \quad \text{Ising model} \qquad (10.0.1)$$

Berlin and Kac [3] relaxed this somewhat and allowed the spin at each site to take on any real value subject to the global constraint

$$\sum_i s_i^2 = N \qquad \text{Spherical model} \qquad (10.0.2)$$

where N is the number of sites in the lattice. The name of the model derives from the observation that, if we imagine the N-dimensional space spanned by the s_i, then the allowed states of the Ising model are the corners of the unit N-dimensional hypercube, whereas the allowed states of the spherical model are all points on the surface of the hypersphere of radius N.

The partition function for the spherical model in an external magnetic field h is

$$Z = \int D[s] \exp\left(J \sum_{\langle ij \rangle} s_i s_j + h \sum_i s_i\right) \delta\left(N - \sum_i s_i^2\right) \qquad (10.0.3)$$

where, as usual,

$$\int D[s] \equiv \prod_{i=1}^{N} \int_{-\infty}^{\infty} ds_i \qquad (10.0.4)$$

The delta function ensures the constraint (10.0.2). Because of this constraint we can add to the Hamiltonian any term of the form $a(N - \sum s_i^2)$, where a is an arbitrary constant, without changing the value of the partition function. The motivation for including this term will be clear in a moment.

If we represent the delta function by its Fourier transform, the partition function becomes

$$Z = \int_{-\infty}^{\infty} \frac{d\omega}{2\pi} \int D[s] \exp\left(J \sum_{\langle ij \rangle} s_i s_j + h \sum_i s_i - (a + i\omega) \sum_i s_i^2\right) \qquad (10.0.5)$$

Note that the integrations over the spin variables are now Gaussian, and the quadratic form can be diagonalized if we introduce the Fourier-transformed fields $s(k)$,

$$s_i = \frac{1}{\sqrt{N}} \sum_k e^{ikx_i} s(k) \qquad (10.0.6)$$

In terms of the $s(k)$, the partition function has the form

$$Z = \int_{-\infty}^{\infty} \frac{d\omega}{2\pi} \int D[s(k)] \exp\left(-\sum [a + i\omega - J\Delta(k)]s(k)s(-k) \right.$$
$$\left. + h\sqrt{N}\, s(0) + (a + i\omega)N\right) \quad (10.0.7)$$

where $\Delta(k)$ is given by

$$\Delta(k) = \sum_{i=1}^{d} \cos k_i \quad (10.0.8)$$

It is important to note that because s_i is real, the Fourier coefficients $s(k)$ and $s(-k)$ are related by the condition

$$s(-k) = s^*(k) \quad (10.0.9)$$

and therefore we are to understand the integrations over the field $s(k)$ with $k \neq 0$ as

$$\int ds(k)\, ds(-k) = \int d\mathcal{R}e[s(k)]\, d\mathcal{I}m[s(k)] \quad (10.0.10)$$

The order of integrations can be reversed, and the Gaussian integrations performed, so long as the real part of the coefficient of the quadratic term is positive. Therefore, we must choose $a > dJ$, and for $k > 0$ we find that

$$\int_{-\infty}^{\infty} d^2s(k) e^{-[a+i\omega-J\Delta(k)]|s(k)|^2} = \frac{1}{a + i\omega - J\Delta(k)} \quad (10.0.11)$$

The integration over the field $s(0)$ can be done separately, and we have

$$\int_{-\infty}^{\infty} ds(0) \exp\left(-[a + i\omega - dJ]s^2(0) + \sqrt{N}\, hs(0)\right)$$
$$= \frac{\exp\left(\frac{1}{4}\frac{Nh^2}{a + i\omega - dJ}\right)}{\sqrt{a + i\omega - dJ}} \quad (10.0.12)$$

Equations (10.0.11) and (10.0.12) can be combined to give, apart from an uninteresting constant factor,

$$Z = \int_{-\infty}^{\infty} \frac{d\omega}{2\pi} \exp\left(N\left[\frac{1}{4}\frac{h^2}{a+i\omega-dJ} + a + i\omega \right.\right.$$
$$\left.\left. - \frac{1}{2N}\sum_k \ln[a+i\omega-\Delta(k)]\right]\right) \quad (10.0.13)$$

Let us now define a new variable z by

$$z = \frac{(a+i\omega+dJ)}{J} \quad (10.0.14)$$

so that the partition function is

$$\int_{-i\infty+c}^{i\infty+c} \frac{dz}{2\pi} e^{N\phi(z)} \quad (10.0.15)$$

where $c = a - dJ$ and we have defined the function $\phi(z)$,

$$\phi(z) = \frac{h^2}{4Jz} + Jz + Jd - \frac{1}{2}g(z) \quad (10.0.16)$$

and $g(z)$ is defined as

$$g(z) = \frac{1}{N}\sum_k \ln[z + d - \Delta(k)] \quad (10.0.17)$$

Thus far, we have succeeded in representing the partition function in terms of a single integral. In the limit $N \to \infty$ we can approximate the integral by the steepest-descents method. We assume that $\phi(z)$ has a maximum for $z = z_0$, and expand about z_0. The conditions that z_0 be a maximum are

$$\frac{d\phi}{dz}(z_0) = 0 \quad (10.0.18a)$$

$$\frac{d^2\phi}{dz^2}(z_0) < 0 \quad (10.0.18b)$$

If we now expand $\phi(z)$ about z_0 and deform the contour as shown in

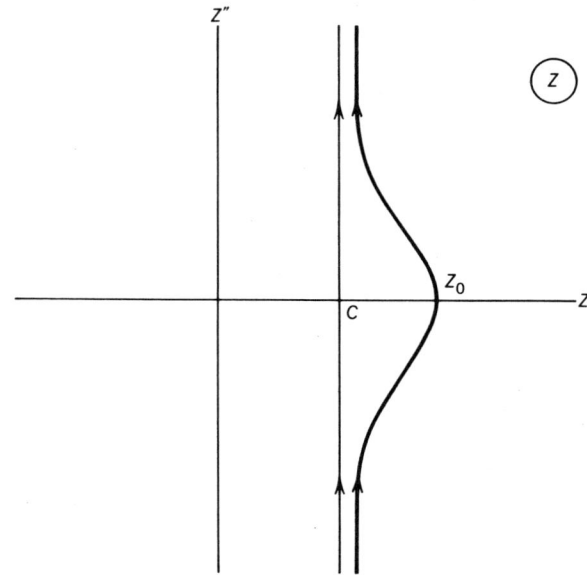

Fig. 10.0.1 Contour of integration for evaluating the free energy in the steepest-descents approximation.

Fig. 10.0.1, we find

$$Z = e^{N\phi(z_0)} \int_{-\infty}^{\infty} dz \, e^{(N/2)g''(z_0)(z-z_0)^2} \qquad (10.0.19)$$

In the limit $N \to \infty$, we only need the leading dependence on N, which is simply $N\phi(z_0)$. Therefore, we finally arrive at the following expression for the free energy per spin:

$$f = -\phi(z_0) \qquad (10.0.20)$$

10.1 THE EQUATION OF STATE AND CRITICAL EXPONENTS FOR THE SPHERICAL MODEL

Equation (10.0.18a) is an implicit relation for z_0, and in general it is rather difficult to evaluate explicitly the function $g(z)$, (10.0.17). We can, however, derive a second relation for z_0 in terms of the magneti-

10.1 THE EQUATION OF STATE AND CRITICAL EXPONENTS

zation by differentiating the free energy with respect to h,

$$\begin{aligned} m &= -\frac{\partial f}{\partial h} \\ &= \frac{\partial \phi}{\partial h}(z_0) + \frac{\partial z_0}{\partial h}\phi'(z_0) \\ &= \frac{h}{2Jz_0} \end{aligned} \qquad (10.1.1)$$

because $\phi'(z_0) = 0$. If we expand (10.0.18a) we can write (10.1.1) in the form

$$1 - m^2 = \frac{g'(z_0)}{2J} \qquad (10.1.2)$$

This is the exact equation of state, and it relates the magnetization to the applied field and the temperature (through J). At the critical point the ratio h/m, which is essentially the inverse of the magnetic susceptibility, vanishes in the limit $h \to 0$, and so the critical temperature is given by

$$J_c = \tfrac{1}{2}g'(0) \qquad (10.1.3)$$

We are by now well acquainted with the fact that close to the critical point the singular behavior of thermodynamic functions arises from the long-wavelength fluctuations of the system. If we approximate the function $\Delta(k)$ for small k by $\Sigma_i(1 - \tfrac{1}{2}k_i^2)$, we find, replacing the sum over k by an integral,

$$\Delta g'(z) = g'(z) - g'(0) = -\int \frac{d^d k}{(2\pi)^d} \frac{z}{\frac{k^2}{2}\left(z + \frac{k^2}{2}\right)} \qquad (10.1.4)$$

If we now let $z \to \lambda z$, we have

$$\Delta g'(\lambda z) = \lambda^{\tfrac{1}{2}d - 1} \Delta g'(z) \qquad (10.1.5)$$

Therefore, $g'(z) - g'(0)$ is a homogeneous function of z of order

$\frac{1}{2}d - 1$, that is,

$$\Delta g'(z) \cong -Az^{d/2-1} \qquad (10.1.6)$$

If we substitute this into the equation of state (10.1.2), we have

$$h = 2mJ_c\left[\frac{2J_c}{A}(t + m^2)\right]^{2/(d-2)} \qquad (10.1.7)$$

where $t = J_c/J - 1$ plays the role of the reduced temperature.

We can now use the equation of state (10.1.7) to derive the critical exponents. First, if we set $T = T_c$, we have

$$m \sim h^{(d-2)/(d+2)} \qquad (10.1.8)$$

From which we see that the critical isotherm exponent is $\delta = (d + 2)/(d - 2)$. Next, in the limit $h \to 0$, the magnetization is given by ($t < 0$)

$$m^2 = -t \qquad (10.1.9)$$

and we see that $\beta = 1/2$. If we differentiate (10.1.7) with respect to h, we find in the limit $m \to 0$,

$$\chi = \frac{\partial m}{\partial h} \sim t^{-2/(d-2)} \qquad (10.1.10)$$

so that $\gamma = 2/(d - 2)$.

The critical exponents for the spherical model are collected in Table 10.1.1.

TABLE 10.1.1 Critical Exponents for the Spherical Model

α	$(d - 4)/(d - 2)$
β	$1/2$
γ	$2/(d - 2)$
δ	$(d + 2)/(d - 2)$
η	0
ν	$1/(d - 2)$

10.2 CRITICAL PROPERTIES OF THE n-VECTOR MODEL IN THE LIMIT $n \to \infty$

Now that we have outlined the solution for the spherical model we must make contact with the physically more interesting n-vector model. As we mentioned in Sec. 10.0, the critical properties of the n-vector model in the limit $n \to \infty$ are identical to those of the spherical model. We begin by writing the free-energy functional in the form

$$L = \int d^d x \left(\tfrac{1}{2} \nabla \phi_\alpha \nabla \phi_\alpha + U[\phi_\alpha \phi_\alpha] \right) \tag{10.2.1}$$

The quantity $\phi^2 = \phi_\alpha \phi_\alpha$ is of $O(n)$, and at any reasonable fixed point we should expect the potential and kinetic parts of the free energy to be of the same order in n, so we assume that $U(\phi^2) \sim O(n)$. We now expand U as a power series in ϕ^2,

$$U(\phi^2) = \sum_{k=1}^{\infty} \frac{u_{2k}}{k} \frac{(\phi^2)^k}{2} \tag{10.2.2}$$

[The definition of the u_{2k} in (10.2.2) is a convention due to Ma [5], which we will follow here.]

If $U(\phi^2)$ is to be of $O(n)$, then it follows that

$$u_{2k} \sim O(n^{1-k}) \tag{10.2.3}$$

This assignment of the n-dependence of the coupling constant u_{2k} is called Ma ordering. Because ϕ^2 is the sum of n terms, where n is large we can expect that the relative fluctuation in ϕ^2 will be small

$$\lim_{n \to \infty} \frac{\langle \phi^4 - \langle \phi^2 \rangle^2 \rangle}{\langle \phi^2 \rangle^2} \sim O\left(\frac{1}{n}\right) \tag{10.2.4}$$

This suggests that we expand the potential about $\langle \phi^2 \rangle$ in powers of the fluctuation $\phi^2 - \langle \phi^2 \rangle$. This procedure is very reminiscent of mean field theory, and indeed, $\langle \phi^2 \rangle$ must be determined self-consistently just as the mean field is. Here, however, we do not introduce symmetry breaking and the method is actually equivalent to the

Hartree approximation. In this and the next section we are following Ma [6]. Expanding U, we have

$$U(\phi^2) = U(\langle\phi^2\rangle) + U'(\langle\phi^2\rangle)(\phi^2 - \langle\phi^2\rangle)$$
$$+ \tfrac{1}{2}U''(\langle\phi^2\rangle)(\phi^2 - \langle\phi^2\rangle)^2 + \cdots \quad (10.2.5)$$

If we keep just the first two terms

$$U(\phi^2) \cong U(\langle\phi^2\rangle) - U'(\langle\phi^2\rangle)\langle\phi^2\rangle + U'(\langle\phi^2\rangle)\phi^2 \quad (10.2.6)$$

the free energy becomes, apart from constant terms,

$$L \cong \tfrac{1}{2}\int d^dx \left(\nabla\phi_\alpha \nabla\phi_\alpha + t(\langle\phi^2\rangle)\phi_\alpha\phi_\alpha\right) \quad (10.2.7)$$

where

$$t(\phi^2) = 2U'(\phi^2) \quad (10.2.8)$$

We must determine $\langle\phi^2\rangle$ self-consistently

$$\langle\phi^2\rangle = \frac{\int D[\phi]\exp\left(-\tfrac{1}{2}\int d^dx[\nabla\phi_\alpha\nabla\phi_\alpha + t(\langle\phi^2\rangle)\phi_\alpha\phi_\alpha]\right)\phi_\alpha\phi_\alpha}{\int D[\phi]\exp\left(-\tfrac{1}{2}\int d^dx[\nabla\phi_\alpha\nabla\phi_\alpha + t(\langle\phi^2\rangle)\phi_\alpha\phi_\alpha]\right)}$$
$$= \frac{n}{N}\sum \frac{1}{k^2 + t(\langle\phi^2\rangle)} \quad (10.2.9)$$

At the critical point $t(\langle\phi^2\rangle)$ vanishes, and converting the sum over k to an integral,

$$\langle\phi^2\rangle_c = n\int_0^\Lambda \frac{d^dk}{(2\pi)^d}\frac{1}{k^2}$$
$$= \frac{nS_d}{(2\pi)^d}\frac{\Lambda^{d-2}}{d-2} \quad (10.2.10)$$

where the subscript c indicates that this is the value at the critical point.

Evidently, $\langle\phi^2\rangle$ is playing the role of r in the usual Landau–Ginzburg theory, and therefore the deviation of $\langle\phi^2\rangle$ from its critical value is a measure of the deviation of the temperature from T_c.

10.2 THE N-VECTOR MODEL IN THE LIMIT $n \to \infty$

Because the susceptibilty is related to the Green function by

$$\chi = \int d^d x\, G(x)$$
$$= G(k = 0) \qquad (10.2.11)$$

we have

$$\chi^{-1} = t(\langle \phi^2 \rangle)$$
$$\sim [\langle \phi^2 \rangle - \langle \phi^2 \rangle_c]^\gamma \qquad (10.2.12)$$

where γ is the usual susceptibility exponent.

By (10.2.9), again going over to an integral,

$$\langle \phi^2 \rangle - \langle \phi^2 \rangle_c = n \int_0^\Lambda \frac{d^d k}{(2\pi)^d} \left[\frac{1}{k^2 + t(\langle \phi^2 \rangle)} - \frac{1}{k^2} \right] \qquad (10.2.13)$$

In the limit $\Lambda^2/t(\langle \phi^2 \rangle) \gg 1$ the upper limit on the integral can be replaced by ∞, and we have

$$\langle \phi^2 \rangle - \langle \phi^2 \rangle_c = -\frac{nS_d}{(2\pi)^d} [t(\langle \phi^2 \rangle)]^{(d-2)/2} \int_0^\infty dx\, \frac{x^{d-3}}{1+x^2} \qquad (10.2.14)$$

The definite integral in (10.2.14) exists for $2 < d < 4$ and it has the value

$$I_d = \int_0^\infty dx\, \frac{x^{d-3}}{1+x^2}$$
$$= \frac{\pi}{2}(d-2) \csc \frac{\pi}{2}(d-2) \qquad (10.2.15)$$

Substitution of (10.2.15) into (10.2.14) gives

$$\gamma = \frac{2}{d-2} \qquad (10.2.16)$$

which is just the result obtained for the spherical model [see (10.1.10)]. Because $\eta = 0$, see Table 10.1.1, the other exponents follow by scaling.

10.3 RENORMALIZATION GROUP IN THE LIMIT $n \to \infty$

Thus far, we have not employed the RG to solve the $n \to \infty$ limit of the n-vector model. This really is not necessary as the model can be solved exactly. However, it is instructive to carry out the construction of the RG equations in this case [6].

As usual, we begin by dividing the fields into $\phi^{(l)}$: $0 \le k \le \Lambda/b$ and $\phi^{(s)}$: $\Lambda/b < p \le \Lambda$. Let us define $\langle \cdots \rangle_s$ to be

$$\langle \cdots \rangle_s = \frac{\int D[\phi_s] e^{-L[\phi^{(s)}, \phi^{(l)}]}(\cdots)}{\int D[\phi_s] e^{-L[\phi^{(s)}, \phi^{(l)}]}} \quad (10.3.1)$$

The partition function is

$$Z = \int D[\phi^{(l)}] \int D[\phi^{(s)}] e^{-L[\phi^{(l)}, \phi^{(s)}]}$$

$$= \int D[\phi^{(l)}] \exp\left(-\tfrac{1}{2} \int d^d x \, \nabla\phi_\alpha^{(l)} \nabla\phi_\alpha^{(l)}\right)$$

$$\times \int D[\phi^{(s)}] \exp\left(-\tfrac{1}{2} \int dx \, \nabla\phi_\alpha^{(s)} \nabla\phi_\alpha^{(s)}\right) e^{-U[(\phi)](l) + \phi^{(s)})^2]} \quad (10.3.2)$$

We can argue that $(\phi^{(l)} + \phi^{(s)})^2 \approx (\phi^{(l)})^2 + (\phi^{(s)})^2$ because $(\phi^{(l)})^2$ and $(\phi^{(s)})^2$ are $O(n)$, whereas $\phi^{(s)}\phi^{(l)}$ is $O(1)$ (see Prob. 10.4). We again make the argument that the fluctuation in ϕ^2 due to the fields $\phi^{(s)}$ is small compared with ϕ^2 itself and so we write

$$U[\phi_\alpha \phi_\alpha] \simeq U[\phi_\alpha^{(l)}\phi_\alpha^{(l)} + n_s] - \frac{t}{2}[\phi_\alpha^{(l)}\phi_\alpha^{(l)} + n_s]n_s$$

$$+ \frac{t}{2}[\phi_\alpha^{(l)}\phi_\alpha^{(l)} + n_s]\phi_\alpha^{(s)}\phi_\alpha^{(s)} \quad (10.3.3)$$

where n_s is given selfconsistently

$$n_s = \langle \phi_\alpha^{(s)} \phi_\alpha^{(s)} \rangle_s$$

$$= \frac{n}{N} \sum_k \left(\frac{1}{k^2 + t[\phi_\alpha^{(l)}\phi_\alpha^{(l)} + n_s]} \right) \quad (10.3.4)$$

Note that the average in (10.3.4) is over the short-wavelength fields only.

10.3 RENORMALIZATON GROUP IN THE LIMIT $n \to \infty$

The partition function now becomes

$$Z = \int D[\phi^{(l)}] \exp\left(-\tfrac{1}{2}\int d^dx\, \nabla\phi_\alpha^{(l)} \nabla\phi_\alpha^{(l)} + U[\phi^{(l)2} + n_s]\right)$$

$$\times \int D[\phi^{(s)}] \exp\left(-\tfrac{1}{2}\int d^dx\, \nabla\phi_\alpha^{(s)} \nabla\phi_\alpha^{(s)} + t[\phi_\alpha^{(l)2} + n_s]\phi_\alpha^{(s)}\phi_\alpha^{(s)}\right) \quad (10.3.5)$$

The integration over the short-wavelength modes is Gaussian and can be carried out with the result

$$\int D[\phi^{(s)}] \exp\left(-\tfrac{1}{2}\int d^dx\, \nabla\phi_\alpha^{(s)} \nabla\phi_\alpha^{(s)} + U[\phi^{(l)2} + n_s]\right)$$
$$= \tfrac{1}{2}\sum \ln[k^2 + t(\phi^{(l)2} + n_s)] \quad (10.3.6)$$

Here we are simply doing the functional integral in little steps, and at each step the self-consistency condition (10.3.4) must be satisfied.

If we now rescale the momenta and the fields, the renormalized potential $U'(\phi^2)$ is

$$U'(\phi^2) = b\Big\{U(b^{2-d}\phi^2 + n_s) - \tfrac{1}{2}b^2 t(b^{2-d}\phi^2 + n_s)n_s$$
$$+ \tfrac{1}{2}\sum_k \ln[k^2 + b^2 t(b^{2-d}\phi^2 + n_s)]\Big\} \quad (10.3.7)$$

It is important to note that n_s minimizes $U'(\phi^2)$,

$$\frac{\partial U'(\phi^2)}{\partial n_s} = 0 \quad (10.3.8)$$

Using this result, a simpler RG equation can be obtained for t,

$$t'(\phi^2) \cong b^2 t(b^{2-d}\phi^2 + n_s) \quad (10.3.9)$$

This, together with (10.3.5) for n_s, constitute the RG equation for the function t. These equations are highly nonlinear, but they can be cast in a much simpler form if we consider an infinitesimal RG transformation

$$b = 1 + \delta \qquad \delta \ll 1 \quad (10.3.10)$$

with

$$x = \frac{(2\pi)^d(d-2)}{nS_d\Lambda^2}\phi^2$$

we find (see Prob. 10.5)

$$\frac{\partial t}{\partial b} = 2t + (2-d)\left(x - \frac{1}{1+t}\right)\frac{dt}{dx} \qquad (10.3.11)$$

At the fixed point $t^*(x)$, the left-hand side of (10.3.11) is 0 and we have

$$2t^* + (2-d)\left(x - \frac{1}{1+t^*}\right)\frac{dt^*}{dx} = 0 \qquad (10.3.12)$$

Rather than trying to find $t^*(x)$, let us change variables [10] and look for an implicit solution $x(t^*)$. By (10.3.12) we have

$$\frac{dx}{dt^*} = \frac{(d-2)}{2t^*}\left(x - \frac{1}{1+t^*}\right) \qquad (10.3.13)$$

The initial condition for this first-order differential equation is

$$x(0) = 1 \qquad (10.3.14)$$

which fixes the coefficient of the term in U^* proportional to x to be 1.
With a little algebra, (10.3.13) can be cast in the form

$$t^*\frac{dx}{dt^*} = \frac{d-2}{2}(x-1) + \frac{d-2}{2}\frac{t^*}{1+t^*} \qquad (10.3.15)$$

Let us "guess" a solution of the form

$$x = 1 + t^{(d-2)/2}f(t) \qquad (10.3.16)$$

Substituting (10.3.16) into (10.3.15) gives

$$t^{*(d-2)/2}\frac{df}{dt^*} = \frac{d-2}{2}\frac{1}{1+t^*} \qquad (10.3.17)$$

which can be integrated to give

$$f(t^*) = \frac{d-2}{2}\int_0^{t^*} ds\, \frac{s^{-(d-2)/2}}{1+s} \qquad (10.3.18)$$

10.3 RENORMALIZATON GROUP IN THE LIMIT $n \to \infty$

Putting these results together, the fixed-point solution $t^*(x)$ is given implicitly by

$$x = 1 + \frac{d-2}{2} t^{*(d-2)/2} \int_0^{t^*} ds \, \frac{s^{-(d-2)/2}}{1+s} \quad (10.3.19)$$

To find the scaling fields, let $t = t^* + t'$. Then t' satisfies

$$\frac{\partial t'}{\partial b} = 2t' + (2-d)\left(x - \frac{1}{1+t^*}\right)\frac{\partial t'}{\partial x}$$

$$+ (2-d)\frac{dt^*}{dx}\frac{1}{(1+t^*)^2}t' \quad (10.3.20)$$

We now again change variables from $x \to t^*$, which gives

$$\frac{\partial t'}{\partial b} = 2t' + (2-d)\left(x - \frac{1}{1+t^*}\right)\frac{dt^*}{dx}\frac{\partial t'}{\partial t^*}$$

$$+ (2-d)\frac{dt^*}{dx}\frac{1}{(1+t^*)^2}t' \quad (10.3.21)$$

By (10.3.12) this can be simplified somewhat to

$$\frac{\partial t'}{\partial x} = 2t' + 2t^*\frac{\partial t'}{\partial t^*} + (2-d)\frac{dt^*}{dx}\frac{1}{(1+t^*)^2}t' \quad (10.3.22)$$

Finally, let us define a new function τ,

$$t' = \frac{dt^*}{dx}\tau \quad (10.3.23)$$

So that (10.3.21) becomes

$$\frac{\partial \tau}{\partial b} = 2\tau - 2t^*\left(\frac{dt^*}{dx}\right)^{-1}\frac{\partial}{\partial t^*}\left(\frac{dt^*}{dx}\tau\right)$$

$$+ (2-d)\left(\frac{dt^*}{dx}\right)^2\frac{1}{(1+t^*)^2}\tau \quad (10.3.24)$$

Using the identity

$$\frac{\partial}{\partial t^*}\left(\frac{dt^*}{dx}\right) = -\left(\frac{dt^*}{dx}\right)^2 \frac{d^2x}{dt^{*2}} \qquad (10.3.25)$$

and (10.3.12), after some algebra we find

$$\frac{d\tau}{db} = (d-2)\tau - 2t^* \frac{\partial \tau}{\partial t^*} \qquad (10.3.26)$$

If we assume that τ is a monomial in t^*,

$$\tau = (t^*)^m \qquad (10.3.27)$$

then (10.3.27) becomes

$$\frac{d\tau}{db} = (d - 2 - 2m)\tau \qquad (10.3.28)$$

from which we see that the eigenvalues of the linearized RG are

$$y_m = d - 2 - 2m \qquad (10.3.29)$$

As usual, the largest ($m = 0$) eigenvalue is y_T,

$$y_T = \frac{1}{\nu}$$

$$= d - 2 \qquad (10.3.30)$$

which agrees with our earlier result (see Table 10.1.1). The other critical exponents can be determined by the scaling laws along with the result $\eta = 0$.

10.4 FIELD-THEORETIC APPROACH TO THE $1/n$ EXPANSION

Just as with the ε expansion, further light can be shed on the $1/n$ expansion by applying field-theoretic methods to the correlation function. The discussion in this section parallels that of Sec. 9.8 rather

10.4 FIELD-THEORETIC APPROACH TO THE 1/n EXPANSION

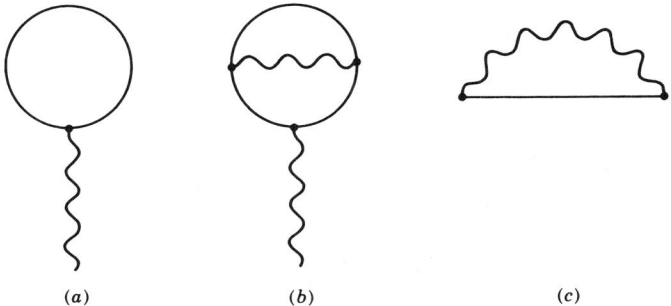

Fig. 10.4.1 First-order proper self-energy insertions.

closely [1]. We begin with the usual free-energy functional, (9.0.4)

$$L = \frac{1}{2} \int d^d x (\nabla \phi_\alpha \nabla \phi_\alpha + r \phi_\alpha \phi_\alpha) + \frac{\lambda}{4!} \int d^d x (\phi_\alpha \phi_\alpha)^2 \quad (10.4.1)$$

where, according to Ma ordering, we take $\lambda \sim (1/n)$.

At first sight this might seem to make our job quite easy because all we need to do is calculate to first order in $1/n$. Unfortunately, as we know from the rules for constructing Feynman diagrams, each closed loop yields a factor n, so we must carefully keep track of all diagrams for G of $O(1)$ and $O(1/n)$.

We begin this discussion with the lowest-order proper self-energy insertion, shown in Fig. 10.4.1. Because Fig. 10.4.1a contains a factor of λ and a closed loop, it is of $O(1)$, whereas Figs. 10.4.1b and c are of $O(1/n)$.

To $O(1)$, the Green function is given by the self-consistent Hartree approximation, shown diagrammatically in Fig. 10.4.2. In the limit $n \to \infty$, the Hartree approximation is exact and yields the spherical model critical exponent $\eta = 0$. The Hartree approximation to the

Fig. 10.4.2 Dyson equation for the Green function in the Hartree approximation.

Green function is given by

$$G_H^{-1}(p) = p^2 + r_H \qquad (10.4.2)$$

where r_H is given self-consistently by

$$r_H = r - \frac{2n\lambda}{4!} \int \frac{d^d q}{(2\pi)^d} \frac{1}{q^2 + r_H} \qquad (10.4.3)$$

It is easy to see that (10.4.2) leads to the spherical model values for the critical exponents. First, by (10.4.3) at $r = r_c$, r_H vanishes and $G_H \approx p^{-2}$, so $\eta = 0$. Near r_c we can write

$$r_H(r) = r - r_c - \frac{2n\lambda}{4!} \int \frac{d^d p}{(2\pi)^d} \left[\frac{1}{p^2 - r_H} - \frac{1}{p^2} \right] \qquad (10.4.4)$$

This equation should be compared with (10.2.13). Note that here r_H^2 is playing the same role as $t(\langle \phi^2 \rangle)$.

Defining $x = p/\sqrt{r_H}$, (10.4.4) can be written in the form

$$r_H = r - r_c - \frac{2n\lambda}{4!} r_H^{(d-2)/2} I_d \qquad (10.4.5)$$

where I_d is given by (10.2.15). To leading order in t we have $r_H \sim t^\gamma$, so by (10.4.5) we must again have

$$\gamma = \frac{2}{d-2} \qquad (10.4.6)$$

If we want to go beyond the $n \to \infty$ limit and calculate the exponents to $O(1/n)$, we need to keep all diagrams of $O(1/n)$. Let Σ^* denote the sum of all proper self-energy insertions of $O(1/n)$. Dyson's equation for the Green function is

$$G^{-1}(k) = G_H^{-1}(k) - \Sigma^*(k) \qquad (10.4.7)$$

As in Sec. 9.8 we split off the $p = 0$ part of the self-energy and define

$$W(k) = \Sigma^*(k) - \Sigma^*(0) \qquad (10.4.8)$$

10.4 FIELD-THEORETIC APPROACH TO THE $1/n$ EXPANSION

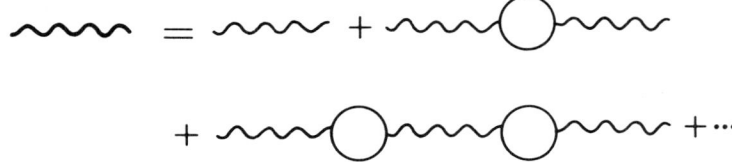

Fig. 10.4.3 Sums of diagrams defining the effective interaction in the ring approximation.

Dyson's equation now becomes

$$G^{-1}(k) = p^2 + \tilde{r} - W(k) \qquad (10.4.9)$$

and $\tilde{r}$ is the inverse susceptibility to $O(1/n)$.

In constructing the self-energy insertion to $O(1/n)$, it is important to realize that in addition to Figs. 10.4.1b and c in which the "bare" coupling λ appears, we can replace the bare coupling by the series of ring diagrams shown in Fig. 10.4.3. This series defines what is called the effective interaction in the ring approximation and is represented by the heavy wavy line in the figure. Because each additional factor of λ is $O(1/n)$ and each additional loop is $O(n)$, each term in the sum is $O(1/n)$. This series can be expressed in the compact diagrammatic equation shown in Fig. 10.4.4.

It is important to keep in mind that the propagator lines in Figs. 10.4.3 and 10.4.4 represent factors of $G_1(p)$, where

$$G_1^{-1}(p) = p^2 + \tilde{r} \qquad (10.4.10)$$

To $O(1/n)$ then, the full set of self-energy insertions are these shown in Figs. 10.4.5a and b. Note that the bare coupling appears in Fig. 10.4.5a along with the effective coupling. The reasons for this is that the Hartree approximation has already summed all diagrams with an arbitrary number of "tadpole" self-energy insertions, Fig. 10.4.1a, so we do not need to take this sum into account again in the interaction

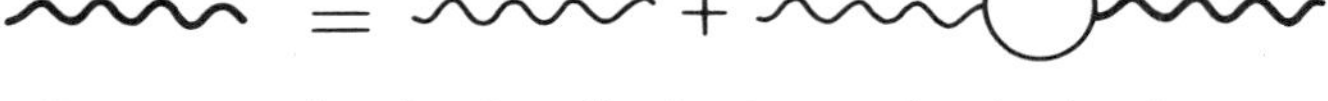

Fig. 10.4.4 Dyson equation for the effective interaction in the ring approximation.

Fig. 10.4.5 (a) Self-energy insertion contributing to $\tilde{r}$. (b) Self-energy insertion contributing to $W(p)$.

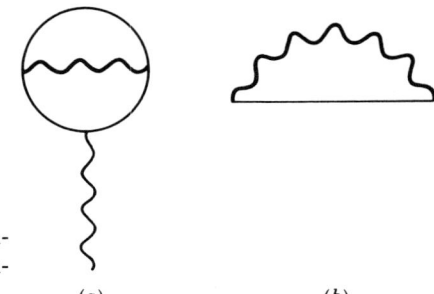

line. Writing out the Dyson equation, Fig. 10.4.4, we have

$$u(p) = \lambda - \frac{2n\lambda}{4!}\Pi(p)u(p) \qquad (10.4.11)$$

where $\Pi(p)$ is the factor coming from the closed loop, and it has the value

$$\Pi(p) = \int d^d q \, G_1(q) G_1(p-q) \qquad (10.4.12)$$

In the language of many-body theory, $\Pi(p)$ is a polarization insertion, and is to the effective interaction what the self-energy is to the Green function.

Equation (10.4.11) is easily solved for the effective interaction

$$u(p) = \frac{\lambda}{1 + \dfrac{2n\lambda}{4!}\Pi(q)} \qquad (10.4.13)$$

The only contribution to $W(p)$ to $O(1/n)$ comes from Fig. 10.4.5b; Fig. 10.4.5a contributes to $\tilde{r}$. We have, to $O(1/n)$,

$$W(p) = -\frac{\lambda}{4!} \int \frac{d^d q}{(2\pi)^d} \frac{1}{\dfrac{2n\lambda}{4!}\Pi(q)} [G_1(p-q) - G_1(q)] \qquad (10.4.14)$$

10.4 FIELD-THEORETIC APPROACH TO THE $1/n$ EXPANSION

At the critical point we can set $\tilde{r} = 0$, and in this case

$$\Pi(p) = \int \frac{d^d q}{(2\pi)^d} \frac{1}{q^2} \frac{1}{|p-q|^2}$$
$$= A_d p^{d-4} \qquad (10.4.15)$$

where

$$A_d = \frac{\Gamma^2\left(\frac{d-2}{2}\right)\Gamma\left(\frac{4-d}{2}\right)}{2^d \pi^{d/2} \Gamma(d-2)} \qquad (10.4.16)$$

and Γ is the usual gamma function.

For $d < 4$, $\Pi(p)$ diverges at small p, so in (10.4.14) we can approximate

$$1 + \frac{2n\lambda}{4!}\Pi(q) \sim \frac{2n\lambda}{4!}\Pi(q) \qquad (10.4.17)$$

which gives

$$W(p) = -\frac{1}{2nA_d} \int \frac{d^d q}{(2\pi)^d} q^{4-d} \left[\frac{1}{|p-q|} - \frac{1}{q^2}\right] \qquad (10.4.18)$$

We want to isolate the small-p behavior of $W(p)$. For small p one finds

$$W(p) \sim \eta p^2 \ln p \qquad (10.4.19)$$

where

$$\eta = 4\left(\frac{4}{d} - 1\right)\frac{S_d}{n} \qquad (10.4.20)$$

The Green function at $T = T_c$ is therefore for large n,

$$G(p) \sim p^2(1 - \eta \ln p)$$
$$\cong p^{2-\eta} \qquad (10.4.21)$$

and so we can identify (10.4.20) with the critical exponent η.

The other critical exponents have been worked out to $O(1/n)$, and they are given in Table 10.4.1.

TABLE 10.4.1 Critical Exponents to $O(1/n)$

α	$\left(\dfrac{d-4}{d-2}\right)\left(1 + 8\dfrac{d-1}{d-4}\dfrac{S_d}{n}\right)$
γ	$\dfrac{2}{d-2}\left(1 - 6\dfrac{S_d}{n}\right)$
ν	$\dfrac{1}{d-2}\left[1 - 8\left(\dfrac{d-1}{d}\right)\dfrac{S_d}{n}\right]$
β	$\dfrac{1}{2} - \left(\dfrac{2}{d-2}\right)(2d-1)\dfrac{S_d}{n}$
η	$\dfrac{4}{d}(4-d)\dfrac{S_d}{n}$

PROBLEMS

10.1 Derive (10.0.12) by explicitly evaluating the integral

$$\int_{-\infty}^{\infty} ds(0) e^{-[a+i\omega - dJ]s^2(0) + \sqrt{N}hs(0)}$$

10.2 Verify that the function $\Delta g'(z)$ given in (10.1.4)

$$\Delta g'(z) = -\int_{-\infty}^{\infty} \frac{d^d k}{(2\pi)^d} \frac{z}{\dfrac{k^2}{2}\left(z + \dfrac{k^2}{2}\right)}$$

is a homogeneous function of z as in (10.1.5).

10.3 Derive the result of (10.2.9) by introducing the Fourier components

$$\phi_\alpha(x) = \frac{1}{\sqrt{N}} \sum_k e^{ikr} \phi_\alpha(k)$$

10.4 For the free-energy functional

$$L_0 = \tfrac{1}{2}\int d^d x \, \nabla\phi_\alpha(x) \nabla\phi_\alpha(x)$$

show explicitly that

$$\sum_\alpha \left\langle [\phi_\alpha^{(l)}(x)\phi_\alpha^{(s)}(x)]^2 \right\rangle_s \simeq O(1)$$

Note that $\phi_\alpha^{(l)}(x)$ can be regarded as a fixed vector.

10.5 Derive the renormalized form for t, (10.3.9) (assume $\eta = 0$).

REFERENCES

1. Barber, M. N. (1977) *Phys. Reports* **29C** 1.
2. Baxter, R. J. (1982) *Exactly Solved Models in Statistical Mechanics.* Academic, New York.
3. Berlin, T. H. and Kac, M. (1952) *Phys. Rev.* **86** 821.
4. Joyce, G. S. (1974) In *Phase Transitions and Critical Phenomena* (C. Domb and M. S. Green, eds.), vol. 2. Academic, New York.
5. Ma, S. K. (1973) *Rev. Mod. Phys.* **45** 589.
6. Ma, S. K. (1976) In *Phase Transitions and Critical Phenomena* (C. Domb and M. S. Green, eds.), vol. 6. Academic, New York.
7. Stanley, H. E. (1968) *Phys. Rev. Lett.* **20** 589.
8. Stanley, H. E. (1968) *Phys. Rev.* **176** 718.
9. Stanley, H. E. (1969) *J. Phys. Soc. Japan* **26** 102.
10. Vvedensky, D. D. and Chang, T. S. private communication.

11

THE TWO-DIMENSIONAL X–Y MODEL AND THE KOSTERLITZ–THOULESS TRANSITION

11.0 INTRODUCTION

The isotropic X–Y model consists of classical two-component spins $S_i = (S_{x,i}, S_{y,i})$ subject to the constraint that the magnitude of the spin at each site is 1,

$$S_{x,i}^2 + S_{y,i}^2 = 1 \qquad (11.0.1)$$

The Hamiltonian for this system is

$$H = -J \sum_{\langle ij \rangle} S_i S_j \qquad (11.0.2)$$

If we define θ_i as the angle made by each spin with respect to some arbitrary direction, then the Hamiltonian can be written as

$$H = -J \sum_{\langle ij \rangle} \cos(\theta_i - \theta_j) \qquad (11.0.3)$$

The X–Y model was originally introduced as a model of superfluid He^4 in which the order parameter is a complex scalar field. If one neglects fluctuations in the magnitude of the order parameter, then the excitations of the system are described in terms of the phase angle of the "condensate wave function." The angle θ plays a role analogous to the phase of the condensate wave function in the theory of superfluid He^4. The low-energy excitations of the X–Y model will consist of arrangements of the spins that vary slowly from point to point. Therefore, it is possible to replace the cosine by a quadratic form, and the discrete difference by a gradient. Apart from uninteresting constant terms, the Hamiltonian can be expressed as

$$H = J \int d^2 r (\nabla \theta)^2 \qquad (11.0.4)$$

The partition function of the X–Y model can then be expressed as a functional integral over all "textures" of the field $\theta(x)$,

$$Z = \int D[\theta(x)] \exp - \beta J \int d^2 r |\nabla \theta|^2 \qquad (11.0.5)$$

The "most probable field" $\phi(x)$ is the function that maximizes the Boltzmann weight, and therefore is a solution of the equation

$$\frac{\delta H}{\delta \theta}[\phi] = 0$$

or

$$\nabla^2 \phi = 0 \qquad (11.0.6)$$

In addition to the rather uninteresting solution $\phi(x) = \phi_0$, a constant, Laplace's equation in two dimensions has singular solutions of the form (see Prob. 11.1)

$$\phi = n \tan^{-1}\left[\frac{y - y_0}{x - x_0}\right] \qquad (11.0.7)$$

The point-like singularity is at (x_0, y_0), and the solution has the property that

$$\int dl \cdot \nabla \phi = 2\pi n \qquad (11.0.8)$$

Such a spin configuration is shown in Fig. 11.0.1 and we refer to it as a "vortex." The vortex solution of Laplace's equation, (11.0.6), is the simplest example of a fascinating class of objects called "topological excitations." In the present situation the symmetry group of the Hamiltonian (11.0.2) or (11.0.4) is the abelian group $O(2)$ and the

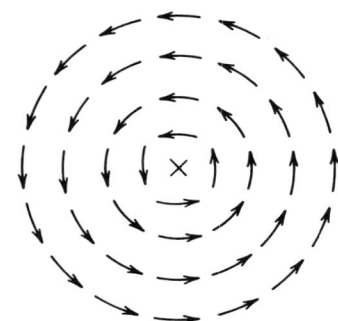

Fig. 11.0.1 Spin configuration of an isolated $n = 1$ vortex in the planar X–Y model.

"topological charge," n of (11.0.8), is the circulation. However, when one considers systems in which the symmetry group of H is nonabelian, then the nature of the topological excitations and the form of the topological invariants, or charges, exhibits a rich variety.

The one-vortex solution can be generalized to a many-vortex solution by virtue of the linearity of (11.0.6), and the general vortex solution has the form

$$\phi(x) = \sum_{i=1}^{N} n_i \tan^{-1}\left[\frac{y - y_i}{x - x_i}\right] \qquad (11.0.9)$$

Because ϕ is a solution of Laplace's equation in two dimensions, one can introduce its harmonic conjugate χ and define a function $f(z)$ of the complex variable $z = x + iy$,

$$f(z) = \phi + i\chi \qquad (11.0.10)$$

which is analytic except at the singular points. In hydrodynamics ϕ is called the velocity potential and χ is the stream function. Because the singularities in f are at isolated points, the Cauchy–Riemann relations hold for the partial derivatives

$$\begin{aligned}\frac{\partial \phi}{\partial x} &= \frac{\partial \chi}{\partial y} \\ \frac{\partial \phi}{\partial y} &= -\frac{\partial \chi}{\partial x}\end{aligned} \qquad (11.0.11)$$

It is easy to see that by (11.0.7) and (11.0.11), χ satisfies Poisson's equation

$$\nabla^2 \chi = 4\pi\rho \qquad (11.0.12)$$

where ρ is given by

$$\rho = \sum_i n_i \delta(r - r_i) \tag{11.0.13}$$

Note that in (11.0.11) and (11.0.12), we have succeeded in mapping the many-vortex system, or "vortex gas," into the two-dimensional Coulomb gas in which the electric charge is replaced by the topological charge, (11.0.8). This equivalence has been used with great success by taking results from one model over to the other.

In dealing with Poisson's equation, it is useful to construct the Green function $g(r)$, such that

$$\nabla^2 g(r - r') = \delta(r - r') \tag{11.0.14}$$

A good approximation for $g(r)$ on the square lattice, for $r > a$ (a is the lattice spacing), is

$$g(r) = \frac{1}{2\pi} \ln \frac{r}{r_0}$$
$$r_0 = \frac{a}{2\sqrt{2}\, e^\gamma} \tag{11.0.15}$$

where $\gamma = 0.577216\ldots$ is Euler's constant. By (11.0.11) the derivatives of ϕ can be expressed in terms of the derivatives of χ, and we have for the energy of the vortex state

$$E_v = J \int d^2 r (\nabla \chi)^2 \tag{11.0.16}$$

Rewriting the integrand as a divergence and integrating by parts gives (see Prob. 11.3)

$$E_v = J \int d^2 r \left[\nabla \cdot (\chi \nabla \chi) - \chi \nabla^2 \chi \right]$$
$$= J \int dl\, \hat{n} \cdot (\chi \nabla \chi) - 4\pi J \int d^2 r\, \rho \chi \tag{11.0.17}$$

where ρ is given by (11.0.12). The boundary term can be evaluated by assuming that all the vortices are far from the boundary, which can be taken to be a circle of radius R. This gives

$$\int dl\, \hat{n} \cdot \chi \nabla \chi = 2\pi \left(\sum_i n_i\right)^2 \ln R \qquad (11.0.18)$$

From this we see that the energy of a state with nonzero net vorticity diverges as the size of the system increases, and such configurations play no role in the thermodynamics of the system. We therefore require strict neutrality of the vortex gas

$$\sum_i n_i = 0 \qquad (11.0.19)$$

The remaining term in the energy can be written as

$$E_v = 16\pi^2 J \sum_{i \neq j} n_i n_j g(r_i - r_j) \qquad (11.0.20)$$

In this double sum, terms with $i = j$ vanish because $g(0) = 0$. Because two vortices occupying the same lattice site are equivalent to a single vortex whose circulation is the sum of the individual circulations, we must also require that the vortices remain at least one lattice spacing apart. In this way we are sure to count each distinct vortex configuration only once, and the approximate form for the Green function, (11.0.15), may be used. The energy for the vortex state is therefore

$$E_v = 2\pi J \sum_{i \neq j} n_i n_j \ln \frac{|\mathbf{r}_i - \mathbf{r}_j|}{r_0} \qquad (11.0.21)$$

Because the lattice spacing a is going to play a central role in the RG calculation, let us rewrite (11.0.21) in terms of a. This gives rise to an additional term, which can be interpreted as the "self-energy" of the vortex, and plays the role of a chemical potential

$$E_v = 2\pi J \sum_{i \neq j} n_i n_j \ln \frac{|\mathbf{r}_i - \mathbf{r}_j|}{r_0}$$
$$= 2\pi J \sum_{i \neq j} n_i n_j \ln \frac{|\mathbf{r}_i - \mathbf{r}_j|}{a} - 2\pi J \sum_i n_i^2 \ln \frac{r_0}{a} \qquad (11.0.22)$$

Note that in (11.0.22), use has been made of the neutrality condition, (11.0.19). It is clear from (11.0.22) that vortices with two or more units of circulation are energetically unfavorable, and therefore we will restrict our attention to vortices with unit circulation

$$n_i^2 = 1 \quad \text{for all } i \tag{11.0.23}$$

In addition to the most probable field $\phi(x)$, we must allow for the field $\theta(x)$ to fluctuate about $\phi(x)$, and we define the "spin wave" field $\psi(x)$ by

$$\theta(x) = \phi(x) + \psi(x) \tag{11.0.24}$$

Because we have explicitly taken into account the topologically interesting configurations of the field $\theta(x)$ in constructing the vortex states, we assume that the spin waves are nonsingular and in the long-wavelength limit, independent of the vortices. The partition function then factorizes into a spin wave and a vortex contribution

$$Z = Z_{\text{sw}} \times Z_v \tag{11.0.25}$$

where the spin wave partition function is

$$Z_{\text{sw}} = \int D[\psi] \exp - \beta J \int d^2 r |\nabla \psi^2| \tag{11.0.26}$$

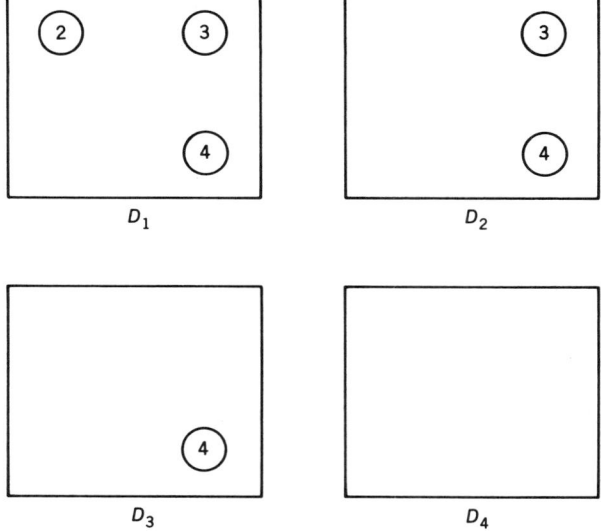

Fig. 11.0.2 Domains of integration for a four-vortex state evaluating the partition function of the vortex gas.

and the partition function for the vortex gas is

$$Z_v = \sum_{n=0}^{\infty} \frac{K^{2n}}{(n!)^2} \int_{D_1} dr^2 \cdots \int_{D_{2n}} d^2_{r_{2n}} \exp - 2\pi\beta J \sum_{i \neq j} n_i n_j \ln \frac{|\mathbf{r}_i - \mathbf{r}_j|}{a}$$

(11.0.27)

The parameter K in (11.0.27) is the fugacity of the vortex gas and is simply related to the chemical potential μ by

$$K = e^{\beta\mu} \qquad \mu = 2\pi J \ln \frac{r_0}{a} \qquad (11.0.28)$$

The factor of $(n!)^2$ is necessary to avoid counting states in which vortices of like sign are simply permuted; each vortex state consists of n positive and n negative vortices. The symbols D_j indicate that the domain of integration of each vortex must take into account the restriction that no two vortices can overlap. Therefore, define D_j to be the whole plane except for circles of radius a centered on the positions of vortices $j+1, j+2, \ldots, 2n$. These domains of integration are shown in Fig. 11.0.2 for a four-vortex state.

11.1 REAL-SPACE RENORMALIZATION FOR THE VORTEX GAS

Before we become involved with the details of the RG for this system, it is useful to get a good physical picture of just what is going on.

We are going to probe the states of the X–Y model by studying the behavior of the spin–spin correlation function

$$G(r) = \langle S(r) \cdot S(0) \rangle \qquad (11.1.1)$$

In the low-temperature state, vortices can exist only as closely bound pairs. The effect of such pairs on the spin–spin correlation function is negligible, and the correlation function is determined solely by the spin waves. As the temperature approaches the transition temperature, the vortices "ionize" and a transition to a vortex plasma state takes place. The spin correlations are screened by the vortex plasma and therefore decay exponentially. The order parameter for this system is not as easily defined as in the case of a ferromagnetic transition and the Mermin, Wagner, Hohenberg theorem [4, 8–9] excludes spontaneous symmetry breaking in $d = 2$. As we will see

11.1 REAL-SPACE RENORMALIZATION FOR THE VORTEX GAS

when we apply the theory to superfluid helium films, the order parameter, which in that case is the superfluid density, is related to a kind of susceptibility.

As with any RG transformation, the first step is to "integrate out" the microscopic details of the model. If one imagines the vortex gas at low temperature as consisting of vortex pairs, some with large separations and some with very small separations, then out first objective is to integrate out the very close pairs. By taking into account the effect of such pairs, the interaction between more distant pairs is renormalized. This is the physical basis for the original work of Kosterlitz and Thouless [7], which motivates the construction of the Kosterlitz renormalization procedure [6].

We assume that the vortex gas is not too dense so that the contribution to the free energy of vortex configurations in which three or more vortices are close together is negligible. The decimation part of the RG then consists of calculating explicitly the contribution from vortex–antivortex pairs whose relative coordinate lies within an annulus of radius da about a,

$$\delta_i(j): a \leq |\mathbf{r}_i - \mathbf{r}_j| \leq a + da \qquad (11.1.2)$$

Contributions from like-signed close pairs are neglected because such states have a much higher energy than close neutral pairs. Once the calculation for each pair has been performed, the allowed minimum separation of vortices has increased from a to $a + da$. In order to restore the problem to its original form, the second part of the RG procedure, rescaling, must be implemented. In this case the scale factor is

$$b = 1 + \frac{da}{a} \qquad (11.1.3)$$

The Kosterlitz RG transformation is an example of an infinitesimal real-space RG.

Define new domains of integration D'_{ij}, which are identical to the D_{ij}, except that the forbidden circular regions about each vortex are of radius $a + da$. Let $\delta_i(j)$ be the thin annular region of radius a and width da centered about vortex j and let $\overline{D}_j$ be the entire plane except for the circular regions of radius a about the $2n - 2$ vortices besides i and j. These domains are illustrated in Fig. 11.1.1, again for a four-vortex configuration. The decimation procedure is to first integrate over r_i within the annular region $\delta_i(j)$ about vortex j, and

352   X-Y MODEL AND THE KOSTERLITZ–THOULESS TRANSITION

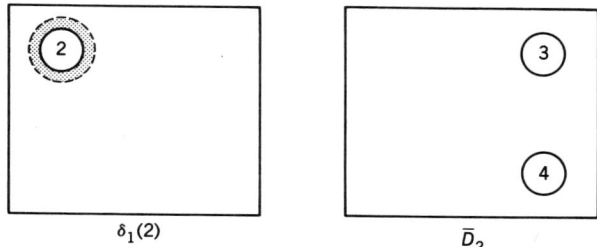

Fig. 11.1.1 Region of integration for the four-vortex state for carrying out decimation of closely bound vortex pairs.

then to integrate r_j over the domain $\overline{D}_j$. In this way the contribution to the partition function from the close-bound pair i, j is taken into account. This procedure must then be carried out for all possible pairs. Schematically, this may be written as

$$\int_{D_{2n}} d^2 r_{2n} \cdots \int_{D_1} d^2 r_1 = \int_{D'_{2n}} d^2 r_{2n} \cdots \int_{D_1} d^2 r_i + \frac{1}{2} \sum_{i \neq j} \int_{D'_{2n}} d^2 r_{2n} \cdots$$

$$\times \int_{D'_{j+1}} d^2 r_{j+1} \int_{D_j} d^2 r_j \cdots \int_{\delta_i(j)} d^2 r_i \cdots \int_{D'_1} d^2 r_1 \qquad (11.1.4)$$

In order to stay as close to the notation of Kosterlitz as possible, let us define

$$p_i = (2\pi J)^{1/2} n_i \qquad (11.1.5)$$

Collecting all terms that involve vortices i and j, we wish to calculate

$$\int_{\delta_i(j)} d^2 r_1 \exp\left[2\beta \sum_k p_i p_k \ln\frac{|\mathbf{r}_i - \mathbf{r}_k|}{a} + 2\beta \sum_k p_j p_k \frac{|\mathbf{r}_i - \mathbf{r}_k|}{a} \right] \qquad (11.1.6)$$

Define a vector $\hat{a} = (\mathbf{r}_j - \mathbf{r}_i)/a$ and $\mathbf{r}_{jk} = \mathbf{r}_j - \mathbf{r}_k$. The logarithm in the first sum in (11.1.6) can be rewritten as

$$\ln|\hat{a} + \mathbf{r}_{jk}| = \ln|\mathbf{r}_{jk}| + \ln\left[1 + \frac{2\hat{a} \cdot \mathbf{r}_{jk}}{|\mathbf{r}_{jk}|^2} + \frac{a^2}{|\mathbf{r}_{jk}|^2}\right] \qquad (11.1.7)$$

The first term in this expression is exactly cancelled by the second sum

11.1 REAL-SPACE RENORMALIZATION FOR THE VORTEX GAS

in (11.1.6) because we require that $p_i = -p_j$. Therefore, we wish to calculate

$$\int_{\delta_i(j)} d^2 r_i \prod_k \left[1 + \frac{2\hat{\mathbf{a}} \cdot \mathbf{r}_{jk}}{|\mathbf{r}_{jk}|^2} + \frac{a^2}{|\mathbf{r}_{jk}|^2} \right]^{\beta p_i p_k} \quad (11.1.8)$$

Assuming that $|r_i - r_k| \gg a$, we have

$$\prod_k \left[1 + \frac{2\hat{\mathbf{a}} \cdot \mathbf{r}_{jk}}{|\mathbf{r}_{jk}|^2} + \frac{a^2}{|\mathbf{r}_{jk}|^2} \right]^{\beta p_i p_k}$$

$$\cong 1 + \sum_k \left\{ \frac{2\beta p_i p_k \hat{\mathbf{a}} \cdot \mathbf{r}_{jk}}{|\mathbf{r}_{jk}|^2} + \frac{\beta a^2 p_i p_k}{|\mathbf{r}_{jk}|^2} + \frac{2\beta p_i p_k (2\beta p_i p_k - 1)(\hat{\mathbf{a}} \cdot \mathbf{r}_{jk})^2}{|\mathbf{r}_{jk}|^4} \right.$$

$$\left. + \sum_{l \neq k} \frac{4\beta^2 p_i^2 p_k p_l (\hat{\mathbf{a}} \cdot \mathbf{r}_{jk})(\hat{\mathbf{a}} \cdot \mathbf{r}_{jl})}{|\mathbf{r}_{jk}|^2 |\mathbf{r}_{jl}|^2} \right\} \quad (11.1.9)$$

The integration over the annular region $\delta_i(j)$ is therefore

$$a\, da \int_0^{2\pi} d\theta \left\{ \left[1 + \sum_k \frac{2a^2}{|\mathbf{r}_{jk}|^2} \beta p_i p_k \cos\theta \right] \right.$$

$$+ \beta p_i p_k \frac{a^2}{|\mathbf{r}_{jk}|^2} + \frac{2\beta p_i p_k (2\beta p_i p_k - 1)}{|\mathbf{r}_{jk}|^2} a^2 \cos^2\theta$$

$$\left. + 4\beta p^2 a^2 \sum_{k \neq l} \left[\frac{p_k p_l \cos(\theta - \theta_k)\cos(\theta - \theta_l)}{|\mathbf{r}_{jk}|^2 |\mathbf{r}_{jl}|^2} \right] \right\} \quad (11.1.10)$$

where we have defined $\hat{\mathbf{a}} \cdot \hat{\mathbf{r}}_{jk} = \cos(\theta - \theta_k)$. The result of the integration is (see Prob. 11.4)

$$2\pi a\, da \left[1 + \beta^4 p^4 \sum_k \frac{a^2}{|\mathbf{r}_j - \mathbf{r}_k|^2} + 2\beta^2 p^2 a^2 \sum_{k \neq l} \frac{(\mathbf{r}_j - \mathbf{r}_k) \cdot (\mathbf{r}_j - \mathbf{r}_l) p_k p_l}{|\mathbf{r}_j - \mathbf{r}_k|^2 |\mathbf{r}_j - \mathbf{r}_l|^2} \right] \quad (11.1.11)$$

We must now integrate this expression over the domain $\overline{D}_{ij}$, which is left as an exercise (Prob. 11.5). The result is

$$2\pi a\, da\left\{A - 2\pi a^2\beta p^2 \sum_{k\neq l} p_k p_l \ln\frac{|\mathbf{r}_k - \mathbf{r}_l|}{a}\right\} \quad (11.1.12)$$

where A is the area of the plane. Note that the sums over k and l in (11.1.12) run over the $2n - 2$ remaining vortices. If we redefine the summation variable $n \to n - 1$, the vortex contribution to the partition can be written as

$$Z_v = \sum_{n=0}^{\infty} \frac{K^{2n}}{(n!)^2} \int_{D'_{2n}} d^2r_{2n} \cdots \int_{D'_1} d^2r_1$$

$$\times \left[1 + \frac{K^2}{(n+1)^2} \sum_{i\neq j}^{n+1} 2a\, da\left(A - 2a^2\beta^2 p^2 \sum_{k\neq l} p_k p_l \ln\frac{|\mathbf{r}_k - \mathbf{r}_l|}{a}\right)\right] e^{-\beta H_v}$$

(11.1.13)

Note that the sums in (11.1.13) over i and j are independent of i and j, which leads to a multiplicative factor $n(n + 1)$. In the thermodynamic limit the number of vortices is quite large, and we may take the limit $1/n \to 0$. Because $da/a \ll 1$, we can exponentiate the second factor in (11.1.12), which gives

$$Z = e^{-dF} \sum_{n=0}^{\infty} \frac{K^{2n}}{(n!)^2} \int_{D'_{2n}} d^2r_{2n} \cdots \int_{D'_1} d^2r_1$$

$$\times \exp -\beta\left[1 - (2\pi)^2\beta p^2(Ka^2)^2\frac{da}{a}\right]\sum_{i\neq j} p_i p_j \ln\frac{|\mathbf{r}_i - \mathbf{r}_j|}{a} \quad (11.1.14)$$

Here dF is the contribution to the free energy from the closely bound pairs. This completes the decimation part of the RG transformation. The minimum separation between oppositely charged vortices is now $a + da$. In order to restore the cutoff to its original value, we rescale by a factor $b = 1 + ba/a$. For example, we can write

$$\ln\frac{|\mathbf{r}_{ij}|}{a} = \ln\left[\frac{r_{ij}}{a + da - da}\right] \cong \ln\left[\frac{|\mathbf{r}_{ij}|}{a + da}\right] + \frac{da}{a} \quad (11.1.15)$$

which leads to a renormalization of the fugacity [see (11.0.28)]

$$K \to \left(1 - \beta p^2\frac{da}{a}\right)K \quad (11.1.16)$$

11.1 REAL-SPACE RENORMALIZATION FOR THE VORTEX GAS

Putting these results together, we have the following RG equations:

$$(\beta p^2)' = \beta p^2 \left[1 - (2\pi)^2 \beta p^2 (Ka^2)^2 \frac{da}{a} \right]$$
$$(Ka^2)' = Ka^2 \left[1 - (\beta p^2 - 2) \frac{da}{a} \right]$$
(11.1.17)

Note that in the RG equation for Ka^2, an extra term arises from the renormalization of a^2. Defining a continuous scale parameter l so that $dl = da/a$, and

$$x = \beta p^2 - 2$$
$$y = 4\pi Ka^2$$
(11.1.18)

the RG equations can be written as

$$\frac{dx}{dl} = -\frac{1}{4}(x+2)^2 y^2 \qquad (11.1.19a)$$

$$\frac{dy}{dl} = -xy \qquad (11.1.19b)$$

These equations have a fixed point at $x^* = y^* = 0$. Expanding around the fixed point, we see that, multiplying (11.1.19a) by x and (11.1.19b) by y,

$$\frac{dx^2}{dl} = -y^2 \qquad (11.1.20a)$$

$$\frac{dy^2}{dl} = -2xy^2 \qquad (11.1.20b)$$

It is easy to show that the solutions of (11.1.20) satisfy

$$x^2 - y^2 = x_0^2 \qquad (11.1.21)$$

RG trajectories for $x_0^2 > 0$ ($T < T_c$) and $x_0^2 < 0$ ($T > T_c$) are shown in Fig. 11.1.12. The dotted line $x = y$ is the critical line and corresponds to $x_0 = 0$, which gives $T_c = \pi J$ by (11.1.18).

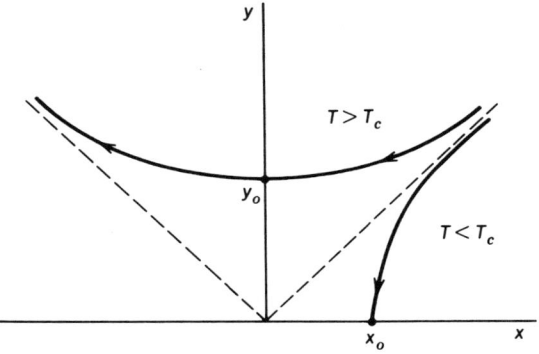

Fig. 11.1.2 Typical RG trajectories for $T > T_c$ and $T < T_c$.

11.2 ANALYSIS OF THE RENORMALIZATION GROUP EQUATIONS

For $T < T_c$ the RG takes y to 0 and x to the real value x_0, which, for small x_0, is related to the reduced temperature $(T_c - T)/T_c$ by

$$x_0^2 \sim \frac{T_c - T}{T_c} \qquad T < T_c \qquad (11.2.1)$$

Because $y \to 0$ for any initial value of (x, y), we see that the RG transformation drives the fugacity of the vortices to 0; the vortices are renormalized right out of the problem. The only remaining degrees of freedom are the spin waves, which determine the spin–spin correlations for $T < T_c$. As we will presently see, the correlation function in the presence of spin waves decays algebraically, and therefore we have for the correlation length

$$\xi = \infty \qquad T \leq T_c \qquad (11.2.2)$$

This is our first indication that the singularity associated with the vortex dissociation transition is of a different character then the usual second-order phase transition we have been studying. There is no way of defining the critical exponent ν for $T < T_c$. The general solution to (11.1.20) in the neighborhood of the fixed point is (Prob. 11.6)

$$x(l) = x_0 \frac{1 + \exp(-2x_0 l)}{1 - \exp(-2x_0 l)} \qquad (11.2.3)$$

For $T > T_c$ the solution for $T < T_c$ can be analytically continued by taking $x_0 \to iy_0$ in (11.2.3). This gives for the solution in the neighborhood of the fixed point

$$x(l) = \frac{y_0}{\tan y_0 l} \qquad (11.2.4)$$

Again, y_0 is a measure of the reduced temperature $(T - T_c)/T_c$. Very close to the critical point, the RG trajectory approaches to within y_0 of the fixed point and then diverges rapidly away from it. Physically, this means that the system looks critical on the original length scale because the correlation length is much greater than a. As we examine the system on larger and larger length scales (increasing l), we finally reach a point where the length scale we are using is of the same order as the correlation length, and for this value of l, l_0, we expect the RG trajectory to deviate substantially from the critical trajectory. It is natural to take this as the point where y attains its minimum value y_0. Solving for l_0 from (11.2.4), we have

$$l_0 = \frac{1}{y_0} \tan^{-1} \frac{y_0}{x(l_0)} \to \frac{\pi}{2y_0} \qquad (11.2.5)$$

The correlation length is then given by

$$\xi \sim e^{l_0}$$
$$= \exp \frac{\pi}{2}\left(\frac{T_c}{T - T_c}\right) \qquad (11.2.6)$$

For $T > T_c$ the critical behavior of the planar X–Y model is governed by an essential singularity; the correlation length diverges faster than any power of $(T - T_c)$ and therefore it is not possible to define the exponent ν in the usual way for $T > T_c$.

11.3 THE SPIN–SPIN CORRELATION FUNCTION

In this section we consider the spin–spin correlation function for $T < T_c$. As we have observed, the RG equations drive K, the fugacity of the vortex gas, to 0. This has the effect of "renormalizing out" the vortices, leaving only the spin waves. The only effect of the closely bound vortex pairs is to renormalize the spin wave coupling constant.

The spin–spin correlation function is

$$G(r) = \langle S(r) \cdot S(0) \rangle = \langle \cos(\theta(r) - \theta(0)) \rangle \quad (11.3.1)$$

If the field θ is decomposed into the vortex (ϕ) and spin wave (ψ) parts, we have

$$G(r) = \langle \cos(\phi(r) - \phi(0))\cos(\psi(r) - \psi(0)) \rangle \quad (11.3.2)$$

Because, within the present approximation, the spin waves are independent of the vortices, we can write

$$\langle \cos(\phi(r) - \phi(0))\cos(\psi(r) - \psi(0)) \rangle$$
$$= \langle \cos(\phi(r) - \phi(0)) \rangle \langle \cos(\psi(r) - \psi(0)) \rangle \quad (11.3.3)$$

By the preceding argument, under repeated renormalizations we should have for $T < T_c$,

$$G(r; x, y) = G(r; x(l), y(l)) = \lim_{l \to \infty} G(r; x_0, 0) \quad (11.3.4)$$

Therefore, if we are interested only in the behavior of the correlation function below T_c, we only need to evaluate the correlation function due to the spin waves. Therefore, we must calculate

$$G_{sw}(r) = \frac{\int D[\psi] \exp\left(-\beta J \int d^2 r (\nabla \psi)^2\right) \cos(\psi(r) - \psi(r))}{\int D[\psi] \exp\left(-\beta J \int d^2 r (\nabla \psi)^2\right)} \quad (11.3.5)$$

This may be rewritten as a Gaussian integral

$$G_{sw}(r) = \frac{\int D[\psi] \exp\left(-\beta J \int d^2 r (\nabla \psi)^2 + i \int d^2 r \cdot [\delta(r - r') - \delta(r')] \psi(r')\right)}{\int D[\psi] \exp\left(-\beta J \int d^2 r (\nabla \psi)^2\right)}$$

$$(11.3.6)$$

Completing the square and noting that $g(0) = 0$, this gives (see Prob. 11.7)

$$G_{sw}(r) = \exp - \frac{g(r)}{2\beta J} \quad (11.3.7)$$

Therefore, the spin wave correlation function is given by

$$G_{\text{sw}}(r) = \left|\frac{r}{r_0}\right|^{1/-4\pi\beta J} = \left|\frac{r}{r_0}\right|^{-\eta}$$

$$\eta(T) = \frac{T}{4\pi J}$$
(11.3.8)

Equation (11.3.8) proves our previous statement that the correlations decay algebraically for all $T < T_c$, and therefore the correlation length exponent is not well defined. Note that if we take $T = T_c = \pi J$ in (11.3.8) we find $\eta(T_c) = 1/4$. If we choose a value of T different from T_c, then the RG trajectory terminates at x_0, and the exponent η calculated in (11.3.8) will depend on x_0 (and therefore T). This should be contrasted with the usual situation for a second-order phase transition in which the exponents do not depend on the temperature.

11.4 UNIVERSAL BEHAVIOR OF SUPERFLUID HELIUM FILMS

We conclude our study of the Kosterlitz–Thouless transition with a discussion of the behavior of a system which should be described by the two-dimensional X–Y model, thin films of superfluid He^4 [11]. In applying the theory to liquid helium, we have come full circle as the X–Y model was originally constructed as a model of helium.

It has long been known that quantized vortices exist in superfluid helium [12, 13, 15], and it can also be shown that the two-dimensional Bose fluid can be described in terms of quantized vortices [2]. The superfluid state is described by a complex scalar order parameter whose phase is equivalent to the field θ in the X–Y model. The superfluid velocity is given by

$$v_s(x) = \frac{\hbar}{m}\nabla\theta$$
(11.4.1)

It is beyond the scope of this text to delve into the details of the microscopic theory of superfluidity, and we will simply quote the result that the superfluid density ρ_s can be calculated from the superfluid velocity autocorrelation function [5]

$$\frac{m^2 kT}{\hbar^2 \rho_s(T)} = \frac{m^2}{\hbar^2}\int d^2r \langle v_s(r) \cdot v(0)\rangle$$
(11.4.2)

Here m is the mass of the helium atom. In order to calculate the preceding correlation function, we consider first the correlation function for the phase field itself.

By an argument completely parallel to that used previously to evaluate the spin correlation function, we can argue that for $T < T_c$ we only need to evaluate the spin wave part of (11.4.2). This is easily done by introducing a generating functional

$$F[h] = \frac{\int D[\psi]\exp\left(-\beta J \int d^2r (\nabla \psi)^2 + \int d^2r h(r)\psi(r)\right)}{\int D[\psi]\exp\left(-\beta J \int d^2r (\nabla \psi)^2\right)} \quad (11.4.3)$$

where $h(r)$ is an arbitrary field.

The desired correlation function is found by taking the appropriate functional derivatives of $F[h]$,

$$\Gamma(r - r') = \lim_{h \to 0} \frac{\delta^2 F[h]}{\delta h(r)\, \delta h(r')} \quad (11.4.4)$$

The integral (11.4.3) is a Gaussian and can be evaluated explicitly. Keeping in mind that $g(0) = 0$, one finds (Prob. 11.8)

$$\Gamma(r - r') = -\frac{1}{2\beta J} g(r - r') \quad (11.4.5)$$

Putting (11.4.1) and (11.4.2) and (11.4.5) together, the superfluid density is given by

$$\frac{m^2 kT}{\hbar^2 \rho_s(T)} = \int d^2r\, \nabla_r \nabla_{r'} \left(-\frac{1}{2\beta J} g(r - r')\right) \quad (11.4.6)$$

The Laplacian acting on $g(r)$ yields a delta function, and the integral over the plane can be performed, yielding

$$\frac{m^2}{\hbar^2} \frac{kT}{\rho_s(T)} = \frac{1}{2\beta J}$$

$$= 2\pi \eta \quad (11.4.7)$$

If we now specialize this result to $T = T_c$, we have by (11.3.8)

$$\frac{m^2}{\hbar^2} \frac{kT_c}{\rho_s(T_c)} = \frac{\pi}{2} \quad (11.4.8)$$

11.4 UNIVERSAL BEHAVIOR OF SUPERFLUID HELIUM FILMS

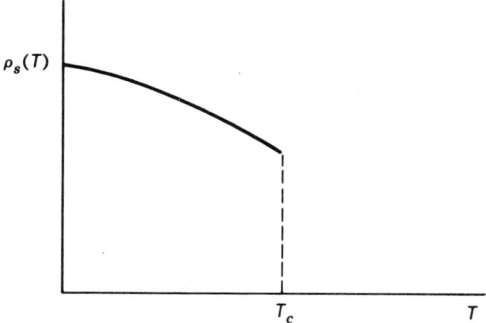

Fig. 11.4.1 Behavior of the superfluid density with temperature. Note the discontinuity at $T = T_c$.

As first shown by Nelson and Kosterlitz [11] and illustrated in Fig. 11.4.1, the superfluid density rises discontinuously from 0 at T_c, but what is even more intriguing, their calculation implies that the ratio of $\rho_s(T_c)/T_c$ should be universal. It should not depend, for example, on the type of substrate used or the areal density of the film.

Experimental results taken from Bishop and Reppy [1] for real helium films are shown in Fig. 11.4.2. The agreement with the result

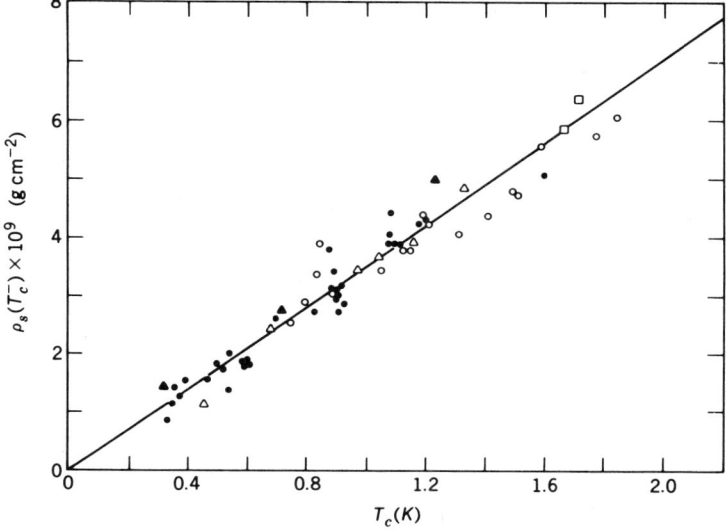

Fig. 11.4.2 Discontinuity in the superfluid density as measured by several methods after Bishop and Reppy [11]. Data are from: ▲ Bishop and Reppy, T = const., ● Bishop and Reppy, T_c = const., □ Hallock [3], △ Mochel [10], and ○ Rudnick [14]. The straight line fit to the data is from the Kosterlitz–Thouless static theory [7].

of Nelson and Kosterlitz is quite satisfying, and illustrates the real power of the RG method.

PROBLEMS

11.1 Verify that (11.0.7) is a solution of Laplace's equation and that

$$\int dl \cdot \nabla \phi = 2\pi n$$

11.2 Construct the harmonic conjugate χ to ϕ given by (11.0.7) and show that it satisfies Poisson's equation.

11.3 Assuming that all the vortices are far from the boundary of the system ($R \gg r_i$), show that the surface term in (11.0.4) is given by (11.0.18).

11.4 To lowest order in $a|r_{jk}|$, show that (11.1.10) follows from (11.1.7). Evaluate the integral and derive (11.1.11).

11.5 Perform the integration of (11.1.10) over the domain $\overline{D}$ and verify (11.1.11).

11.6 Show that (11.2.3) is the most general solution to the fixed-point RG equations (11.1.19) subject to the condition

$$\lim_{l \to \infty} x(l) \to x_0$$

11.7 Evaluate the functional integral in (11.3.6) and derive the expression (11.3.8) for $G_{sw}(r)$.

11.8 Evaluate the functional integral (11.4.3) and derive the result (11.4.5) for $\Gamma(r)$.

REFERENCES

1. Bishop, D. J. and Reppy, J. D. (1978) *Phys. Rev.* **L40** 1727.
2. Creswick, R. J. (1981) Ph.D. thesis (unpublished), Univ. California, Berkeley.
3. Hallock, R., see [1].
4. Hohenberg, P. C. (1967) *Phys. Rev.* **158** 383.
5. Hohenberg, P. C. and Martin, P. C. (1965) *Ann. Phys.* **34** 291.

6. Kosterlitz, J. M. (1974) *J. Phys. C.* **7** 1046.
7. Kosterlitz, J. M. and Thouless, D. J. (1973) *J. Phys. C* **6** 1181.
8. Mermin, N. D. (1966) *Phys. Rev.* **176** 250.
9. Mermin, N. D. and Wagner, H. (1966) *Phys. Rev.* **L22** 1133.
10. Mochel, J., see [1].
11. Nelson, D. R. and Kosterlitz, J. M. (1977) *Phys. Rev.* **L39** 1201.
12. Rayfield, G. W. and Reif, F. (1963) *Phys. Rev.* **L11** 305.
13. Rayfield, G. W. and Reif, F. (1964) *Phys. Rev.* **136** 1194.
14. Rudnick, L. (1978) *Phys. Rev. Lett.* **40** 1454.
15. Yarmchuck, E. J., Gordon, M. J. V., and Packard, R. E. (1979) *Phys. Rev.* **L43** 214.

APPENDIX A

THE CUMULANT EXPANSION

In many places in physics, but especially in statistical mechanics, we run across averages of the exponential of an operator

$$Z(\lambda) = \langle e^{\lambda A} \rangle \tag{A.1}$$

The parameter λ has been included to facilitate power counting, and in some cases (e.g., high-temperature expansions) λ is a real parameter which is, in some sense, small.

One approach to evaluating $Z(\lambda)$ is to expand the exponential in a power series and average term by term

$$Z(\lambda) = \sum_{n=0}^{\infty} \frac{\lambda^n}{n!} \langle A^n \rangle \tag{A.2}$$

It is possible to derive a more accurate expression for $Z(\lambda)$ by expanding what is essentially the free energy

$$F(\lambda) = \ln Z(\lambda) \tag{A.3}$$

rather than $Z(\lambda)$ directly. Thus we write

$$F(\lambda) = \sum_{n=1}^{\infty} \frac{\lambda^n}{n!} C_n \tag{A.4}$$

where C_n is called the nth cumulant and the series (A.4) is the cumulant expansion.

The cumulant expansion is equivalent to the sum of connected graphs in field theory. The advantage of the cumulant expansion over the direct series (A.2) is that in many cases the cumulants are much easier to evaluate for a given order in λ, and because the cumulants give an approximation to F, they represent some of the terms to all orders in λ that appear in the direct expansion for $Z(\lambda)$.

By (A.4) we have

$$C_n = \left.\frac{\partial^n F(\lambda)}{\partial \lambda^n}\right|_{\lambda=0}$$

$$= \left.\frac{\partial^n}{\partial \lambda^n} \ln\langle e^{\lambda A}\rangle\right|_{\lambda=0} \tag{A.5}$$

Note that the cumulants themselves are independent of the parameter λ, but we will find it useful to define

$$C_n(\lambda) = \frac{\partial^n}{\partial \lambda^n} \ln\langle e^{\lambda A}\rangle \tag{A.6}$$

With this definition we have

$$C_1(\lambda) = \frac{\langle A e^{\lambda A}\rangle}{\langle e^{\lambda A}\rangle} \tag{A.7}$$

and

$$C_2(\lambda) = \frac{\langle A^2 e^{\lambda A}\rangle}{\langle e^{\lambda A}\rangle} - \frac{\langle A e^{\lambda A}\rangle^2}{\langle e^{\lambda A}\rangle^2}$$

$$= \langle (A - \langle A\rangle_\lambda)^2\rangle_\lambda \tag{A.8}$$

where we have defined the average $\langle \cdots \rangle_\lambda$ as

$$\langle B\rangle_\lambda = \frac{\langle e^{\lambda A} B\rangle}{\langle e^{\lambda A}\rangle} \tag{A.9}$$

Furthermore, let us define

$$B_n(\lambda) = \langle (A - \langle A\rangle_\lambda)^n\rangle_\lambda \tag{A.10}$$

Then

$$\frac{\partial}{\partial \lambda} B_n(\lambda) = B_{n+1}(\lambda) - nB_{n-1}(\lambda)B_2(\lambda) \qquad (A.11)$$

[Note that $B_1(\lambda) \equiv 0$.] By (A.11) we can generate all the cumulants recursively:

$$\begin{aligned}
C_3 &= B_3 \\
C_4 &= B_4 - 3B_2^2 \\
C_5 &= B_5 - 10B_3 B_2 \\
C_6 &= B_6 - 15B_4 B_2 - 10B_3^2 + 30B_2^3
\end{aligned} \qquad (A.12)$$

and so on.

In general, the relationship between the B's and C's is

$$C_n = \sum_{k=1}^{n} \frac{1}{k} \sum_{m_1,\ldots,m_k} \frac{n!}{m_1! m_2! \cdots m_k!} \delta_{\Sigma m_i, n} B_{m_1} \cdots B_{m_k} \qquad (A.13)$$

The reciprocal relation is also easy to write down

$$B_n = \sum_{k=1}^{n} \frac{1}{k!} \sum_{m_1,\ldots,m_k} \frac{n!}{m_1! m_2! \cdots m_k!} \delta_{\Sigma m_i, n} C_{m_1} C_{m_2} \cdots C_{m_k} \qquad (A.14)$$

As an example of the advantage of the cumulant expansion over the direct power-series expansion, we consider the Ising model on a one-dimensional chain. The Hamiltonian is

$$H = -J \sum_i S_i S_{i+1} \qquad (A.15)$$

and we want to find

$$Z(J) = \mathrm{Tr}\, e^{-H} \qquad (A.16)$$

First, let us expand the exponential to sixth order in H (the trace of all the odd powers of H give 0). The trace vanishes unless the indices are equal in pairs, groups of four, and so on. The second cumulant is

$$C_2 = J^2 \sum_{ij} \text{Tr}(S_i S_{i+1} S_j S_{j+1}) = NJ^2 \tag{A.17}$$

The trace of H^4 is

$$J^4 \sum_{ijkl} \text{Tr}(S_i S_{i+1} S_j S_{j+1} S_k S_{k+1} S_l S_{l+1}) \tag{A.18}$$

First, we count all terms in which the indices are paired. There are three ways of pairing all the terms, which overcounts the number of terms where all four indices are equal, so

$$B_4 = J^4(3N^2 - 2N) \tag{A.19}$$

By (A.12)

$$\begin{aligned} C_4 &= J^4(3N^2 - 2N) - 3J^4N^2 \\ &= -2NJ^4 \end{aligned} \tag{A.20}$$

Note that the fourth cumulant, like the second cumulant, is proportional to N, which reflects the extensive property of the free energy. If you are not yet convinced, let us calculate the sixth-order contribution to each of the two series. First, we will consider all possible ways to pair six terms; there are 15 ways to do this, which gives a contribution J^6N^3. Each of these includes three terms in which the indices are grouped with one pair and one quartet of equal indices. There should only be 15 such terms, so we have overcounted by an amount $30J^6N^2$. Finally, there is one single term in which all six indices are equal, but we have undercounted by $15J^6N$, so we need to add $16J^6N$. Altogether we have

$$B_6 = (15N^3 - 30N^2 + 16N)J^6 \tag{A.21}$$

The sixth cumulant is, by (A.12),

$$\begin{aligned} C_6 &= [15N^3 - 30N^2 + 16N - 15(3N^2 - 2N)N + 30N^3]J^6 \\ &= 16NJ^6 \end{aligned} \tag{A.22}$$

The series (A-16) for the partition function is

$$Z = 2^N \left[1 + N\frac{J^2}{2} + \frac{1}{4!}(3N^2 - 2N)J^4 \right.$$
$$\left. + \frac{1}{6!}(15N^3 - 30N^2 + 16N)J^6 + \cdots \right] \quad (A.23)$$

whereas the series for the free energy is

$$F = N \ln 2 + N \left[\frac{J^2}{2!} - \frac{2J^4}{4!} + \frac{16J^6}{6!} \cdots \right] \quad (A.24)$$

A few comments are in order here. First, we will consider the all-important question of convergence of the two series. For large N the "most divergent" (with N) terms in the series for Z, (A.21), are of the form

$$z_n \sim \frac{N^{n/2} J^n}{2^{n/2} \left(\frac{n}{2}\right)!} \quad (A.25)$$

If we apply the ratio test we find

$$\frac{z_{n+2}}{z_n} = \frac{NJ^2}{2\left(\frac{n}{2} + 2\right)\left(\frac{n}{2} + 1\right)} \quad (A.26)$$

For large N series for Z must be carried out to very high order, $n \sim \sqrt{N}J$, before the size of the terms begins to decrease and the partial sums begin to settle down. On the other hand, the series for the free energy looks perfectly convergent even at low order.

There is, however, a more compelling reason to prefer the cumulant expansion over the direct-series expansion for Z, and this is reflected in the extensive property of F. The fact that each cumulant is proportional to N means that in summing over all terms there is only one index that runs over the whole lattice. All the other summed indices are tied to a particular lattice site; the other terms in the cumulant expansion are from sites that are connected. Very often it is possible to formulate simple graphical rules for constructing these terms, which makes the evaluation of the cumulants much easier than the coefficients in the series for the partition function.

Let us consider the fourth cumulant, which we can write as

$$\sum_{ijkl} \text{Tr}(H_i H_j H_k H_l) - 3\,\text{Tr}(H_i H_j)\text{Tr}(H_k H_l) \qquad (A.27)$$

where we have introduced the notation

$$H_i = -JS_i S_{i+1} \qquad (A.28)$$

There are terms in the sum where, for example $i = j$, $k = l$, and $i \neq k$. It is easy to see that in this case no contribution to (A.27) is made because the first term factorizes

$$\text{Tr}(H_i^2 H_k^2) = (\text{Tr}\,H_i^2)(\text{Tr}\,H_k^2) \qquad i \neq k \qquad (A.29)$$

The only term that survives is the one for which $i = j = k = l$, that is,

$$\sum_i \text{Tr}(H_i^4) - 3(\text{Tr}(H_i^2))^2 = -2NJ^4 \qquad (A.30)$$

The same holds for the sixth cumulant. The coefficient appearing in the cumulant expansion is found simply by summing the coefficients for C_6, (A.12),

$$1 - 15 - 10 + 30 = 16 \qquad (A.31)$$

We come to the conclusion that to find the coefficients appearing in the cumulant expansion all we need to consider are those terms in which every index is the same. On lattices of higher dimension the rules are not so simple because there are closed paths on the lattice. Nevertheless, the effort in calculating the cumulants is much less than calculating the corresponding coefficient in the direct expansion of Z.

PROBLEMS

A.1 The partition function for the Ising chain can be calculated exactly, and one finds

$$Z = 2^N (\cosh J)^N$$

The free energy is therefore (apart from sign)

$$F = N \ln \cosh J$$

Expand $\ln \cosh J$ to sixth order and compare with the results of the cumulant expansion, (A.24).

A.2 Using the relation (A.13), show that

$$\sum_{n=1}^{\infty} \frac{\lambda^n C_n}{n!} = \ln\left(1 + \sum_{n=1}^{\infty} \frac{\lambda^n B_n}{n!}\right)$$

and similarly, by (A.14), that

$$1 + \sum_{n=1}^{\infty} \frac{\lambda^n B_n}{n!} = \exp \sum_{m=1}^{\infty} \frac{\lambda^m C_m}{m!}$$

APPENDIX B

FEYNMAN DIAGRAMS

B.1 RULES FOR CONSTRUCTING FEYNMAN DIAGRAMS

In this appendix we derive the rules for constructing Feynman graphs for the free energy and two-point correlation function. We use only the mathematical tools developed in the text. Those of you who are interested in seeing similar derivations in the context of quantum field theory are referred to Fetter and Walecka [3], Abrikosov, Gorkov, and Dzyloshinski [1], and Amit [2]. In notation we most closely follow Amit.

We begin with the functional integral for the partition function

$$Z = \int D[\phi] + e^{-(L_0[\phi] + L_{\text{int}}[\phi])} \tag{B.1}$$

where

$$L_0[\phi] = \frac{1}{2} \int d^d x (\nabla \phi_\alpha \nabla \phi_\alpha + r \phi_\alpha \phi_\alpha) \tag{B.2}$$

and, for example,

$$L_{\text{int}}[\phi] = \frac{\lambda}{4!} \int d^d x (\phi_\alpha(x) \phi_\alpha(x))^2 \tag{B.3}$$

The generating functional $Z[\eta]$ is given by

$$Z_0[\eta] = \int D[\phi] e^{-L_0[\phi] + \int d^d x\, \eta_\alpha(x)\phi_\alpha(x)} \qquad (B.4)$$

With the aid of the generating functional, it is possible to rewrite (B.1) as

$$Z = \lim_{\eta \to 0} e^{-L_{\text{int}}[\delta/\delta\eta]} Z_0[\eta] \qquad (B.5)$$

Because the functional integral (B.4) is Gaussian, it can be evaluated explicitly with the result

$$Z_0[\eta] = \exp\left(\frac{1}{2} \int d^d x \int d^d y\, \eta_\alpha(x) G^{(0)}_{\alpha\beta}(x-y) \eta_\beta(y)\right) \qquad (B.6)$$

where

$$G^{(0)}_{\alpha\beta}(x-y) = \delta_{\alpha\beta} \int \frac{d^d k}{(2\pi)^d} e^{ik(x-y)} \frac{1}{k^2 + r} \qquad (B.7)$$

is the "free" Green function.

We now establish the rule for evaluating derivatives of the Gaussian functional (B.6) of the form[†]

$$\lim_{\eta \to 0} \frac{\delta}{\delta\eta(x_{2n})} \cdots \frac{\delta}{\delta\eta(x_1)} Z_0[\eta]$$

$$= \sum_{\text{all pairings}} G^{(0)}(x_1 - x_2) G^{(0)}(x_3 - x_4) \cdots G^{(0)}(x_{2n-1} - x_{2n})$$

$$(B.8)$$

where the sum is over all possible pairs of points. We prove this recursively as follows. The second functional derivative of $Z_0[\eta]$ is, to

---

[†]In what follows we will suppress the internal indices in the interest of compactness of notation.

B.1 RULES FOR CONSTRUCTING FEYNMAN DIAGRAMS

$O(\eta^2)$,

$$\frac{\delta}{\delta\eta(x_1)}\frac{\delta}{\delta\eta(x_2)}Z_0[\eta]$$
$$= \left[G^{(0)}(x_1 - x_2) + \int dz_1 \int dz_2\, G^{(0)}(x_1 - z_1) \right.$$
$$\left. \times G^{(0)}(x_2 - z_2)\eta(z_1)\eta(z_2) \right] Z_0[\eta] \quad \text{(B.9)}$$

Taking two more functional derivatives, we have

$$\frac{\delta}{\delta\eta(x_4)} \cdots \frac{\delta}{\delta\eta(x_1)} Z_0[\eta]$$
$$= [G^{(0)}(x_1 - x_2)G^{(0)}(x_3 - x_4) + G^{(0)}(x_1 - x_4)G^{(0)}(x_2 - x_3)$$
$$+ G^{(0)}(x_1 - x_2)G^{(0)}(x_3 - x_4) + G^{(0)}(x_1 - x_4)G^{(0)}(x_2 - x_3)]Z_0[\eta]$$
$$+ \left[G^{(0)}(x_1 - x_2)\int dz_2\, G^{(0)}(x_3 - z_2)\int dz_4\, G^{(0)}(x_4 - z_4)\eta(z_2)\eta(z_4) \right.$$
$$+ G^{(0)}(x_1 - x_4)\int dz_2\, G^{(0)}(x_2 - z_2)\int dz_3\, G^{(0)}(x_3 - z_3)\eta(z_2)\eta(z_3)$$
$$+ G^{(0)}(x_2 - x_3)\int dz_1\, G^{(0)}(x_1 - z_2)\int dz_4\, G^{(0)}(x_4 - z_4)\eta(z_1)\eta(z_4)$$
$$+ G^{(0)}(x_2 - x_4)\int dz_1\, G^{(0)}(x_1 - z_1)\int dz_3\, G^{(0)}(x_3 - z_3)\eta(z_1)\eta(z_3)$$
$$\left. + G^{(0)}(x_1 - x_3)\int dz_3\, G^{(0)}(x_2 - z_3)\int dz_4\, G^{(0)}(x_4 - z_4)\eta(z_3)\eta(z_4) \right] Z_0[\eta]$$
$$+ \text{terms fourth order in } \eta \quad \text{(B.10)}$$

Let us denote the sum over all pairs of n products of G's, as in (B.8), by $S_{2n}(x_1, x_2, \ldots, x_{2n})$. We assume that the pattern that emerges from (B.9) and (B.10) continues so that we may write for the $2n$th functional derivative of Z_0,

$$\frac{\delta}{\delta\eta(x_{2n})} \cdots \frac{\delta}{\delta\eta(x_1)} Z_0[\eta]$$
$$= S_{2n}(x_1 - x_{2n})Z_0[\eta]$$
$$+ \sum_{i \neq j} S_{2n-2}(x_1, \ldots, x_{2n}) \int dz_i \int dz_j\, G^{(0)}(x_i - z_i)$$
$$\times G^{(0)}(x_j - z_j)\eta(z_i)(z_j)Z_0[\eta]$$
$$+ \text{terms of order } \eta^4 \quad \text{(B.11)}$$

In the second term of (B.11) it is understood that the arguments x_i and x_j are excluded from the function S_{2n-2}. We now show that by differentiating (B.11) two more times we arrive again at (B.11) with $n \to n+1$,

$$\frac{\delta}{\delta\eta(x_{2n+2})} \frac{\delta}{\delta\eta(x_{2n+1})} \frac{\delta}{\delta\eta(x_{2n})} \cdots \frac{\delta}{\delta\eta(x_1)} Z_0[\eta]$$

$$= S_{2n}(x_1,\ldots,x_{2n})\Big[G^{(0)}(x_{2n+2} - x_{2n+1})$$

$$+ \int dz_{2n+2} \int dz_{2n+1} G^{(0)}(x_{2n+2} - z_{2n+2})$$

$$\times G^{(0)}(x_{2n+1} - z_{2n+1})\eta(z_{2n+2})\eta(z_{2n+1})\Big]Z_0[\eta]$$

$$+ \sum_{i \neq j} S_{2n-2}(x_1,\ldots,x_{2n})\Big[G^{(0)}(x_i - x_{2n+2})$$

$$\times G^{(0)}(x_j - x_{2n+1}) + G^{(0)}(x_i - x_{2n+1})$$

$$\times G^{(0)}(x_j - x_{2n+2})Z_0(\eta)\Big]$$

$$+ \Big[G^{(0)}(x_i - x_{2n+2})\int dz_j \int dz_{2n+1} G^{(0)}(x_j - z_j)$$

$$\times G^{(0)}(x_{2n+1} - z_{2n+1})\eta(z_j)\eta(z_{2n+1})$$

$$+ G^{(0)}(x_i - x_{2n+1})\int dz_j \int dz_{2n+2} G^{(0)}(x_j - z_j)$$

$$\times G^{(0)}(x_{2n+2} - z_{2n+2})\eta(z_j)\eta(z_{2n+2})$$

$$+ G^{(0)}(x_j - x_{2n+2})\int dz_i \int dz_{2n+1} G^{(0)}(x_i - z_i)$$

$$\times G^{(0)}(x_{2n+1} - z_{2n+1})\eta(z_i)\eta(z_{2n+1})$$

$$+ G^{(0)}(x_j - x_{2n+1})\int dz_i \int dz_{2n+1} G^{(0)}(x_i - z_i)$$

$$\times G^{(0)}(x_{2n+2} - z_{2n+2})\eta(z_i)\eta(z_{2n+2})\Big]Z_0[\eta]$$

$+$ terms fourth order in η \hfill (B.12)

Combining terms, we have

$$\frac{\delta}{\delta\eta(x_{2n+2})} \cdots \frac{\delta}{\delta\eta(x_1)} Z_0[\eta]$$

$$= \Bigg[S_{2n+2}(x_1, x_2, \ldots, x_{2n+2}) Z_0[\eta]$$

$$+ \sum_{i \neq j} S_{2n}(x_1, x_2, \ldots, x_{2n+2}) \int dz_i \int dz_j \, G^{(0)}(x_i - z_i)$$

$$\times G^{(0)}(x_i - z_i) G^{(0)}(x_i - z_j) \eta(z_i) \eta(z_j) \Bigg] Z_0[\eta] \quad \text{(B.13)}$$

We have proved that if (B.11) is true for n, then it is true for $n + 1$. By construction we know that (B.11) is true for $n = 1$ and 2, and so by induction it is true for all n. Now the final step in the proof of (B.8) is simply to set $\eta = 0$ in (B.11). For those with a field-theory background, this proof is essentially equivalent to Wick's theorem.

The first element of the diagrammatic method is that for each factor $G^{(0)}(x - y)$ we draw two points labeled x and y and connect

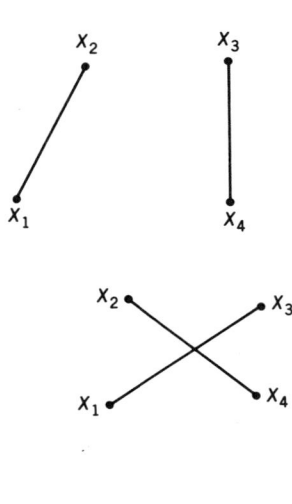

Fig. B.1 The three graphs corresponding to the fourth functional derivative of $Z_0[\eta]$.

them with a solid line. The expression (B.11), in the limit $\eta \to 0$, is therefore the sum of graphs constructed in the following way: Indicate each of the $2n$ arguments $x_1, x_2, \ldots, x_{2n}$ by a point, then connect the points in pairs in all possible ways, as shown in Fig. B.1.

B.2 FEYNMAN GRAPHS FOR THE PARTITION FUNCTION

If the exponential in (B.5) is expanded in a power series, we have

$$Z = \lim_{\eta \to 0} \sum_{\eta \to 0}^{\infty} \frac{(-1)^n}{n!} \left(L_{\text{int}}\left[\frac{\delta}{\delta \eta}\right] \right)^n Z_0[\eta] \qquad (B.14)$$

For the special case where $L_{\text{int}}[\phi]$ is given by (B.3), we may represent L_{int} by a squiggly line with four "terminals," two at each end, as shown in Fig. B.2. There are two types of diagrams that occur in the expression (B.14). Connected diagrams are diagrams in which every part is connected to the whole by at least one line. A disconnected diagram consists of two or more parts with no line connecting the parts. Evidently, disconnected diagrams are simply products of connected diagrams.

Generally, the nth-order contribution to the sum in (B.14) will consist of a sum over all possible products of connected diagrams whose separate orders add up to n. Let us denote the sum of all connected diagrams of order k by Φ_k. Exactly which of the n factors of H_{int} go into a given factor Φ_k is immaterial. All permutations among the various connected parts give identical contributions so, if we can determine the number of connected parts, we can sum all of these together and represent the entire sum by a single diagram.

Let us suppose that there are exactly l connected parts in a particular set of diagrams of order $m_1, m_2, \ldots, m_l$, and of course

$$\sum_{j=1}^{l} m_j = n \qquad (B.15)$$

The number of ways of partitioning n factors of H_{int} into l different

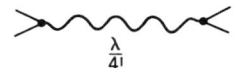

Fig. B.2 Graphical representation of $L_{\text{int}}[\phi] = (\lambda/4!)\int d^d x (\phi_\alpha(x)\phi_\alpha(x))^2$.

terms is

$$f(n, m_1, m_2, \ldots, m_l) = \frac{n!}{m_1! m_2! \cdots m_l!} \tag{B.16}$$

Now suppose that there are several of the m_i that are equal, so that interchanging a factor H_{int} between them does not produce a new term, but one that has already been counted. Obviously, (B.16) overcounts in this case by a factor of $p_{m_i}!$, where p_{m_i} is the multiplicity, and we must divide by $p_{m_i}!$. Putting all this together, we have

$$Z = \sum_{n=0}^{\infty} (-1)^n \sum_{p_1, p_2 \ldots} \frac{1}{p_1!} \frac{1}{p_2!} \left(\frac{\Phi_1}{1!}\right)^{p_1} \left(\frac{\Phi_2}{2!}\right)^{p_2} \cdots \delta\left(\sum_j j p_j, n\right) \tag{B.17}$$

The sum on n can be performed with the result

$$Z = \sum_{p_1} \frac{1}{p_1!} \left(\frac{(-1)^1 \Phi_1}{1!}\right)^{p_1} \sum_{p_2} \frac{1}{p_2!} \left(\frac{(-1)^2 \Phi_2}{2!}\right)^{p_2} \cdots$$

$$= \exp \sum_{j=1} \frac{(-1)^j \Phi_j}{j!} \tag{B.18}$$

A comparison of the expression (B.18) with the cumulant expansion shows that Φ_j is in fact the jth cumulant.

We have enormously simplified the task of evaluating the diagrams for the partition function, but we can go even further. Consider a connected diagram of order m. Because there are no external points, that is, all arguments $x_1, x_2 \ldots$ are integrated over, the diagram forms a closed loop of n links. Which of the m factors of L_{int} goes with which link is completely arbitrary, and in fact there are $(n-1)!$ ways of permuting the m factors. This allows us to represent the entire sum of equivalent diagrams by just one, and we have

$$Z = \exp \sum_{j=1}^{\infty} \frac{(-1)^j}{j} \tilde{\Phi}_j \tag{B.19}$$

where $\tilde{\Phi}_j$ is the sum of all "topologically distinct" connected, closed diagrams of order j. Topologically distinct just means that one diagram cannot be taken into another by a simple permutation of the factors L_{int}.

With the particular choice of interaction, (B.3), there is a further symmetry that always comes up. At each end of the interaction line, there are two terminals for propagator lines with the same internal index. If we connect these in a particular way, there exists another diagram in which the terminals are switched, giving rise to an overall factor of $2 \cdot 2 = 4$ for each interaction line. However, if the graph contains a closed loop this equivalence is lost and we must correct with a factor $1/2$. In addition, we pick up a factor of n, the number of components of ϕ_α, from the trace of the identity associated with the sum over internal indices around a closed loop. Finally, the wiggly interaction line can be flipped end over end without changing the value of the graph, and this leads to another factor of 2.

A particular graph may have symmetries in addition to those described previously, and one must take these explicitly into account graph by graph. In this way we only need to consider "topologically distinct" graphs.

We may now list the rules for constructing Feynman graphs of order m for the free energy in real space:

1. Draw all topologically distinct closed, connected graphs with m interaction lines.
2. With each interaction line associate a factor of $8\lambda/4! = \lambda/3$.
3. With each closed loop associate a factor $n/2$.
4. Multiply by $(-1)^m/m$ and by the symmetry factor for the graph (if any).

B.3 GRAPHICAL REPRESENTATION OF THE TWO-POINT GREEN FUNCTION

From a practical standpoint, the Green function

$$G_{\alpha\beta}(x - y) = \langle \phi_\alpha(x)\phi_\beta(y) \rangle \qquad (B.20)$$

is generally more convenient than the free energy for the following reason. In the sum over graphs for the free energy, the order of the graph appears (see rule 4) explicitly in the denominator. This factor makes it impossible to decompose a complicated graph into simpler subgraphs of lower order. There is a trick, called coupling-constant integration, which does allow some progress.

B.3 GRAPHICAL TWO-POINT GREEN FUNCTION

We introduce a parameter $0 \leq \alpha \leq 1$ multiplying L_{int}. The free energy is now

$$F(\alpha) = -\sum_{n=1}^{\infty} \frac{(-1)^n}{n} \alpha^n \tilde{\Phi}_n \tag{B.21}$$

To remove the pesky factor of $1/n$, we integrate (B.21) from 0 to 1,

$$F = \int_0^1 \frac{d\alpha}{\alpha} \sum_n (-1)^n \tilde{\Phi}_n \alpha^n \tag{B.22}$$

The sum in (B.22) is free of the factor $1/n$, and we can now sum the contributions from parts of the graph. Unfortunately, in order to finally arrive at the free energy we must perform the integration over α.

Graphs for the Green function, (B.20), are constructed by the same rules as apply to the free energy except that now the graphs are rooted in the points x and y.

A given graph will consist of a connected part rooted at x and y, and a disconnected part that factorizes. If the overall order of the graph is n and the connected part is of order p, there are $(n!/p!q!)\delta_{p+q,n}$ equivalent permutations of the n factors L_{int} between the connected and disconnected parts. Therefore, we can write

$$G(x, y) = \frac{1}{Z} \sum_{n=0}^{\infty} \frac{(-1)^n}{n!} \sum_{p=0} \frac{n!}{p!q!} C_p(x-y) D_q \delta_{p+q,n} \tag{B.23}$$

where $C_p(x-y)$ means "the sum of all connected graphs of order p rooted in x and y," and D_q is "the sum of all disconnected graphs of order q."

The sum over n can be performed, and we have

$$G(x, y) = \frac{1}{Z} \left(\sum_{q=0}^{\infty} \frac{(-1)^q}{q!} D_q \right) \left(\sum_{p=0}^{\infty} \frac{(-1)^p}{p!} C_p(x-y) \right) \tag{B.24}$$

The sum over all closed disconnected diagrams is just the partition function, and so the factor of $1/Z$ is cancelled and the Green function is equal to the sum over connected diagrams only

$$G(x, y) = \sum_{p=0}^{\infty} \frac{(-1)^p}{p!} C_p(x-y) \tag{B.25}$$

Now consider a connected graph of order p rooted at the points x and y. We can permute the p factors of L_{int} without changing the value of the graph, and so if we allow a single graph to represent all $p!$ of these, the factor of $p!$ cancels and we have

$$G_{\alpha\beta}(x-y) = \sum(\text{connected graphs rooted at } x \text{ and } y) \quad \text{(B.26)}$$

The rules for assigning an algebraic expression to a graph need to be modified only slightly:

1. Draw all topologically distinct graphs rooted in the points x and y with m interaction lines.
2. With each interaction line associate a factor $\lambda/3$.
3. With each closed loop associate a factor $n/2$.
4. Multiply by $(-1)^m$ and by the symmetry factor for the graph (if any).

REFERENCES

1. Abrikosov, A. A., Gorkov, L. P., and Dzyaloshinski, I. E. (1963) *Methods of Quantum Field Theory in Statistical Physics*. Dover, New York.
2. Amit, D. J. (1978) *Field Theory, the Renormalization Group, and Critical Phenomena*. McGraw-Hill, New York.
3. Fetter, A. L. and Walecka, J. D. (1971) *Quantum Theory of Many-Particle Systems*. McGraw-Hill, New York.

APPENDIX C

COMBINATORIAL SOLUTION TO THE ISING MODEL

In this appendix we present the details of the combinatorial method of solving the two-dimensional Ising model presented in Chap. 5. The solution was first given by Burgoyne [1], and this appendix is essentially an expanded version of his 1962 paper, see also Stanley [2].

The partition function for the Ising model on the square lattice is

$$Z = \text{Tr} \exp\left(J \sum_{\langle ij \rangle} S_i S_j\right) = (\cosh J)^{N_b} \text{Tr} \prod_{\langle ij \rangle} (1 + v S_i S_j) \quad \text{(C.1)}$$

where N_b is the number of bonds on the lattice and $v = \tanh J$. The parameter J is equal to the exchange coupling divided by kT.

If the product in (C.1) is expanded and the trace performed term by term, only those factors that are multiples of the identity contribute to Z. Because $S_i^2 = 1$, this means only terms in which a given spin variable appears $0, 2, 4, \ldots$ times need be considered. To each such term we can associate a graph G in which a line connecting two nearest-neighbor sites represents a factor $v S_i S_j$. Evidently, only closed graphs with an even number (0, 2, or 4) of lines at a given vertex contribute to Z. The trace in (C.1) can therefore be written as

$$\text{Tr} \prod_{\langle ij \rangle} (1 + v S_i S_j) = 1 + \sum_G l(G) \quad \text{(C.2)}$$

COMBINATORIAL SOLUTION TO THE ISING MODEL

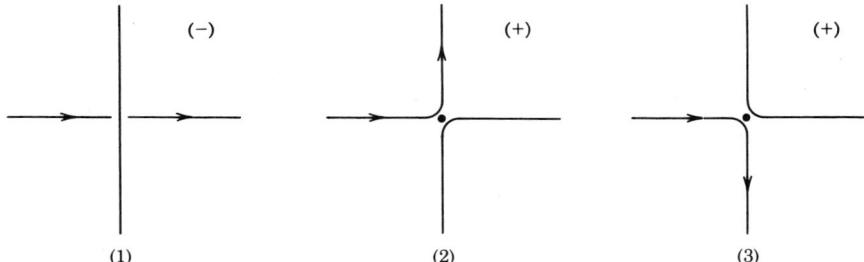

Fig. C.1 Three topologically distinct paths at a self-intersection of a graph. Note (-1) factor associated with type (1).

where $I(G)$ is the weight associated with each graph, which in our case is simply v^b, where b is the number of bonds in the graph.

We now establish a many-to-one correspondence between graphs G and paths, which we will label by P, on the lattice. A "path" is a directed graph on the lattice.

If the graph has no self-intersections, then it is clear that there is just one possible path apart from the trivial multiplicity associated with the sense in which the path is transversed. However, in the case of self-intersection, there are three topologically distinct choices for the path, as shown in Fig. C.1. If a graph has n self-intersections, then there are 3^n topologically distinct paths that "cover" the graph. We assign a phase factor to the three types of intersection, as shown in Fig. C.1, and define the weight function for a given path to be

$$W(P) = (-1)^{n_1} I(G) \tag{C.3}$$

where n_1 is the number of self-intersections of type 1 in the path. The number of paths with n_1 self-intersections of type 1, n_2 of type 2, and so on, is $n!/(n_1!n_2!n_3!)$, so we have

$$\sum_P W(P) = \sum_{n_1, n_2, n_3} \delta_{n_1+n_2+n_3, n} \frac{n!}{n_1!n_2!n_3!} (-1)^{n_1} I(G)$$

$$= (-1 + 1 + 1)^n I(G)$$

$$= I(G) \tag{C.4}$$

Therefore, the sum of $W(P)$ over all paths covering a given graph is just $I(G)$.

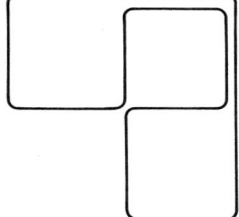

Fig. C.2 An improper path on the lattice in which some bonds are traversed more than once.

We now wish to prove the following theorem:

$$1 + \sum_G I(G) = \prod_P [1 + W(P)] \tag{C.5}$$

where $W(P)$ now includes all nonperiodic paths, which may or may not correspond to an allowed graph. For example, in $W(P)$ we allow paths that traverse a given bond more than once, as shown in Fig. C.2. Such terms never appear in the sum (C.2) because each factor vS_iS_j occurs at most once in the expansion on the left-hand side of the product in (C.2). The improper path of Fig. C.2 can result from the product of two proper paths, Fig. C.3. A periodic path is one in which each bond is traversed in exactly the same way more than once, as for example Fig. C.4, and these paths are, for the moment, forbidden.

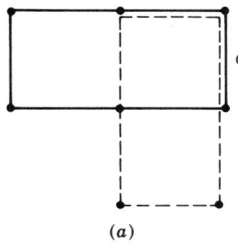

(a)

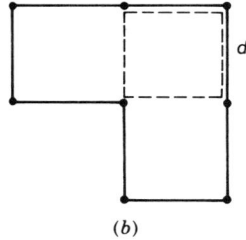

(b)

Fig. C.3 Products of two paths, a and b, giving rise to the improper path in Fig. C.2. The bonds in the first path are shown by solid lines, those in the second path are shown by dotted lines.

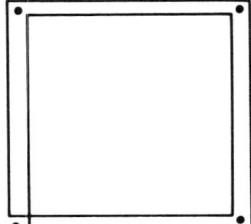

Fig. C.4 A periodic path of period 2.

We have already proved that the sum over all paths in which no bond is covered twice, and which therefore corresponds to an allowed graph, reproduces $\Sigma_G I(G)$. We now want to show that the sum of all nonperiodic paths, (C.5), again reproduces the sum over graphs. In the final step we will show that the sum over all paths, including periodic paths, is equal to the sum over graphs.

Let us group together all terms in the expansion of the product in (C.5) in which the same bonds are covered the same number of times. The weights for these paths will differ only in sign.

Now choose a particular bond d which is covered $N > 1$ times. In the product $W(P_1)W(P_2)\ldots$, the paths P_i either (i) contain the bond d or (ii) they do not. In the example shown in Figs. C.2 and C.3, both factors contain the bond d. Imagine removing the bond d from each of the factors, giving N path segments which begin and end at the ends of bond d. We call the N end points at either end of the bond d "terminals." We group all terms that lead to the same set of path segments into a subgroup. Each subgroup will share identical paths of type (ii), and the terms arising from each subgroup differ only in the way in which the N bonds through d are connected. In Fig. C.5 we show all six paths that belong to the group of the path shown in Fig. C.2 together with the corresponding path segments. Note that each subgroup in this case decomposes into two path segments. In general, a subgroup will decompose into N path segments $\{Q_1, Q_2, \ldots, Q_N\}$, and all terms in the subgroup are formed by connecting the N "top" terminals with the N bottom terminals. For $N = 3$ this is illustrated in Fig. C.6.

The relative phase for each of the $N!$ ways of attaching the terminals is just the number of self-intersections of the N lines drawn between the upper and lower sets of terminals. As one can see in Fig. C.6, half of the terms have an even number of self-intersections and half have an odd number of self-intersections. Therefore, the sum over all terms in a subgroup gives 0. The see this in general, we note that the number of intersections is equal to the minimum number

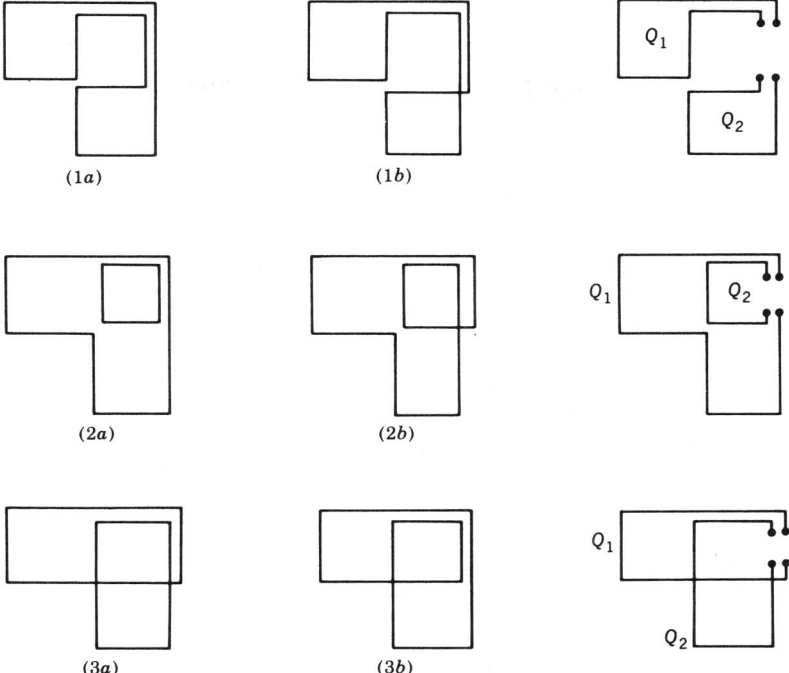

Fig. C.5 The six terms in the group of Fig. C.2. To the right are the path segments, which when joined in the two possible ways yield the two corresponding diagrams to the left.

of permutations of $1, 2, 3, \ldots, N$ to obtain the sequence $S(1)$, $S(2), \ldots, S(N)$. Exactly half of the permutations are even, and contribute with a $(+1)$ phase, and half are odd.

We assumed that the path segments $\{Q_1, Q_2, \ldots, Q_N\}$ in the subgroups were all different because if two are the same then it is possible to join them in a way which leads to a periodic or repeated path. We now consider a subgroup S in which we allow path segment Q_1 to occur r_1 times, Q_2 occurs r_2 times, and so on, and

$$\sum_i r_i = N \tag{C.6}$$

The notation used in Fig. C.7 needs a little explanation. We have labeled each path by the sequence of subpaths of which it is composed, and the order of the subpaths in the label follows the actual sequence in which the path is traversed. There is an arbitrariness as to the sense in which the entire path is traversed, but inequivalent paths

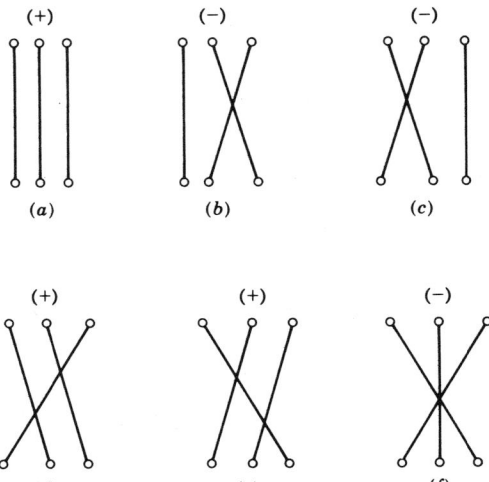

Fig. C.6 The six possible ways of linking three path segments through a single bond. The sign attached to each term is the phase, relative to path (a), and is simply $(-1)^c$ where c is the number of self-intersections.

can be obtained by reversing the sense in which a subpath is traversed relative to the rest of the diagram. We indicate that a path is traversed in the opposite sense by a bar over the label.

The parenthesis around a particular sequence indicates that this subpath is closed. Thus, for example, the path $(Q_1 Q_2)(Q_1 \bar{Q}_2)$ is allowed, whereas the periodic path $(Q_1 Q_2)(Q_1 Q_2)$ of Fig. C.8 is forbidden.

We now wish to prove that the contribution to the product (C.5) from paths such as those in Fig. C.7 also vanishes. If we decompose such a subgroup, which we denote by Σ, by removing bond d, and then recombine the path segments in all possible ways, we will generate some forbidden paths in addition to the allowed paths that make up the subgroup. Therefore, we must show that (i) the forbidden graphs so formed all cancel out and (ii) the allowed paths also all cancel out as they did in the case where none of the path segments were identical.

The proof is by induction. We first assume that it is true that all the paths for the group

$$\Sigma_0 = \{Q_0; Q_1, \ldots, Q_1; Q_2, \ldots, Q_2, \ldots\} \qquad \text{(C.7)}$$

cancel. This group is identical to Σ except that one path segment of

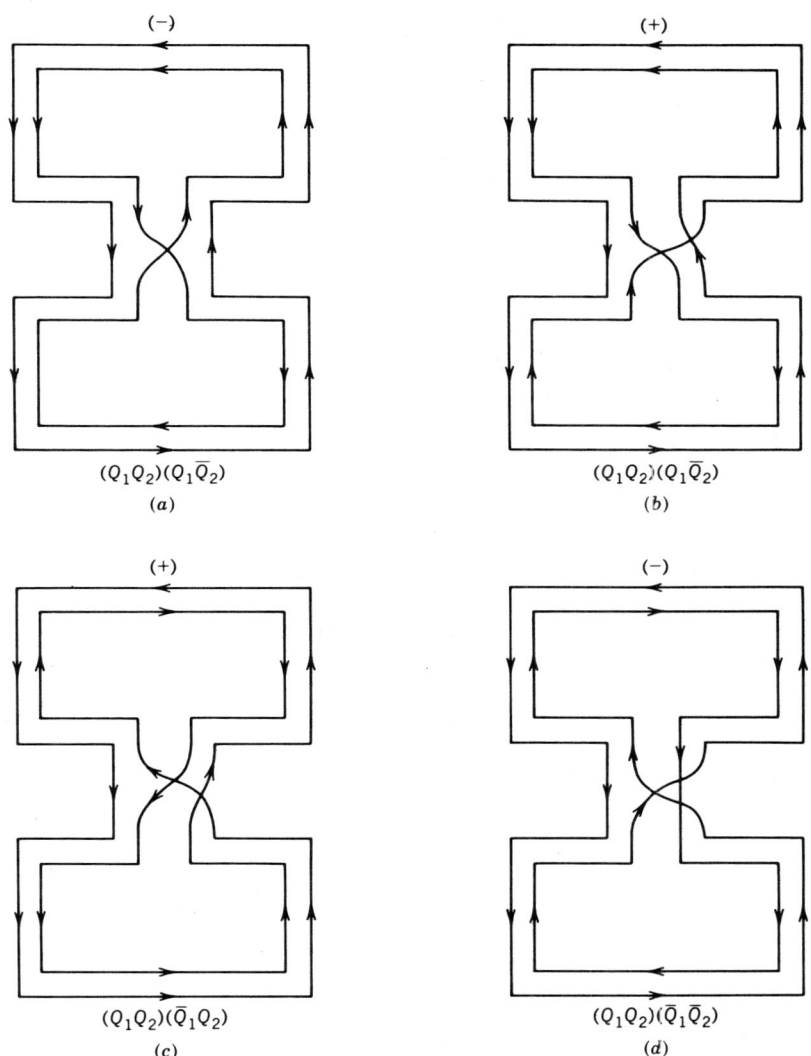

Fig. C.7 A subgroup, (a) through (d), consisting of four paths, which when decomposed lead to identical path segments. In this case there are two sets of identical path segments, which we denote by $\{Q_1, Q_1, Q_2, Q_2\}$. Note the paths (a) and (d) have sign (-1), whereas (b) and (c) are $(+1)$. In (e) we see the path segments formed when the repeated bond is removed.

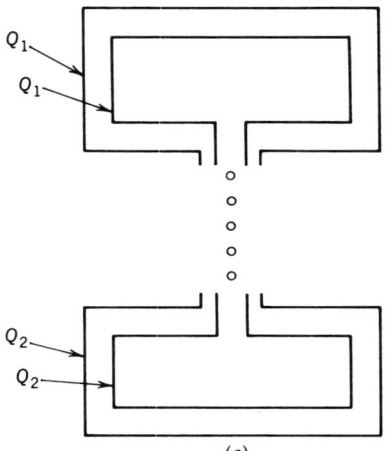

Fig. C.7 *(Continued)* (e)

the type Q_1, which we have labeled Q_0, is assumed to be different, leaving $r_1 - 1$ identical path segments of type Q_1.

Each path in Σ corresponds to r_1 paths in Σ_0, because there are r_1 places in which one can make the replacement $Q_0 \to Q_1$. Because none of the paths in Σ are forbidden, replacing Q_1 by Q_0 cannot lead to a forbidden diagram.

Now if we link the path segments in all possible ways and then identify $Q_0 = Q_1$, we will generate forbidden paths. That is, the path containing Q_0 will either become periodic or will become equal to another subpath.

In the first case the relevant subpath has the form

$$P = (Q_0 X Q_1 X \cdots Q_1 X)_{\omega - 1} \tag{C.8}$$

$Q_1 X$ is a nonperiodic subpath and, as the subscript indicates, occurs $\omega - 1 \geq 1$ times in P. On equating $Q_0 = Q_1$, P becomes periodic with period $\omega \geq 2$.

Now consider the path

$$P' = (Q_0 X Q_1 X \cdots Q_1 X)_{\omega - 2} \tag{C.9}$$

Such a path occurs in a term of the form $W(P')W(Q_1 X)$. This term also belongs in Σ_0 and leads to a forbidden path when $Q_0 = Q_1$ of the second type, that is, one in which a subpath is repeated more than once. Furthermore, the sign of the term $W(P)$ is opposite to that of $W(P')W(QX)$ and therefore all the forbidden graphs cancel in pairs. The remaining terms in Σ_0 that correspond to the allowed paths in Σ

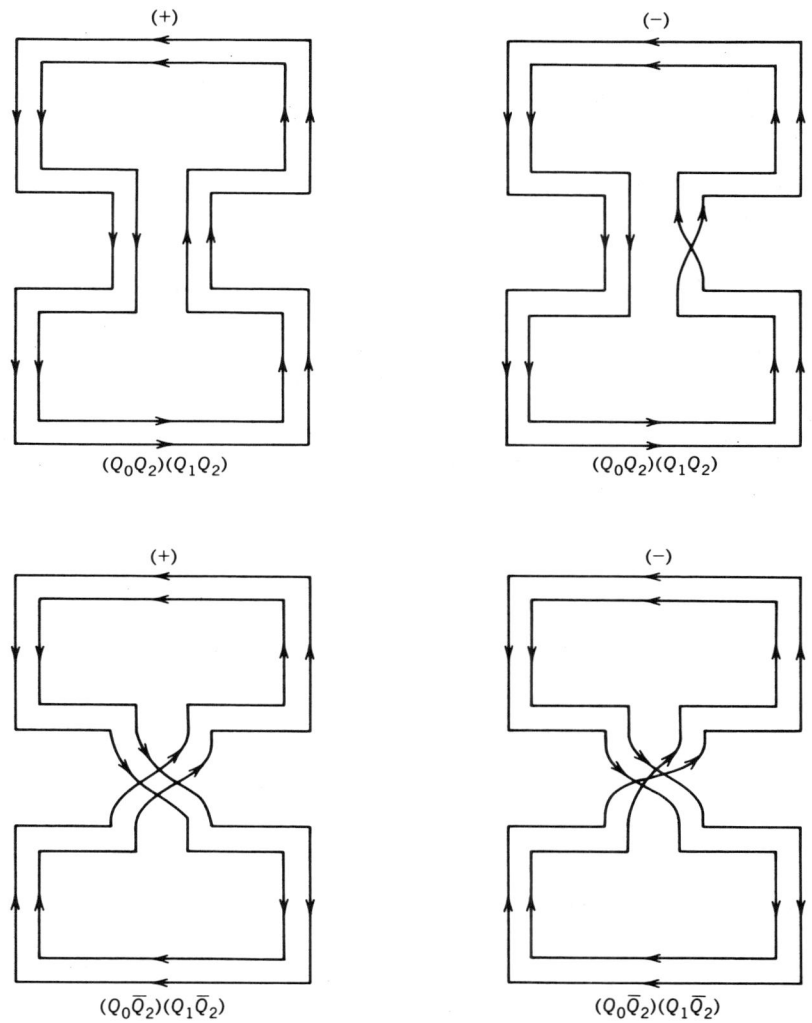

Fig. C.8 Forbidden graphs which cancel pairwise when $Q_0 = Q_1$.

have already been assumed to cancel, and therefore all the terms in Σ_0 cancel.

For the example of Fig. C.7, on taking one of the path segments Q_1 for Q_0, we get 12 paths, eight of which are equivalent to the paths in Figs. C.7a–d, and the remaining four are forbidden. These are shown in Fig. C.8.

We have thus far proved that the product (C.3) over all nonperiodic paths reproduces the sum over all allowed graphs in the partition function for the Ising model. We now show that we can write the free energy as a sum over all paths, including periodic paths.

Taking the logarithm of the partition function, apart from uninteresting factors, we have

$$\ln\left[1 + \sum_G I(G)\right] = \sum_P \ln[1 + W(P)] \qquad \text{(C.10)}$$

Expanding the logarithm on the right-hand side of (C.9),

$$\sum_P \ln[1 + W(P)] = \sum_P \left[W(P) - \frac{1}{2}W(P)^2 + \frac{1}{3}W(P)^3 \cdots\right] \qquad \text{(C.11)}$$

The first term in the sum is over all nonperiodic paths, the second is over period-2 paths, the third over period-3 paths, and so on.

In order to actually evaluate the sum in (C.11), we now show that the terms in (C.11) can be represented as a Markov process; that is, each step in a path (now including periodic paths) is, apart from the restriction that the step cannot go backwards, independent of all the previous steps. Let us label the possible directions in which a step can enter and leave in site as shown in Fig. C.9. For each of the four incoming directions, there are three possible outgoing directions. We indicate the site by a Latin index, for example, i, and the outgoing direction from site i by a Greek index, α. We denote this pair of indices by

$$p = (i, \alpha) \qquad \text{(C.12)}$$

The sign we attach to a path is $(-1)^{n_p}$, where n_p is the number of self-intersections of the path. By a theorem of Whitney [3]

$$(-1)^{n_p} = -(-1)^{t_p} \qquad \text{(C.13)}$$

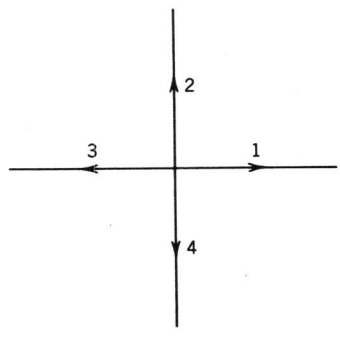

Fig. C.9 Labeling of the four directions by which a path may leave a vertex.

where t_p is the winding number of the path, and is equal to the number of counterclockwise rotations minus the number of clockwise rotations. If we assign a phase $e^{i\theta/2}$ at each vertex, where θ is the angle through which the path turns, then

$$(-1)^{t_p} = \prod_j \exp\left(\frac{i\theta_j}{2}\right) \tag{C.14}$$

This, together with Whitney's result (C.13), gives a convenient way of counting self-intersections. Let us now define a $4N \times 4N$ "transition matrix" $M_1(p, p')$, which assigns the proper phase for making a step from $p' = (j, \beta)$ to $p = (i, \alpha)$. Obviously, unless sites i and j are nearest neighbors the corresponding matrix element will vanish. In fact, there are only 12 nonvanishing matrix elements, which are illustrated in Fig. C.10.

The sum of all paths of l bonds connecting p and p' is simply the matrix product

$$M_l(p, p') = \sum_{p_1} \cdots \sum_{p_{l-1}} M(p, p_1) \cdots M_1(p_{l-1}, p')$$

$$= M_1^l(p, p') \tag{C.15}$$

Because M_1 is translationally invariant we can write

$$M_l(p, p') = \frac{1}{N} \sum_q e^{i\mathbf{q}\cdot(\mathbf{r}-\mathbf{r}')} M_1(\mathbf{q}; \alpha, \beta) \tag{C.16}$$

where $M_1(\mathbf{q}; \alpha, \beta)$ is the Fourier transform of $M_1(p, p'')$. Introducing this into the product (C.14) gives

$$M_l(p, p') = \frac{1}{N} \sum_q e^{i\mathbf{q}\cdot(\mathbf{r}-\mathbf{r}')} M_1^l(q; \alpha, \beta) \tag{C.17}$$

The Fourier transformed matrix $M_1(\mathbf{q}; \alpha, \beta)$ is only 4×4 and has the form [C.2]

$$M_1(q) = \begin{bmatrix} e^{-iq_1} & e^{-i\pi/4}e^{-iq_2} & 0 & e^{i\pi/4}e^{iq_2} \\ e^{i\pi/4}e^{-iq_1} & e^{-iq_2} & e^{-i\pi/4}e^{iq_1} & 0 \\ 0 & e^{i\pi/4}e^{-iq_2} & e^{iq_1} & e^{-i\pi/4}e^{iq_2} \\ e^{-i\pi/4}e^{-iq_1} & 0 & e^{i\pi/4}e^{iq_1} & e^{iq_2} \end{bmatrix} \tag{C.18}$$

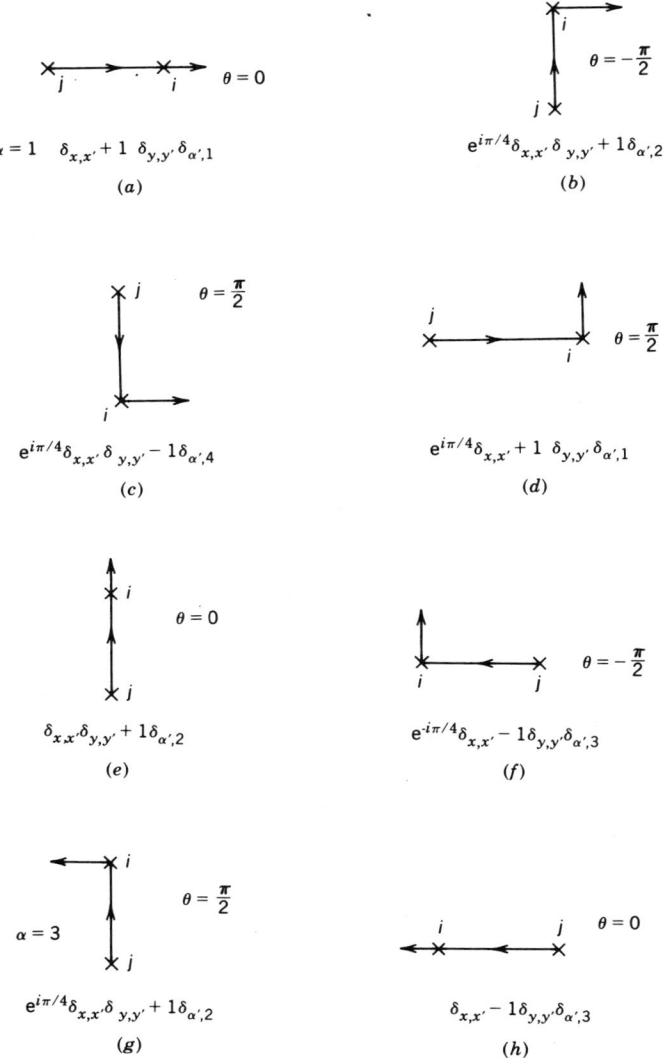

Fig. C.10 The nonzero matrix elements, (a) through (l), of $M_1(p, p')$.

A closed path is obtained by setting $q = q'$, so the sum over all closed paths of l bonds Ω_l can be written as

$$\Omega_l = -\frac{1}{l} \text{Tr} \sum_q M_1^l(q) \tag{C.19}$$

The factor of $1/l$ arises because there are l possible points along the closed path at which to begin, all of which are equivalent. The overall

(i) $e^{i\pi/4}\delta_{x,x'}\delta_{y,y'} - 1\delta_{\alpha',4}$

(j) $e^{-i\pi/4}\delta_{x,x'} + 1\delta_{y,y'}\delta_{\alpha',1}$

(k) $e^{i\pi/4}\delta_{x,x'} - 1\delta_{y,y'}\delta_{\alpha',3}$

(l) $\delta_{x,x'}\delta_{y,y'} - 1\delta_{\alpha',4}$

Fig. C.10 *(Continued)*

minus sign comes from Whitney's theorem, and we divide by 2 because a path can be traversed in two possible directions. If we finally sum over all l and replace the sum over wave vectors by an integral, we have

$$\sum_l \Omega_l = -\frac{1}{2}\text{Tr}\int_0^{2\pi}\frac{dq_1}{2\pi}\int_0^{2\pi}\frac{dq_2}{2\pi}\sum_l \frac{M_1(l)}{l}$$

$$= \frac{1}{2}\text{Tr}\int_0^{2\pi}\frac{dq_1}{2\pi}\int\frac{dq_2}{2\pi}\ln(1 - M_1) \quad (C.20)$$

Using the identity $\text{tr}\ln A = \ln\text{Det } A$, we finally have

$$\frac{1}{N}\sum_P \ln[1 + W(P)] = \frac{1}{2}\int_0^{2\pi}\frac{dq_1}{2\pi}\int\frac{dq_2}{2\pi}\ln\text{Det}[1 - M_1(\mathbf{q})] \quad (C.21)$$

REFERENCES

1. Burgoyne, P. N. (1962) *J. Math. Phys.* **4** 1320.
2. Stanley, H. E. (1971) *Phase Transitions and Critical Phenomena*. Clarendon, Oxford.
3. Whitney, H. (1937) *Comp. Math.* **4** 276.

AUTHOR INDEX

Boldface numbers refer to pages on which complete references appear.

Abrikosov, A. A., 371, **380**
Adler, J., 78, 83, **107**
Alexander, S., 206, **216**
Amit, D. J., xiv, 259, **266**, 269, **321**, 371, **380**
Anderson, P. W., 219, **237**
Arfken, G., 214, **216**
Arnold, V. I., 37, **68**

Barber, M. N., xiv, 307, **321**, **343**
Baxter, R. J., 144, **181**, 213, 214, **216**, 322, **343**
Berker, A. N., 128, **143**
Berlin, T. H., 322, 323, **343**
Bernoulli, D., 22
Bessel, F. W., 282
Bethe, H. A., 83
Binder, K., 231, **237**
Bishop, D. J., 361, **362**
Blumenthal, L. M., 31, **32**
Boltzmann, L. E., 39, 146, 159, 345
Bose, S. N., 137, 359
Braathen, H. J., 204, **216**
Bravais, A., 86, 90, 95, 145
Brown, R., 18, 19, 22
Burgoyne, P. N., 162, **181**, 381, **393**
Burkhardt, T. W., 26, **33**, 223, **237**

Cantor, G. F. L. P., 3, 8, 30, 55
Cauchy, A. L., 346
Cayley, A., 80, 83, 85, 105

Chang, T. S., 334, **343**
Coulomb, C. A., 347
Creswick, R. J., 133, **143**, 359, **362**
Cvitanovic, P., 40, 50, 51, **68**

Den Nijs, M. P. M., 93, **107**
Derrida, B., **80**
Deutscher, G., 78, 83, **107**
Dirac, P. A. M., 20, 302
Domb, C., **107**, **216**, **237**, **343**
Dyson, F. J., 312, 339
Dzyaloshinski, I. E., 371, **380**

Edwards, S. F., 219, **237**
Eminhizer, C. R., 38, **68**
Essam, J. W., 84, **107**
Euler, L., 347

Family, F., 26, **32**, **33**, 69, **107**
Farach, H. A., 133, **143**
Feigenbaum, M. J., 34, 40, 46, 47, 50, 67, **68**
Fetter, A. L., 371, **380**
Feynman, R. P., 272, 307, 337, 371, **376**
Fisher, M. E., 128, **143**, 266, 267, 290, **321**
Flory, P. J., 84, **107**
Fourier, J. B. J., 15, 112, 140, 167, 251, 271, 323
Freed, K. F., 22, **32**

Gaunt, D. S., **80**

AUTHOR INDEX

Gibbs, J. W., 146
Ginzburg, V. L., 137, 257, 281, 291, 322, 330
Glen, M., **80**
Goldstone, J., 137
Gordon, M. J. V., 359, 361, **363**
Gorkov, L. P., 371, **380**
Grassberger, P., 55, 67, **68**
Grebander, U., 176, **181**
Green, G., 140, 272, 347, 372, 378
Green, M. S., **107, 216, 343**
Gunton, J. D., 204, 205, **216**
Gustavson, F., 36, 37, **68**

Hallock, R., 361, **362**
Halsey, T. C., 8, **32**
Hamilton, W. R., 35
Hao, B.-L., 34, **68**
Hartree, D. R., 330, 337
Hauge, E. H., 27, **33**
Hausdorff, F., 2
Heiles, C., 36, 37, 39, 40
Heisenberg, W. K., 287
Helleman, R. H. G., 34, 36, 37, 56, 57, **68**
Hemmer, P. C., xiv, 27, **33**, 204, **216**
Henon, M., 36, 37, 39, 40
Hermann, D. W., **80**
Hohenberg, P. C., 137, 141, **143**, 350, 359, **362**
Houghton, A., 291, 296, **321**
Hsu, S.-C., 204, 205, **216**
Hu, B., 184, **216**
Hubbard, J., 239, 247, 255

Ising, E., 127, 144, **181**, 183, 239, 255, 287, 366, **381**

Jacobi, K. G. J., 35, 58
Jensen, M. H., 8, **32**, 40, **68**
Joyce, G. S., 322, **343**

Kac, M., 322, 323, **343**
Kadanoff, L. P., 8, **32**, 112, 217, 223, 227, 228, **237**
Klein, W., **80**, 93, 94, 95, **107**
Koch, H. von, 5–7, 30
Kogut, J., xiv, 90, **107**, 269, 275, **321**
Kolmogorov, A. N., 37, **68**
Kosterlitz, J. M., 141, **143**, 151, 351, 359, 361, 362, **363**
Kronecker, L., 113, 164, 207, 261, 300

Lagrange, J. L., 259
Landau, L., 137, 257, 281, 291, 322, 330
Lane, J., xiv
Laplace, P. S., 345
Lebowitz, J. L., **237**

Lenz, H. F. E., 144
Liouville, J., 56

Ma, S. K., xiv, 269, **321**, 329, 332, 337, **343**
Mandelbrot, B. B., 1, 2, 5, 6, 31, **32**
Markov, 390
Marro, J., 90, **107**
Martin, P. C., 359, **362**
McCoy, B. M., 144, 176, 178, **181**
McKenzie, D. S., 25, **33**
Menger, K., 31, **32**
Mermin, N. D., 137, 141, **143**, 350, **363**
Mesa, M., xiv
Metropolis, N., 40, **68**, 231, 234, **237**
Migdal, A. A., 217, 223, 227, 228, **237**
Mochel, J., **363**
Moser, J., 36, 37, **68**

Nakanishi, H., 99, **107**
Napiorkowski, M., 27, **33**
Nelson, D. R., 141, **143**, 359, 361, **363**
Newton, I., 34, 192
Niemeijer, Th., 183, 184, **216**

Onsager, L., 144, 145, 155, **182**

Packard, R. E., 359, 361, **363**
Pauli, W., 146
Pawley, G. S., 180, **182**, 235, **237**, 255, **266**
Peierls, R., 144, **182**
Pietronero, L., 1, **33**
Poincaré, J. H., 34
Poisson, S. D., 346, 362
Poole, C. P., Jr., 133, **143**
Potts, R. B., 206, 211, 214
Procaccia, I., 8, **32**, 34, **68**

Rayfield, G. W., 359, **363**
Rayleigh, J. W. S., 40
Redner, S., 26, 28, 29, **33**, 69, **80**, **107**
Reichl, L., 34, 36, 37, **68**
Reif, F., 359, **363**
Reppy, J. D., 361, **362**
Reynolds, O., 40
Reynolds, P. J., 26, 28, 29, **33**, 69, **80**, 93, 94, 95, 99, **107**
Riemann, G. F. B., 346
Rosenbluth, A. W., 231, **237**
Rosenbluth, M. N., 231, **237**
Rudnick, L., 361, **363**
Rushbrooke, G. S., 142

Schick, M., 208, 209, **216**
Shapiro, B., 26, 29, **33**
Shraiman, B.I., 8, **32**

Sierpinski, W., 7, 31
Sisson, C., xiv
Stanley, H. E., 26, **33**, 69, **80**, 93, 94, 95, **107**, 162, **182**, 322, **343**, 381, **393**
Stauffer, D., 69, **80**, 84, 85, **107**
Stein, M. L., 40, **68**
Stein, P. R., 40, **68**
Stinchcombe, R. B., 219, 220, **237**
Stirling, J., 150, 155
Stratonovich, R. L., 239, 247, 255
Sudbo, A. S., 204, **216**
Swendsen, R. H., 180, **182**, 232, 235, **237**, 255, **266**
Sykes, M. F., **80**
Szegö, G., 176, **181**

Teller, E., 231, **237**
Teller, M. N., 231, **237**
Thouless, D. J., 141, **143**, 151, 351, 359, 361, 362, **363**
Tosatti, E., 1, **33**

Van der Waals, J. D., 238
Van Leeuwen, J. M. J., 26, **33**, 183, 184, **216**, **237**

Vdovichenko, N. V., 162, 168, **182**
Verlarde, M. G., 204, **216**
Vvedensky, D. D., 334, **343**

Wagner, H., 137, 141, **143**, 350, **363**
Walecka, J. D., 371, **380**
Walker, J. S., 208, **216**
Wallace, D. J., 180, **182**, 235, **237**, 255, **266**
Wannier, G., 206, **216**
Wegner, F. J., 291, 296, **321**
Weiss, P., 144, 238
Whitney, H., 390, 391, **393**
Wick, G. C., 375
Wilks, S., **80**
Wilson, K. G., xiv, 90, **107**, **182**, **237**, 239, 255, **266**, 267, 269, 275, 282, 290, **321**
Wortis, M., 208, **216**
Wu, F. Y., 213, 214, **216**
Wu, T. T., 144, 176, 178, **181**

Yarmchuck, E. J., 359, 361, **363**
Young, A. P., 220, **237**

Zallen, R., 78, 83, **107**

SUBJECT INDEX

Boldface numbers designate the main discussion of that subject.

Abelian group, 139
AF *denotes* antiferromagnetic
Algorithm, Metropolis, 231, 234
Antiferromagnet, 219
Antiferromagnetic Ising model, **205**
Antiferromagnetism, 145
Antikink, 150
Aperiodic attractor, **54**
Approximation:
    cluster, 246
    Hartree, 330, 337
    mean field, 240
    Migdal-Kadanoff, **223**
    ring, 339
    self consistent, 337
    Stirling's, 150, 155
Asymmetric universality class, 130
Asymptotic series, 205
Attractor, **41**, 44, 48
    aperiodic, 42, **54**, 55
    chaotic, 42

Ball, d-dimensional, 2
Basin of attraction, 12, 43, **120**
Baxter-Wu model, 213, 214
BCS theory, 238
Bernard, *see* Rayleigh-Bernard cell
Bessel function, 282
Bethe lattice, **83**

Bifurcation, 34, 42–46, 60, 63, 64
    pitchfork, 42
Binomial, 220
    coefficient, 150
Block:
    structure, matrix, 167, 171, 173
    transformation, 94
Block spin, 112, 113, 183, 198, 208, 209
    nearest neighbor coupling, 185
    triangular lattice, 184
Boltzmann:
    factor, 159
    weight, 345
Bond:
    closed, 78
    moving, **217**, 225, 226
    open, 78
    percolation, 69, **86**, **96**, 127
    renormalized, 88
    renormalized horizontal open, 98
    terminal, 101
Bose fluid, 359
Boundary:
    conditions, periodic, 70, 209
    site, 72
Brownian:
    fixed point, 18
    motion, 9, 22

SUBJECT INDEX

Calculation, gory details, 270
Cantor set, 3–9, 31, 55
Carpet, Sierpinski, 7, 31
Cauchy-Riemann relations, 346
Cayley tree, 86
  critical probability, 90
  order parameter, 84
  paths, 85
  percolation, **83**, 105
  percolation threshold, 80
Cell:
  nine spin, 208
  seven spin, 204
  three spin, 198
Chain:
  Ising, 369
  rule, 135
Chaos, **34**, 59
  behaviour, 38, 39
  large scale, 37
  onset, 65
  regions, 37, 39
  transition to, 39
Characteristic function, 21
Charge, topological, 346
Chemical potential, 24, 350
Circulating square, 48
Circulation, 348
Cluster:
  approximation, **246**
  average size, 74
  bond, 87
  connected, 70, 88
  distribution function, 105
  fractal dimension, 76
  infinite, 70, 73, 76
  isolated, 70
  largest, 77
  mean size, 85
  number, 71, 73, 74, 76, 82
  number of sites, 75
  three-spin, 146, 246
Coarse grained, 10, 11, 15
  Hamiltonian, 114, 116
  lattice, 26, 89
  parameter, 18
  variable, 111
Combinatorial solution, Ising model, **381**
Completeness relation, 193
Composition rule, 118
Condensate wave function, 344
Conjugate parameter, 119
Connectivity, 69, 70, 94
Conservative map, **65**
Constraint, global, 323

Contour integration, 253, 326
Control parameter, 40, 53, 59, 64
Coordination number, 240
Corner rule, 26
Correlation:
  exponent, 105, 106, 254
  fixed point, 236
  length, 26, 74, 88, 109, 121, 147, 254
    divergence, 122, 126
    exponent, 74, 89, 90, 122
    at fixed point, **120**
  Ising, 149, 152
Correlation function, **110**, 139, 178, 360
  cumulant part, 125
  decay, 141
  determinant, 175
  exponent, 154
  Gaussian, 260, 262
  Ising, 151, 152, **168**
  longitudinal, 269
  order parameter, 120, 125
  scaling form, **124**
  spin–spin, 147, **357**
  spin wave, 358
  transverse, 269
  two point, 371
Coulomb gas, 347
Coupling:
  critical, 181
  renormalized, 209
  three spin, 198, 199
Covering lattice, 77
Critical:
  behaviour $d = 4$, **317**
  coupling, 181
  exponent, 25, 74, 76
    determination, 285
    epsilon, 287
  invariant, **134**
  Ising, 180, 255
  spherical model, **326**, 328, 338
  table, 80, **110**, 180, 287, 328, 342
  fluctuation, 109
  index, 203
  index table, 203
  isotherm, 123, 242
  line, 103
  opalescence, 109
  phenomena, **108**
  point, 1, 195, 357
    liquid–gas, 109
    signal, 121
  probability, 80
  surface, 12, 119, 195, 286
  temperature, Ising model, 241

SUBJECT INDEX

table, 161, 241
trajectory, 357
Crossover:
  exponent, 130
  phenomena, **129**
Cumulant, **21**
  connected, 125
  evaluation, 292
  expansion, **186**, 209, 271, **364**
  expansion asymptotic, 205
  Feynman diagram, 274
  first, **187**, 197, 201
    table of contributions, 202
    with h, 197
  revisited, **191**
  second, **188**, 199, 245
    evaluation, **277**
    table of contributions, 200
  table for Ising model, 205
Curve, 2
  continuous, 2
Cutoff:
  momentum, 260
  restore, 271

Decimation, 11, 116, 351
  diamond lattice, **218**
  fields, 282
  hierarchical lattice, **218**
  honeycomb lattice, 161
  transformation, 114
Degrees of freedom, integrating out, 262
Delta function, Dirac, 15, 20, 113, 302
Detailed balance, 231
Determinant:
  correlation function, 175
  Toeplitz, 176, 178
DHL, see Diamond, hierarchical lattice
Diagram:
  Feynman, **272**, 308, **371**
  Green function, 310
  self energy, 308
Diagrammatic method, 375
Diamond:
  hierarchical lattice, 87, 107
  lattice, 80, 241
    decimation, **218**
Dielectric breakdown, 69
Diffusion, 9
Dilution, site-bond, **99**
Dimension:
  Euclidean, 1, 9, 10, 90
  fractal, 2, 3, 10, 30
  Hausdorff, 2
  map, 34

three, 20
topological, 2
Dirac, see Delta function, Dirac
Divergence, exponential, 35
Domain:
  Ising, 158
  of attraction, 194
Duality, 78
  transformation, 167
Dyson equation, 312, 338, 339

Edwards–Anderson spin glass model, 219
Eigenvalue, 58
  equation, 118
  thermal, 286
Eigenvector, 13
Elastic properties, 99
Energy:
  density, 124
  internal, 146
  self, 308, 348
Ensemble:
  canonical, 24
  grand canonical, 24
Entropy, 146, 206
ε:
  critical exponent, 287
  expansion, 239, 266, **267**
  expansion field theoretic approach, **307**
  expansion renormalization, 282
Equation:
  difference, 59
  of state, 123
  spherical model, **326**, 327
Ergodic, 39
Euclidean, 1, 9
  dimension, 90
  random walk, 10
Euler's constant, 347
Exchange:
  constant, 145, 149
  coupling, 148, 153
Excitation, topological, 151, 345
Expansion:
  $1/n$, **322**, 336
  cumulant, 186, 209, 271
  cumulant asymptotic, 205
  ε, 239, 266, **267**
  gradient, 281, 282
Exponent:
  correlation, 254
  critical:
    determination, 285
    ε, 287
    spherical model, **326**, 328

Exponent, critical (*Continued*)
  table, 342
  maximal, 129
  specific heat, 255
  susceptibility, 331

Factorial, 150
Feigenbaum-Cvitanovic:
  equation, 51
  universal function, 51
Ferroelectricity, 144
Ferromagnet, 14, 219
Ferromagnetic, fixed point, 149
Ferromagnetism, 113, 144
Feynman:
  diagram, **272**, 278, 308, **371**
    cumulant, 274
  graph, **376**
Field:
  arbitrary, 360
  internal space, 268
  interpolation, 282
  local, 239
  momentum space, 265
  product, 273
  renormalized, 198
  rescaling, 294
  scalar product, 268
  source, 124
  symmetry breaking, 122, 131
  theory, 248, 375
    ε expansion, **307**
    $1/n$ expansion, **336**
Fig tree, 40
Finite size scaling, **131**
Fixed point, **13**, 40, 299
  analysis, **192**
  Brownian, 18, 19
  conservative map, 66
  correlation length, **120**
  critical surface, 194
  eigenvector, 194
  ferromagnetic, 149, 153, 154
  Gaussian, 263, **264**, 276, 290, 318
  Hamiltonian, 120
  hierarchical lattice, 89
  isolated, 126
  logistic map, 43
  neighborhood, **117**
  neighborhood RG, 134
  nontrivial, 27, 89
  paramagnetic, 154
  parameter space, 193
  phase transition, first order, **128**
  renormalization group, **117**

sol-gel, 103
standard map, 65
table, 203
trivial, 27, 89, 121
Wilson-Fisher, 290, 318
Fluctuation, 1, 109, 258, 319
  critical, 109
  dissipation theorem, 127, 143
  free energy density, 254
  long wavelength, 327
  order parameter, 109, 121, 344
  quadratic, 244
  thermal, 259
  thermodynamic average, 127
FM *denotes* ferromagnetic
Fourier transform, 20, 112, 114, 140, 167,
    251, 260, 323
Fractal, 2, **5**, **8**, 55
Fractal dimension:
  Cantor set, 5, 9
  cluster, 76
  Koch island, 6, 30
  random walk, 10
  Sierpinski carpet, 7
Free energy, 125, 138
  correlation function, 260
  density, 117, 122, 128, 131
    fluctuations, 254
  functional, 342
  Gaussian, 264
  Gibbs, 146
  homogeneity, **121**
  invariance, 112, 116
  Landau-Ginzburg, **255**, 257, 268, 291
  minimizing, 244
Frustration, **206**
Fugacity, 24, 28, 350, 354, 356
Functional:
  derivative, 372
  integral, 115, 362, 372

Gamma function, 341
  incomplete, 214
Gas:
  Coulomb, 347
  vortex, 348, 350
Gaussian, 21, 31
  distribution, 22
  fixed point, 263, **264**, 276, 318
  fixed point stability, **264**
  free energy, 264
  functional, 372
  integral, 140, 248, 358
  model, 17, **259**
  model RG analysis, 261

model susceptibility, 261
probability, 14
weight, 273
Gellation, 69
Generating:
  function, 24, 124
  functional, 372
  potential, 125
Generator:
  group, 291
  hierarchical lattice, 86
  rotation, 138
Global:
  constraint, 323
  structure, aperiodic attractor, 54
Goldstone:
  bosons, 137
  modes, 137
Gradient, expansion, 281, 282
Graph, 26
  cancel pairwise, 389
  cover, 382
  decomposition, 163
  forbidden, 157
  loop, 164
  partition function, 157, 163
  path, 162, 163
  rooted, 168, 169
  self intersection, 382
  subloop, 162, 164
  sum over, 162
  vertex, 162
Green function, 140, 272, 281, 341
  diagram, 310
  Dyson equation, 312, 337
  free, 372
  two point, **378**
Group, xiii
  abelian, 139
Gyration radius, 74, 75

Hamiltonian, 13
  coarse grained, 111, 114, 116
  fixed point, 120
  Ising, 145, 148, 184
  operator, 110
  Potts model, 206
  RG transformation, 110
  universality class, 129
Hamilton-Jacobi theory, 35
Hamilton's equations, 35
Hartree approximation, 330, 337
Hausdorff dimension, 2
Heisenberg, order parameter, 287
Helium:
  superfluid, 345, **359**
  superfluid film, 359, 361
Henon-Heiles Hamiltonian, 36, 37, 39
Hierarchical lattice, **86**, 87
  decimation, **218**
  generator, 86
Honeycomb, lattice, 78, 80, 159, 160, 241
Hubbard-Stratonovich:
  partition function, 255
  transformation, 239, **247**
Hyperbolic, 152, 156
Hypercube, 226, 323
Hyperscaling, **126**, 127
Hypersphere, 323
Hypersurface, 37
Hypothesis:
  scaling, 75, 319
  universality, 91, 127, 256, 287

Image point, 57
Incommensurate, 145
Insertion:
  polarization, 340
  self energy, 310, 337, 340
Integral, functional, 115, 362, 372
Interaction, three spin, 201
Internal energy, 146
Intersecting storage ring model, 39
Invariance:
  critical exponent, **134**
  scale, 1
Irrelevant, 299
  scaling field, 121
  variable, 13
Ising:
  chain, 148, 369
  critical exponent, 255
  order parameter, 287
Ising model, 27, 78, 127, **144**, **183**, **217**
  antiferromagnetic, 183, **205**, 207
  bond dilute, **219**
  combinatorial solution, **381**
  correlation function, **168**
  critical temperature, 241
  exact solution, **162**, **168**
  ferromagnetic, **155**, 183, 205, 206
  Hamiltonian, 184
  magnetic field, **152**, 155
  magnetization, 240
  mean field theory, **239**
  Metropolis algorithm, 231
  one dimension, **147**, **152**
  one dimensional chain, 366
  order parameter, 147, 240, 287
  partition function, **162**, 247, 250, 255, 381

Ising model (*Continued*)
   phase diagram, 222
   random bond, **219**
   renormalization group, **183**
   renormalization group equation, 136
   specific heat, 132
   square lattice, **162, 168**
   table:
      of critical indices, 203
      of cumulant expansion, 205
      of fixed points, 203
   triangular lattice, **183**
   two dimensional, **155**, 161
Isotherm, critical, 123, 242, 328
Iteration, 86

Jacks, 31
Jacobian, 58

Kagomé lattice, 181
   dual, 105
KAM *denotes* Kolmogorov, Arnold, and Moser
KAM theorem, **37**
Kink, 149, 150, 152, 158
Koch Island, 30
   quartic, 6, 7
   triadic, 5, 6
Kosterlitz, renormalization, 351
Kosterlitz–Thouless, transition, 141, 151, **344**, 359
Kronecker delta, 113, 164, 165, 214, 302

Lagrangian, 259
Landau–Ginzburg:
   form, 281
   free energy, **255**, 257, 268, 291
   model, 322
   theory, 330
Laplace's equation, 345, 362
Lattice:
   animal, 72, 73, 105
   artificial, 86
   Bethe, 83
   body centered cubic, 80, 241
   Bravais, 86, 90, 95, 145
   chain, 80, 241
   coarse grained, 89
   covering, 77, 90
   diamond, 80, 86, 241
   dual, **78**, 79
   face centered cubic, 80, 241
   gas, 144
   hierarchical, 86
   hierarchical self-similar, 89

honeycomb, 78, 80, 159, 160, 241
hypercubic, 226
isolated site, 71
Kagomé, 105, 181
matching, 78
self dual, 159
self-matching, 79
simple cubic, 80, 241
site, 69
square, 71, 78, 79, 80, 241
square self dual, 160
square site percolation, **94**
square transformed, 114
three dimensional, 77
triangular, 78, **91**
triangular block spin, 113
triangular decomposition, 207
triangular dual, 159
triangular lattice, 92
triangular majority rule, 93
triangular three site cell, 93
Law:
   large numbers, 22
   scaling, **110**, 122, **126, 127**
Left-to-right rule, 106
Length:
   correlation, 109
   large scale, 285
   rescaling, 276
Liouville's theorem, 56, 57
Logistic map, 40, 42–47, 63–66

Ma ordering, 329, 337
Macromolecule, 69
Magnetic field:
   Migdal–Kadanoff transformation, **227**
   renormalization group, **196**
Magnetization, 124, 146, 147, 151, 178
   Ising, 179, 240
   spherical model, 327
   spontaneous, 179
Majority rule, **91**, 112, 207, 208
   triangular lattice, 93
Map, 26
   area preserving, 56, 58
   conservative, 56, 57, **65**
   dissipative, 56, 65
   dynamical, 39
   linearizing, 57
   logistic, 40, 42–47, 63–66
   nonconservative, 56, 57
   nontrivial, 40
   one dimensional, **39, 47**
   Poincaré, 36, 37
   renormalized, 49, 50, 60, 64, 66

sine, 41
standard, 58–61
stroboscopic, 38
two dimensional, 34, **56**, 57
universal behaviour, 56
universal properties, **65**
Marginal, variable, 13, 16
Markov process, 390
Master equation, 231
Matrix:
  block structure, 171, 173
  elements, nonzero, 392
MC *denotes* Monte Carlo
MCRG *denotes* Monte Carlo renormalization group, 231, 236
MCS *denotes* Monte Carlo steps per spin
Mean field, **238**
  Ising model, **239**
  magnetization, 240
  variational, **243**
Melting, 108
Menger sponge, 32
Metropolis algorithm, 231, 234
MFT *denotes* mean field theory
Microscopic details, 112
Migdal–Kadanoff:
  approximation, **223**
  method, **217**, **223**
  potential, **224**
  transformation, 223
  transformation, magnetic field, **227**
MK *denotes* Migdal–Kadanoff
Mode:
  long wavelength, 262
  short wavelength, 262, 333
Model, *see also* Ising model
  Baxter–Wu, 213, 214
  Gaussian, 17, **259**
    RG analysis, **261**
    susceptibility, 261
  Landau–Ginzburg, 322
  n-vector, 322, **329**
  spherical, **322**
  X–Y, 151, **344**, 359
    planar, 357
Molecular field, 144
  Weiss, 238
Moment, 21
Momentum:
  cutoff, 111, 114, 260, 262
  shell, 293, 303
  space, 112, 114
    field, 265
    variables, 115
Monte Carlo:

calculation, 93, 94, 95, 96
  Ising model, 133
  renormalization group, **230**

Neighbors, number of nearest, 80
Newton's method, 192
Nonanalytic, 108
Nonintegrability, 34
Numerical solution, 35
n-vector model, **329**

Occupation probability, 75
Oil flow, 69
Opalescence, critical, 109
Operator:
  Hamiltonian, 110
  irrelevant, **119**
  marginal, **119**
  parameter space, 118
  projection, 111, 193
  relevant, **119**
Orbit, 35, 40
  Hamiltonian, 117
  period-1, 41, 45, 46
  period-2, 41, 45, 46
  period-4, 63, 64
  superstable, 48, 49, 53, 67
  unstable, 41
Order:
  long range, 141
  short range, 141
Ordering, Ma, 329, 337
Order parameter, 109
  Cayley tree, 84
  correlation function, 120, 125
  fluctuation, 121, 344
  Heisenberg, 287
  Ising model, 147, 240, 287
  percolation, 73
  renormalized, 123
  scaling relations, 110
Overcounting, 209

Paramecium, 72
Parameter:
  physically adjustable, 14
  relevant, 130
Partition function, 24, 78, 115, 124, 270
  coarse grained, 112
  Feynman graph, 376
  graph, 156, 157, 162
  honeycomb lattice, 161
  Hubbard–Stratonovich, 255
  Ising, 146, 148, 156, **162**, 165, **168**, 170, 247, 250, 255, 381

Partition function (*Continued*)
  spherical model, 325
  spin wave, 349
Partitioning, 115
Path:
  forbidden, 388
  graph, 162, 163
  rooted, 169
  segments, 385
Percolation, **69**
  applications, 83
  bond, **69**, 77, 79, 80, 84, **86**, 90, **96**, 127
  bond-cell for, 97
  Cayley tree, **83**
  critical exponent table, 80
  one dimension, **81**
  order parameter, 73
  site, **69**, 77, 80, 84, 90, 91
  site on square lattice, **94**
  site RSRG, 97
  threshold, 70, 73, **80**
Period doubling, 34, **39**, 41, 45–47, 51, **59**
Perturbation theory, 1
Phase:
  disordered, 108
  ordered, 108
  space, 35
  thermodynamic, 108
Phase diagram:
  Ising model, 222
  sol–gel, 103
Phase transition:
  first order, 108, 128
  fixed point, 128
  liquid–gas, 238
  second order, 108, 118, 128
  zero temperature, 141
Physical subspace, 120
Pitchfork bifurcation, 42
Plaquette:
  square lattice, 79
  triangular lattice, 206
Plasma, 350
Poisson equation, 362
Polarization insertion, 340
Polymer, 22
Potential, generating, 125
Potts model, 206
  three state, 211
  two state, 207, 214
Power law, 320
Probability:
  bond, 85
  critical, 80

distribution, 15, 20
  renormalized, 16
  rescaled, 16
  Gaussian, 14
  occupation, 73, 75
  renormalized bond, 88
  renormalized site, 95
  site, 78
Projection operator, 111, 193
Propagator line, 378

Quadrature, solving coupled equations, 288
Quasicontinuous, 220

Random bond Ising model, **219**
Random walk, **9**, 16
  coarse grained, 10
  with drift term, 19
  renormalization group, **14**
  self avoiding, 17, **22**
  self-similar, 16
  universality class, **20**
  unrestricted, 14
Rayleigh–Bernard cell, 40
Real space renormalization group, **22**, **96**
Relevant:
  parameter, 130
  scaling field, 121
  variable, 13
Renormalization, 12
  cell, 94
  equations, 27
    solution, **296**, **300**
    to order ε, **282**
  group:
    fixed point, **117**
    Gaussian model, **261**
    Monte Carlo, **230**
  group equation, 12, 149, 192, 196, 210, 355, **356**
  group generator, 291
  group in limit $n \to \infty$, **332**
  group Ising, **183**
  group linearized, 135
  group majority rule, 91
  group momentum space, 112
  group real space, **22**, **96**
  group transformation, **269**
  Kosterlitz, 351
  transformation, **11**, 49, **110**
    on Bravais lattice, 95
    on square cells, 96
Renormalized:
  away, 88

SUBJECT INDEX 407

coupling, 148, 209
field, 153, 198
horizontal bond, 102
order parameter, 123
site probability, 95, 100
trajectories, 130
Rescaling, 11, 15, 18, 50
field, 294
length, 276
variable, 263
Resistor network, 69, 98
Reynolds number, 40
RG *denotes* renormalization group
RG equation:
 magnetic field, **196**
 second order, **198**
Ring approximation, 339
RMS *denotes* root mean square
 distance, 9, 10, 16
Rotation:
 generator, 138
 operator, 138
 symmetry, 137, 281
RSRG *denotes* real space renormalization group
Rule:
 chain, 135
 composition, 118
 corner, 26
 left-to-right, 106
 majority, **91**, 112, 207, 208
 majority triangular lattice, 93
 symmetric, 99, 100
Rushbrooke scaling law, 127, 142
RW *denotes* random walk

SAW *denotes* self avoiding walk
Scale:
 factor, 8, 11, 118
 factor renormalization group, 270
 invariance, 1
 invariant, 4
 large length, 285
 loss of, 1, 109, 121
 small, 14
 transformation, 62
Scaled temperature, 134
Scaling, 47
 behaviour, 109
 equation, 116
 exponent, 17, 18, 124
 field, **13**, 117, 121
  irrelevant, 121
  one dimensional map, 52

 relevant, 119, 121
 sol-gel, 104
finite size, **131**
form of correlation function, 124
hypothesis, 75, 319
law, 76, **110**, 122, **126**, **127**
law, Rushbrooke, 127, 142
linear, 290
order parameter, 110
relation, 3, 110
source, 125
Self:
 avoiding walk, **22**, 96
 consistent approximation, 337
 dual, 78, 159
 energy, 308, 348
 energy diagram, 308
 energy insertion, 310, 337, 340
 intersection, 384
Self-similar:
 aperiodic attractor, 55
 Cantor set, 5
 Cayley tree, 83
 hierarchical lattice, 89
 orbit, 65
 random walk, 16
Self-similarity, **1**, **3**, 4
Semiconvergent, 205
Series, asymptotic, 205
Sgn *denotes* the sign of, 186, 187
Shear modulus, 99
Sierpinski carpet, 7, 31
Sine map, 41
Site:
 bond dilution, **99**
 bond problem, 69
 percolation, 69
 terminal, 86
Sol-gel:
 phase diagram, 103
 renormalization group trajectories, 104
 transition, **99**
Source:
 field, 124
 scaling, 125
Space, internal of fields, 268
Specific heat, 122, 131, 146, 243
 critical exponent, 255
 discontinuity, 245
 figure, 132, 133
 Ising, 132, 168
 scaled, 133, 134
Spherical model, **322**
 critical exponent, **326**, 328, 338

Spherical model (*Continued*)
  equation of state, **326**, 328
  magnetization, 327
  partition function, 325
  susceptibility, 327
Spin:
  glass, 145
  -spin correlation function, **357**
  wave, 139, 349, 356, 358, 360
  wave partition function, 349
Sponge, Menger, 32
Standard map, 58–61
Star-triangle transformation, 160, 181
Statistical:
  independence, 84
  weight, 75, 87, 91
Steepest descent, 250, 325
Step function, 303
Stirling's approximation, 150, 155
Stochastic, 37
Subbranch, 84, 85
Subspace, physical, 120
Superconductivity, 238
Superfluid:
  density, 361
  helium, 345, **359**
  transition, 14
  velocity, 359
Superstable, 48
Surface, 2
  critical, 12, 119, 195, 286
Susceptibility, 109, 127, 146
  exponent, 331
  Gaussian model, 261
  noninteracting system, 316
  spherical model, 327
  zero field, 123
Symmetric rule, 99, 100
Symmetry:
  breaking, 122, 126, 127, 156, 241, 258
    spontaneous, **255**
  breaking field, 131
  rotational, 281
  spontaneously broken, 269
    continuous, 141
Szegö's theorem, 176

Tadpole, 339
Temperature:
  parameter, 154
  relevant parameter, 121
  rounding, 131
  scaled, 134
Texture, 345

Thermal:
  eigenvalue, 286
  fluctuation, 259
Thermodynamic:
  action, 247
  density, 128, 129
  fluctuation average, 127
  function, 110
  limit, 131, 354
Third-order calculations, 204
Three-body problem, 35
Threshold, percolation, 70, 73, **80**, 86, 96
  triangular lattice, 91
  RG transformation, 106
Toeplitz determinant, 176, 178
Topological:
  charge, 346
  distinct, 377
  excitation, 151
  excitations, 345
  invariant, 346
Trajectory:
  critical, 357
  of Hamiltonian, 117
  renormalized, 130
  RG of sol-gel transformation, 104
Transformation:
  approximate renormalization group, 86
  block, 94
  block spin, **112**
  decimation, 114
  duality, 78, 167
  Hubbard-Stratonovich, 239, **247**
  renormalization group, 11, **110**, **269**
  RG on Bravais lattice, 95
  star-triangle, 160, 181
Translation invariance, 170
Triangular lattice:
  decomposition, 207
  Ising, 183
  plaquette, 206
Truncation, 191, 209
  parameter space, 111

Uncorrelated site and bond problems, 69
Universal:
  behaviour superfluid, 359
  function, 50
  parameter, 53
Universality, 40, 47, **126**, **127**
  class, **20**, 58, 127
    asymmetric, 130
  hypothesis, 91, 127, 256, 287

Variable:
   coarse grained, 111
   irrelevant, **13**
   marginal, **13**, 16
   relevant, **13**
   rescaled, 263
   separation of, 52
Variational theory, mean field theory, **243**
Vertex, function, 273
Vortex, 345
   bound, 357
   bound pair, 352
   circulation, 348
   dissociation, 356
   fugacity, 356
   gas, 348, **350**
   ionized, 350
   plasma, 151, 350
   quantized, 359
   state, 349
Vorticity, 348

Wavefunction, condensate, 344
Wegner–Houghton:
   generator, 296
   infinitesimal renormalization, **291**
Weight function, 382
Weiss molecular field, 238
WF *denotes* Wilson–Fisher
Whitney's theorem, 390
Wick's theorem, 375
Winding number, 169

X–Y model, 151, **344**, 359
   planar, 357